DEVELOPMENTS
IN
APPLIED
SPECTROSCOPY
Volume 10

DEVELOPMENTS IN
APPLIED SPECTROSCOPY

Selected papers from the Annual Mid-America Spectroscopy Symposia

A Publication of the Chicago Section of the Society for Applied Spectroscopy

DEVELOPMENTS IN APPLIED SPECTROSCOPY
Volume 10

edited by

A. J. Perkins
University of Illinois
College of Pharmacy
Chicago, Illinois

E. L. Grove
Freeman Laboratories, Inc.
Rosemont, Illinois

Emmett F. Kaelble
Monsanto Company
St. Louis, Missouri

Joan E. Westermeyer
Howmet Corporation
Misco Division
Whitehall, Michigan

Selected papers from the Tenth National Meeting of the Society for Applied Spectroscopy and the Twenty-Second Annual Mid-America Spectroscopy Symposium held in St. Louis, Missouri, October 18-22, 1971

Ⱋ **PLENUM PRESS · NEW YORK – LONDON · 1972**

Library of Congress Catalog Card Number 61-17720
ISBN 978-1-4684-3134-6   ISBN 978-1-4684-3132-2 (eBook)
DOI 10.1007/978-1-4684-3132-2

© 1972 Plenum Press, New York
Softcover reprint of the hardcover 1st edition 1972

A Division of Plenum Publishing Company, Ltd.
227 West 17th Street, New York, N.Y. 10011

United Kingdom edition published by Plenum Press, London
Davis House (4th Floor), 8 Scrubs Lane, Harlesden,
London, NW10 6SE, England

PREFACE

Volume 10 of Developments in Applied Spectroscopy presents a collection of twenty selected papers presented during the 10th National Meeting of the Society of Applied Spectroscopy held in St. Louis, Oct. 18-22, 1971.

The 10th National Meeting was sponsored by the St. Louis Section, Society of Applied Spectroscopy in cooperation with Baltimore-Washington, Chicago, Cleveland, Cincinnati, Houston, Intermountain (Idaho Falls), Kansas City, New England, Niagra Frontier, Northern California, North Texas, Pittsburgh and Rocky Mountain (Denver) Sections of the Society and combined with the 22nd Mid-American Symposium. Both theoretical and applied principles were presented in sessions on emmision, atomic absorption, molecular, nuclear, mass, x-ray, flame, Massbauer, and magnetic resonance spectroscopy. In addition special symposia were held on on-line computers, spectra data processing and retrival, application of spectrographic techniques for the museum, in environmental control, in space, in biomedicinal, sampling, the analysis and characterization of electronic materials and gas chromatography.

The members of the Program Committee, I. Adler, L. Beaver, J. E. Delmore Jr., J. Eichelberger, J. R. Ferraro, M. Fishman, C. L. Grant, X. W. Heitsch, W. M. Hickam, M. T. Jones, S. R. Koistyohann, J. Kopp, D. W. Larsen, F. Pogge, W. Ritchey, M. E. Salmon, R. K. Skogerboe, J. J. Spijkerman, P. Urone, C. Veillon, M. S. Wang, and J. S. Ziomek, and the other meeting committee members are to be commended for the excellent program. Thanks are also extended to the exhibitors for their part in making the meeting a success.

A. J. Perkins
E. L. Grove
J. E. Westermeyer
E. F. Kaeble

v

CONTENTS

SPECTROSCOPIC TECHNIQUES
FOR THE MUSEUM

MOLECULAR SPECTROSCOPY

ATMOSPHERIC AND SPACE SPECTROSCOPY

SPECTROSCOPY IN BIOMEDICINE

SPECTROSCOPIC WEAR METAL
ANALYSIS IN LUBRICATING OILS

X-RAY AND EMISSION SPECTROSCOPY

SPECTROSCOPIC TECHNIQUES FOR THE MUSEUM

RECENT INTERFACES BETWEEN MASS SPECTROMETRY

AND ART

Thomas Cairns and Ben B. Johnson

Conservation Center, Los Angeles County Museum of Art
5905 Wilshire Boulevard,
Los Angeles, California 90036, U.S.A.

CHEMISTS have long been aware of the limitations imposed on
their studies by the limits of sensitivity of many of those
currently available analytical methods which permit a number of
elements to be determined in one sample, e.g. optical emission
spectrographic techniques, x-ray fluorescence analysis. This
has led to the adoption of more sensitive methods for a restricted
number of elements, e.g. neutron activation analysis. More
recently, spark source mass spectrometry has provided the
analytical chemist with the ability to cover the full range of
elements in any sample in a single determination and the ability
to detect those elements down to very low concentrations, i.e.
parts in 10^9. The advent of this technique was the product of
semi-conductor technology where 10 parts per billion (ppb) of
copper in germanium or 2 ppb of gold in silicon is reflected in
the electrical properties of those semi-conductors.

EXPERIMENTAL

The AEI MS702 Spark Source Mass Spectrometer in the
Conservation Center of the Los Angeles County Museum of Art
was purchased with a grant from the Samuel H. Kress Foundation.
This instrument like most other mass spectrometers consists
essentially of three parts which are housed in a highly evacuated
system (Figure 1):
    (a) the source for ionizing the sample - a radio frequency

spark;

(b)   The analyzer to separate these ions according to their
       mass to charge ratio (m/e) - a combination of both
       electrostatic and magnetic;

(c)   the detector to record the position and intensity of the
       ions - a photographic plate.

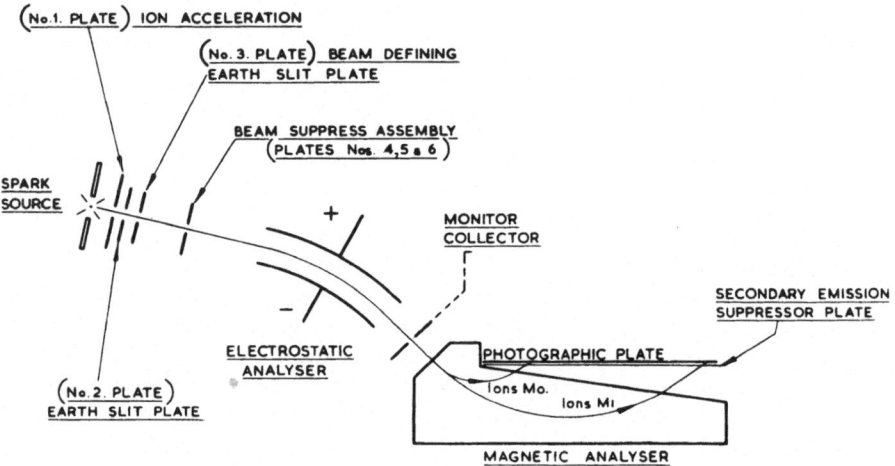

Figure 1.   The AEI MS702 Spark Source Mass Spectrometer
            Schematic

A radio frequency (about one megacycle) voltage of several
tens of kilovolts is applied in pulses across a small gap (a few
thousandths of a centimeter) between the two sample electrodes
in the source under high vacuum.   The vacuum breakdown that
occurs in this gap initiates an electrical vacuum discharge.   The
positive ions from this discharge which is representative of the
sample under study are then accelerated through to the
electrostatic analyzer, monitored and finally separated according
to their m/e ratios.   Usually a series of about fourteen graded

exposures (0. 0001 nanocolumbs to 100 nC) per sample are recorded photographically on an Ilford Q2 plate.

With the radio frequency spark source, all elements are supposedly ionized with approximately equal sensitivity. Two factors must be taken into consideration; first that an ion beam with a large energy spread is produced, and secondly that the output from such a spark is somewhat variable. The large energy spread factor is overcome by using a double focusing instrument which gives the required resolution (3000) and the output variation from the mass spectrometer is corrected by using an integrator to measure the ion beam intensities. Advantages of the photo-plate method of detection are its simple integrating properties and the convenient record of the spectrum obtained.

SAMPLE PREPARATION

In the case of drillings taken from a cast bronze object or any other metallic artifact, electrodes (20 x 2 mm) can easily be formed by compressing such drillings in a specially designed polytetrafluoroethtylene (PTFE) slug. Lack of sufficient material to form a whole electrode can be overcome by "topping up" with the necessary amount of ultra-pure silver powder. In many cases tipped electrodes such as these containing only a few milligrams can be prepared suitable for qualitative analysis.

Electrodes made by compressing pigment samples cannot be used because of their low electrical conductivity. The technique of incorporating a conducting powder must therefore be adopted. Silver powder and the sample are mixed together in a PTFE capsule, containing a pestle of the same material by means of a vibratory mixing machine. The resulting mixture is then com- pressed in the specially designed PTFE slug to form a solid electrode. Electrodes produced by this method have satisfactory mechanical strength and can be handled without difficulty.

More recently, in order to achieve very homogeneous electrodes the bronze drillings are dissolved in acid and then added to ultra-pure graphite powder. The resulting slurry is then taken to dryness and electrodes prepared in the normal manner. This new and novel technique can be used to prepare any concentration of the major components desired and so estimate them as if they were trace elements in the carbon matrix.

Futhermore the use of enriched isotopes as internal standards
has greatly improved the accuracy of the analysis - also added
to the carbon as an acidic solution.   These new methods of
sample preparation have greatly increased the accuracy of both
the major and trace elements in a bronze sample.

SPECTRUM INTERPRETATION

Densitometer traces at the 3nC level of exposure of a bronze
sample are illustrated in Figures 2a, 2b, and 2c.   This spectrum
will be used to illustrate the salient features of a spark source
mass spectrum.

The spectral lines in a spark source mass spectrum are
caused by the following types of ions:
  (a) Singly-charged ions.   The most intense lines in the
       spectrum of a sample are caused by singly-charged ions
       of the major component isotopes.   Thus in Figures 2a
       and 2b, the most intense lines (i.e. peak heights or areas)
       are lines due to $^{56}Fe^+$, $^{63}Cu^+$, $^{65}Cu^+$ and $^{120}Sn^+$.

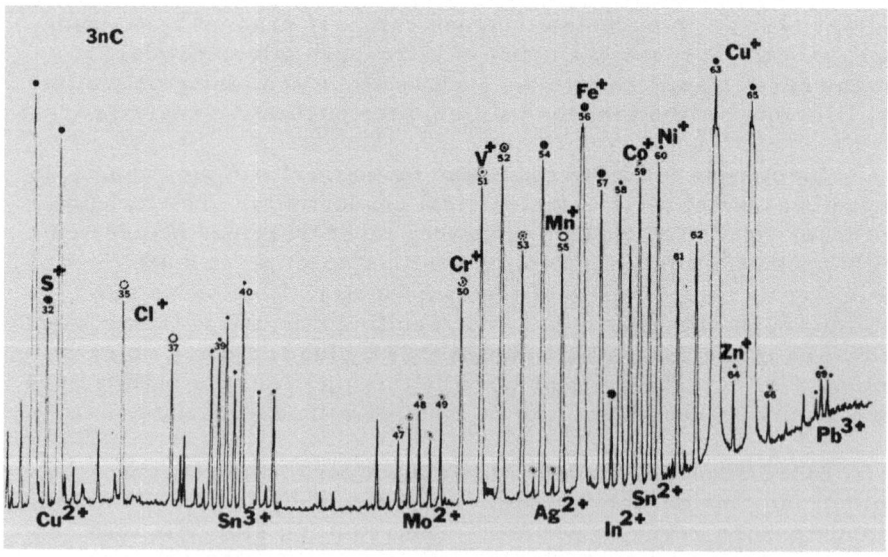

Figure 2a.   Densitometer trace of 3nC exposure of a bronze
             sample over m/e 30 to m/e 70.

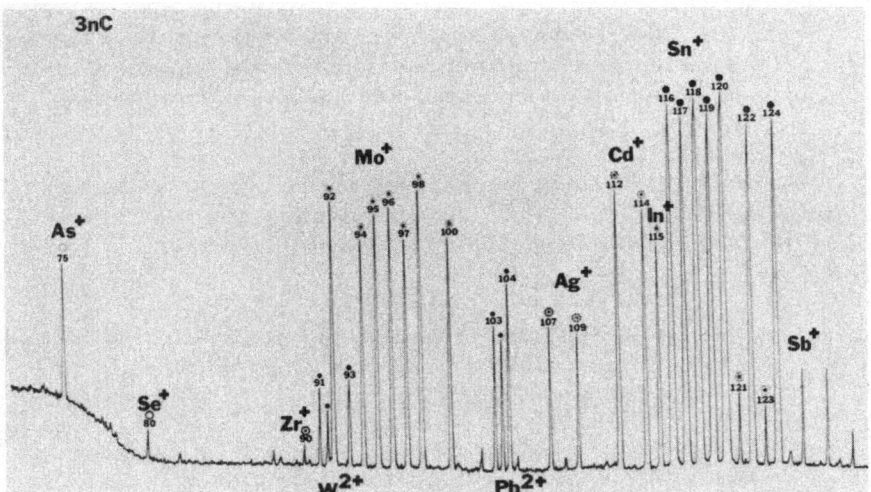

Figure 2b.   Densitometer trace of 3nC exposure of a bronze
              sample over m/e 71 to m/e 130.

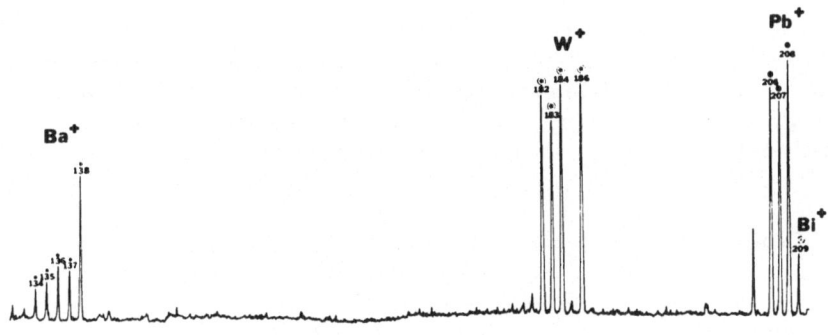

Figure 2c.   Densitometer trace of 3nC exposure of a bronze
              sample over m/e 131 to m/e 210.

However, as the exposure is increased lines due to the lesser
abundant components appear.   At the 3nC level of exposure trace
elements at the concentration level of 10 parts per million (ppm)
are clearly visible, e.g. Bismuth at m/e 209 (Figure 2c).

    (b)   Multiply-charged ions.   Lines caused by multiply-
          charged ions are of much lesser intensity than those due

to the singly-charged species. In Figures 2a and 2b lines caused by the undermentioned multiply-charged ions can be seen - Cu^{2+} at m/e 31.5 and 32.5: Pb^{2+} at m/e 103 104.

The intensity of multiply-charged ions decreases with each degree of ionization (1.e. Pb^{3+} of less intensity than Pb^{2+}) and the ratio of intensities depends on the prevailing spark conditions.

(c) Polyatomic ions. Polyatomic ions are usually a minor feature in most mass spectra, but carbon and silicon are exceptions to the rule.

(d) Complex ions. Complex ions, composed of two or more different elements, are occasionally observed, in particular the types XO^+ and XOH^+.

QUALITATIVE IDENTIFICATION OF ELEMENTS

Elements can usually be identified by their characteristic masses and isotopic patterns. In Figures 2b and 2c the elements Lead, Tin and Molybdenum may be identified by their characteristic groups of isotopes (Table 1) while the relative abundances of naturally occuring isotopes of these elements are given in Figure 3.

Table 1.   Characteristic Isotopic Masses for Lead, Tin
and Molybdenum

Element	Mass of isotopes
Lead	204, 206, 207, and 208
Tin	112, 114, 115, 116, 117, 118, 119, 120, 122, and 124
Molybdenum	92, 94, 95, 96, 97, 98, and 100

An element is reported as being present only when at least one of the following conditions is satisfied:

(a)   lines corresponding to ions of at least two isotopes of
       the element, and in their correct relative intensity ratio,
       are observed;

(b)   lines due to multiply-charged ions are observed at one-
       half or one-third of the masses of the major isotope.
       This condition is particularly stringent when lines due
       to such multiply-charged ions occur at fractional masses.

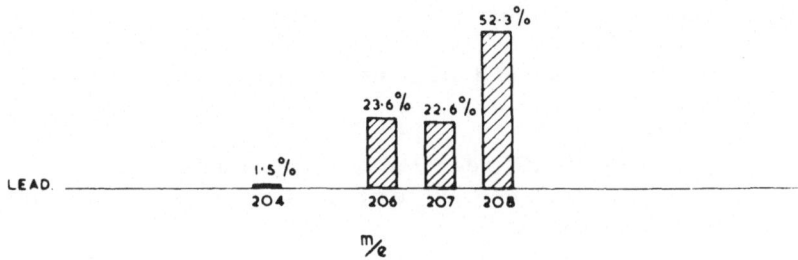

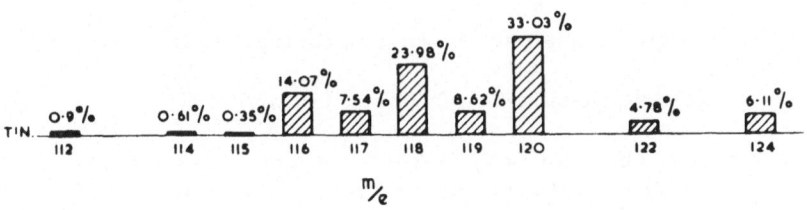

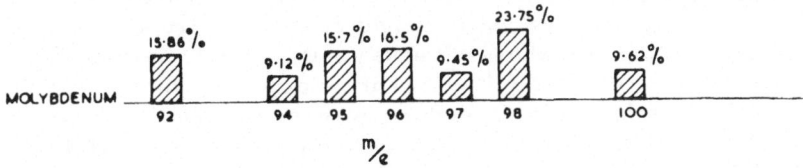

Figure 3.   Relative abundances of naturally occuring isotopes
              of Lead, Tin and Molybdenum.

QUANTITATIVE ANALYSIS

The approach adopted can be summarized in the following basic equation:

$$C_E = C_S \times \frac{Exp_S}{Exp_E} \times \frac{I_S}{I_E} \times \frac{1}{R}$$

where

C_E = content of element E in electrode analyzed (ppm atomic);

C_S = content of a second element S, e. g. internal standard (ppm);

Exp_S = exposure (in nC) required to give a line of chosen predetermined density for a chosen isotope of element S on a photoplate;

Exp_E = exposure (in nC) required to give a line of same density for a chosen isotope of element E on the same photoplate;

I_S = isotopic abundance of chosen isotope of S;

I_E = isotopic abundance of chosen isotope of E;

R = relative sensitivity factor - introduced as a measure of sensitivity of total recording procedure to line of element E used compared with sensitivity to line of element S used, which is arbitrarily assigned a value of unity.

Exposures are obtained from the developed photographic plate and a plot of exposure vs. peak height drawn (Figure 4). The predetermined density chosen (about 0. 8D) for calculations corresponds to a peak height of 90 mm - exposure values are then read off the graph. This procedure ensures that any variables due to line width and shape are cancelled out since essentially the two elements or lines representing these elements are being compared at exactly the same density on the plate. Isotopic abundances can be obtained for all elements from published tables of isotopic abundances. Relative sensitivity factors are predetermined using known standards.

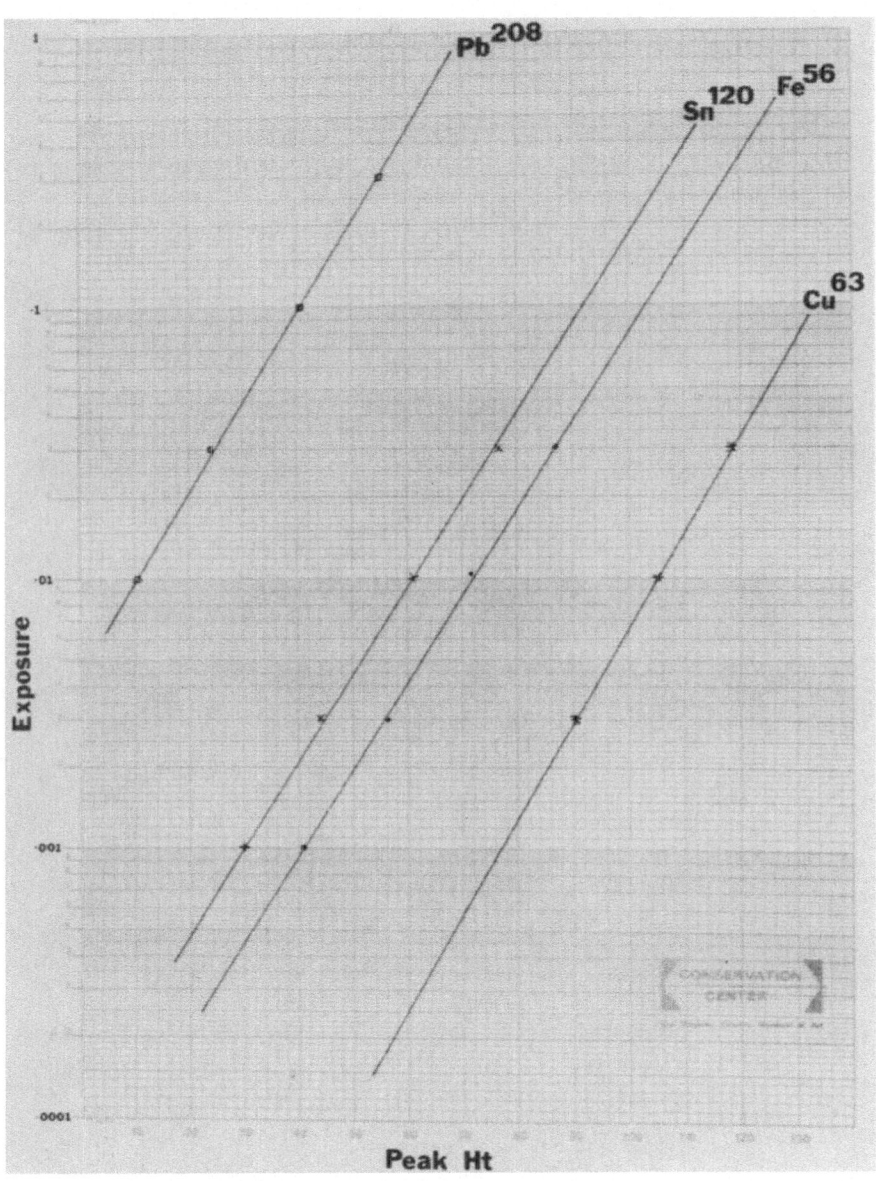

Figure 4.   Plot of exposure vs. peak height (i. e. density)
            for various ions.

APPLICATIONS

South Indian Bronzes

Quite recently an interesting set of bronzes of the early
12th century A.D. depicting Krishna and his two wives, Rukmini
and Satyabhama, together with his messenger, Garuda (Figure 5)
were acquired by the Los Angeles County Museum of Art through
the generosity of Mr. and Mrs. Hal Wallis.

Figure 5. Krishna Rajamannar Bronzes, South Indian, early
           12th century. From left to right, Rukmini, Krishna,
           Satyabhama, and Garuda.

The occurrence of such a group of high quality and importance
is unique. On religious grounds they were probably cast and
fabricated (cold worked after casting the rough shape) as a group
with Sutras dictating proportions, attributions, etc. This

prerequisite, therefore, permitted a detailed study of the chemical composition from piece to piece in an attempt to ascertain if an exact science was operative or was each composition purely by chance. It turned out that all four pieces resembled each other fairly closely. Drillings were taken from various locations in each piece and analyzed several times, then averaged. Table 2 illustrates the typical results obtained for Krishna. Many other South Indian bronzes are now under investigation to determine whether any further correlations can be established.

Table 2. Spark Source Mass Spectrographic Analysis of Krishna (Wt.: 125 lbs.)

Element	Sample Location				
	Hip	Neck	Heel	Average	Base
Cu	94.1 %	93.8 %	93.9 %	93.9 %	95.27 %
Sn	2.42	2.75	3.71	2.96	2.09
Pb	2.81	2.31	1.64	2.25	1.92
Bi	.010	.013	.04	.02	.019
Sb	.092	.098	.09	.09	.064
Ag	.05	.06	.07	.06	.05
As	.09	.14	.09	.10	.08
Zn	.11	.33	.20	.21	.17
Ni	.04	.07	.04	.05	.04
Co	.02	.02	.02	.02	.02
Fe	.20	.31	.17	.23	.27
TOTAL	99.94	99.90	99.97	99.89	99.99

Mughal Indian Miniature Painting

Although there are several excellent studies of the technique of Indian miniature painting which are based purely on literary sources and tradition passed from generation to generation, until present day very little has been published which deals with the technical examination of actual paintings.

The common appearance of blue pigment in Mughal paintings (16th and 17th centuries A. D.) raises a natural question as to

whether lapis lazuli ($3Na_2O. 3Al_2O_3.6SiO_2. 2NaS$), a costly stone
mineral, yielded the blue or whether it was obtained from azurite
($2CuCO_3. Cu(OH)_2$), a much cheaper material. To answer some of
these questions regarding pigments, a study was undertaken of the
miniature paintings from the collection of Nasli and Alice
Heeramaneck and the Los Angeles County Museum of Art.

The limited number of pigment analyses has precluded any
final conclusions at this stage, but some interesting patterns have
evolved. The analysis of the blues during the early Mughal
period (mid-15th to mid-16th century) has revealed that genuine
lapis lazuli was the most preferred and most frequently used blue,
i. e. lack of evidence for the suggested use of azurite as a cheap
substitute. In addition Mughal artists preferred to use their
pigments in pure form in overlapping layers or in mixtures with
a second pigment such as lead white.

For white pigments Mughal artists of the first half of the 17th
century preferred lead white whereas in earlier paintings belonging
to the 15th and 16th centuries kaolin ($Al_2O_3. 2SiO_2. 2H_2O$) was
dominant.

Vermilion (HgS) was the most widely used red pigment. Second
to vermilion, minium (Pb_3O_4) was employed, especially in paintings
of the 16th and 17th centuries. Orpiment (As_2S_3) and realgar
(As_2S_2), yellow and orange, respectively, were identified on
16th and 17th century paintings but not in significant patterns to
be of diagnostic value. In a study of the green pigments, both
malachite and copper resinate were found.

A number of organic pigments were encountered and research
is in progress via combined Gas Chromatography - Mass
Spectrometry to elucidate their molecular structures.

CONCLUSION

Conservation-chemistry represents a new and fascinating
applied scientific discipline harnessing chemical knowledge
to unlock the secrets of the history of technology.

THE USE OF EDX FOR FRAGILE AND REFRACTORY OBJECTS

J. F. Hanlan

National Conservation Research Laboratory

National Gallery of Canada, Ottawa, Canada

INTRODUCTION

The Chairman and other speakers of this session have discussed the problems of adequate sampling of cultural objects. Serious restrictions are placed on the analyst as regards both sample size and location since the object cannot be defaced or altered in any irreversible way. The objects are unique, frequently of uncertain provenance, usually badly documented in any technical sense and almost always seriously altered by time. As a result, the museum scientist must approach each problem with caution and with as much background as possible. There are no routine analyses. Problems must be carefully defined and the optimum available technique selected. Frequently, due to the restrictions imposed, the scientist and scholar must settle for what is possible rather than whatever might have been most desired.

One saving feature is that, in many cases, a relatively imprecise analysis may solve the problem. Simple elemental identification may be sufficient, for example, the occurrence of zinc in a white passage would indicate a nineteenth century origin; a proteinaceous media can be differentiated from a drying oil on the basis of simple staining reactions. When accurate detailed analysis is necessary (1) great ingenuity is required and the analyst must analyze as many samples as possible, taking a statistical view of the results. In some instances, the inhomogeneities which bedevil the analyst, contain the information sought and techniques such as electron microprobe analysis to study diffusion of metals (2) and the X-ray "Macroprobe" to study the layer structure of paintings (3) have been used. Another point that should be made is that methods such as neutron activation analysis (4), spark source mass spectro-

metry (5), and emission spectroscopy with the laser probe (6) require so little sample that if a micro-sample can be considered
truly representative, these methods can be applied to nearly any
type of museum object.

The refinement, in recent years, of high resolution lithium-
drifted silicon X-ray detectors, has permitted the development of
X-ray fluorescent instrumentation capable of rapid, safe and convenient elemental analysis in situ on objects of any size or shape.
The potential of this technique for the analysis of museum objects
was immediately obvious and several workers undertook preliminary
studies (7-11). The possibility of field use was tested successfully by Frierman et al (12).

INSTRUMENTATION & PROCEDURES

The principles of X-ray fluorescence spectrography are familiar
to most and, in any event, are given in many text books on the subject. A good survey of the operational considerations relevant to
the use of silicon detectors for X-ray spectroscopy was given by
various authors at the ASTM workshop on Energy Dispersion Analysis
held in Toronto, 1970 (13).

The electronic signal processing used in our laboratory is
somewhat different than the usual pulse-height analyzer instrumentation commonly employed and should be discussed briefly. A "black-
box" schematic is given in Figure 1. In the schematic power supplies
and the detector bias are not shown.

Pulses from the preamplifier are shaped by the amplifier which
is equipped with pole zero cancellation and base-line restoration.
Coarse and fine gain controls can be preadjusted to any of 4 positions so that specific energies, i.e., pulse heights, can be set to
fall into the desired channels of the encoder for full scale values
between 10-100 Kev. A 5th position is variable over the entire range
and can be used for unusual problems or preliminary testing.

The heart of the system is the encoder. For digital counts
data, each pulse is converted to a digital number which is sorted
into 8 (in two sets of 4) preselected "windows". The channel number
and width is selected by program cards which are plugged into the
encoder. These cards can also be programmed to produce marker bars
for the channel selected. 128 channels are available. For spectrum
display each pulse is used to drive the X-axis of a Hewlett-Packard
181A storage oscilloscope. At the appropriate position a line is
written whose intensity is modulated to zero in the Y axis. As
identical events occur, the trace on the oscilloscope screen inten-
sifies into a vertical wedge. Visually this presentation is equivalent to several thousand channels of a multi-channel analyser.

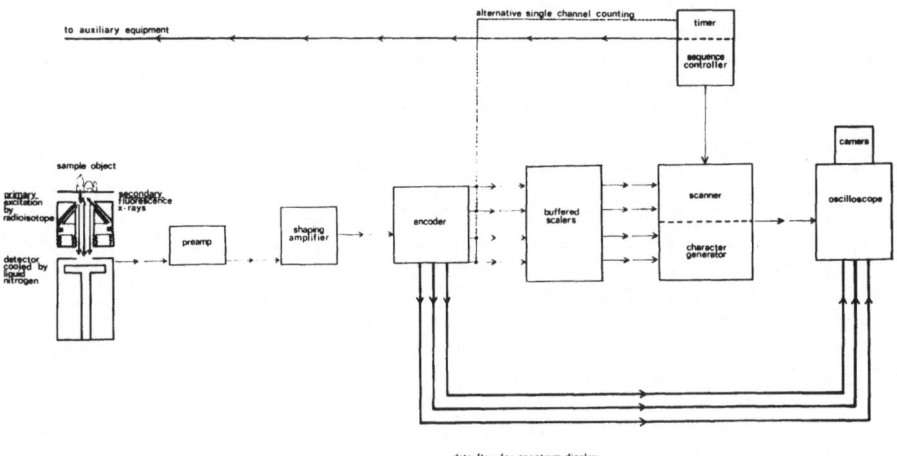

Figure 1.   Schematic drawing of EDX spectrometer.

The pattern on the oscilloscope is recorded by a Tektronix C-12
camera equipped with a projected graticule accessory. A 15 KeV and
30 KeV graticule has been prepared and fine gain and position controls on the encoder permit accurate placement of the spectrum
relative to the energy scale.

    Digital data from the encoder are accumulated for a pre-set time
(or pre-set count) by the set of 4 buffered scalers. A scanner controlled by the timer and sequence controller reads the data in the
scalers and a character generator writes this information on the
oscilloscope screen for up to 4 elements along with the live-time
of the analysis. This can be repeated for a second set of 4 elements programmed on the other card in the encoder so that a total
of eight elements can be determined per experiment.

    For single channel data accumulation, the signal can go
directly to the timer in its counter or ratemeter mode. The ratio
of the count rates for two windows can also be obtained directly
using the external clock mode of the timer.

    The detector (187eV resolution FWHM at 5.9 Kev) and preamplifier

were obtained from Kevex Corporation[*] and the spectrometer from Inax
Instruments Ltd.[**]

The radio-isotope used for excitation is I^{125} adsorbed on char-
coal beads and arranged symmetrically in an annular configuration.
The source spectrum is Te Kα and Te Kβ at 27.5 and 31. KeV respectively with a γ at 35 KeV and has a half life of about 60 days.
This inconveniently short half-life is largely compensated for by
the fact that the material is cheap, very pure, can be obtained in
high specific activities (100 - 300 mCi initially) and has a convenient energy. The main Compton peak is near Kα energy of Sn and for
this reason, the ring source is housed in a conical collimator
fabricated from Sn so that the small amount of fluorescence from
the collimator does not cause interferences in the remainder of the
energy region. The source collimator assembly is about 7/8" thick
by 2" in diameter. The high initial activity permits a useful
source life of several half-lives of I^{125}. The collimator can be
closed down to irradiate a spot about 3 mm in diameter at a distance
of 5 mm but is usually used with a spot size of 1 cm.

For light elements a lower energy isotope such as Fe^{55} or
bremsstrahlung from tritium-zirconium sources or a source-target
assembly of the type described by Giaque (14) would, of course,
improve the signal relative to the background. The use of X-ray
tubes for excitation offers the advantages of higher fluxes and/or
finer collimation and selection of maximum excitation energy but has
associated hazards when used for in situ analysis and loses the
geometric advantages of a compact, annular isotope source as well as
requiring a relatively large tube and power supply.

An overall view of the instrument is given in Figure 2 with the
detector positioned to study a 300 lbs. carved and painted "Madonna
and Child" (Florentine, early 16th century). Neglecting the knobs
and buttons of the oscilloscope which are not changed after the
initial set-up, the operating controls are few and simple. Full
scale energy ranges of about 10 - 100 KeV can be selected and calibrated for correct channel identification. We usually operate at
30 KeV full scale. Gain and position controls are used to provide
correct positioning relative to the camera graticule. Once calibrated, the settings are stable for weeks, with only the oscilloscope
requiring a lengthy warm-up. Unless it is to be unused for several
days, the instrument is left on continually except that the high
voltage bias is reduced to about 100 volts when not in use in order
to prevent damage in case of failure of the cryostat vacuum.

---

[*] Kevex Corporation, 898 Mahler Rd., Burlingame, Calif., U.S.A.
[**] Inax Instruments Ltd., P. O. Box 6044 Stn. J, Ottawa, Canada.

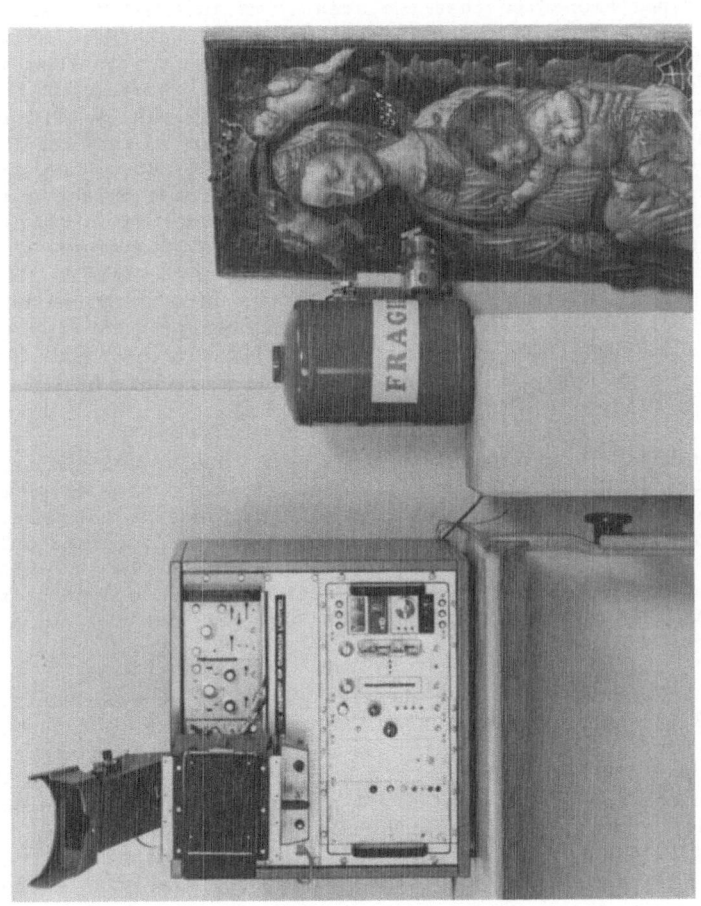

Figure 2. EDX spectrometer in use at the National Gallery of Canada.

Our operating procedures have, to date, been extremely simple.
The spectrum display is checked with a calibration standard of 80%
Cr_2O_3, 20% SrO and 1% Ag_2O (all percentages approximate). If the
positions are slightly off, it is frequently faster to check an
uncertain assignment by briefly presenting the standard for that
element than to set the controls precisely. Alternatively, one can
programme in the channel marker for that element. The object to be
studied is positioned so that the area(s) of interest is centrally
located relative to the source and detector at a distance of about
5 mm from the front of the collimator. The spectrum is then accumulated for a period which may be as short as 1 minute, for the
major elements in a painting or sculpture or as long as 10 minutes
for trace elements in a mineral or ceramic or for detection of the
small amounts of pigment in a drawing or watercolour. When a series
of similar objects or a number of areas on the same object are studied, a uniform accumulation time is used in order to facilitate
comparison. The scaler has typically been used as a simple clock,
or to obtain the backscatter integral, or, in the external clock
mode, to obtain the ratio of two peaks. The data is collected and
stored on Polaroid photographs of the oscilloscope screen. It is
planned to install the apparatus in a wheeled carriage with the
detector on a movable arm to increase mobility and convenience in
the study of large objects.

Nearly all of our work so far has been purely qualitative with
only element identification being required. The character generator
plus buffered scalers modules are recent additions to the instrument
capability and without them integral data could be acquired only one
region at a time. However, for reference purposes, Table I lists
the count rate in counts per sec (averaged over 10 seconds) for 22
elements in a 24 element standard with each element at 1% in a
silicate matrix.

The background count rate is also given for several energies.
Of other 2 elements in the standard Be is, of course, not seen, and
Au is obscured. All overlapping has not been taken into account so
the data are only approximate. The sample was presented as a
slightly compacted powder behind a 1 mil Mylar film. Note also
that no direct count rate comparison is possible between the
15 KeV and 30 KeV full scale portions of the table since each
channel is 117 eV wide in the former case and 234 eV wide in the
latter.

TABLE I

15 KeV full scale.

Line	K∝KeV	Counts/sec.	Overlap
V	4.95	--------	
Cr	5.41	4.8	
Bkgd	5.7	3.8	
Mn	5.90	4.8	
Fe(1)	6.40	7.9	
Bkgd	6.6	4.2	
Co	6.93	8.0	
Ni	7.48	9.0	
Cu	8.05	11.	
Bkgd	8.2	5.1	
Zn	8.64	14.	
Ga	9.25	16	
Ge	9.88	25	Au L∝
As	10.54	38	Pb L∝
Bkgd	10.9	17	Bi L∝
Se	11.22	30	Au Lβ

30 Kev full scale.

Line	KeV	Counts/sec.[3]	Overlap
Tl Lβ	12.21	29	
Pb Lβ	12.61	32	
Bi Lβ	13.02	27	
Tl lγ	14.29	5.6	
Pb Lγ	14.76	7.4	
Bi Lγ	15.24	7.6	
Mo L_α	17.48	140	
Bkgd	18.7	2.2	
Ag K_α	22.16	150	
Bkgd	22.8	16	
Cd K_α	23.17	160	
In K_α	24.21	43	
Sn K_α	25.27	90	Compton Scatter
Sb K_α	26.36	55	
*Te K_α [2]	27.47	33	Coherent Scatter

(1) Fe ≠ 1%, later addition
(2) Te = 0.916%
(3) Total count rate 2900 counts/sec.

RESULTS AND DISCUSSION

Most of the objects to be discussed fall into category of those which cannot be satisfactorily sampled. With others the intent was to provide a rapid preliminary survey so that more detailed analyses could be performed wherever anomalies might appear.

I. Drawings

In the past, examination and analysis of works of art on paper have been severely handicapped due to the impossibility of obtaining representative samples. Limitations have, thereby, been imposed on both the description and the scholarship of these works. Analysis of the inorganic constituents, e.g., pigments, grounds, impurities, of such objects would provide a valuable factual basis for scholarship and, in some case, conservation. Organic inks such as bistre, sepia and carbon black cannot, of course, be tested this way.

Figure 3.   Metal point drawing on prepared ground "Four Heads" by Gheerart David, full size.

To initiate this project, four metalpoint drawings and one ink
and wash drawing by Durer were selected. An example is shown full
size in Figure 3: "Four Heads" by Gheerart David. With the exception
of lead, the metals used for metalpoint drawings do not make a
satisfactory line on paper and the surface is first prepared by the
application of a white or coloured ground. The main question to be
answered at this point was whether or not there was enough material
in the lines to detect and determine the metalpoint used. Once this
was established, however, certain unexpected observations were made.
The data are given in Table II. The 4th column labelled "Drawing"
gives the elements found which are associated with the drawing.
Elements in clear areas are given in the fifth column.

TABLE II

Artist	Dates	Title	Drawing	Paper and Ground
Memling	1430-1494	Virgin & Child, etc.	Ag	Ca,(As),(Pb),(Sr)
David	1460-1523	Four Heads	Ag	Ca,As,(Pb),(Sr)
Raphael	1483-1520	Youthful Saint, etc.	Ag,Fe,Pb	Ca,(As),(Pb),(Sr)
Legros	1837-1911	Jeune Paysanne	--------	Ba,Sr,(Fe),(Pb),(Rb)
Durer	1471-1528	Nude Woman and Staff (ink and wash)	Fe,Zn,Pb,(As)	--------------------

( ) indicates minor constituent.

The first three drawings are on a calcium ground, probably
either carbonate or phosphate. Strontium is an impurity associated
with calcium and the calcium-strontium ratio may provide a method of
specifying the ground more precisely. The relatively modern Legros
drawing is on a barium sulfate ground - again strontium is a normal
impurity. The presence of arsenic in the first three may indicate
fungicidal treatment in the past or may be merely a common impurity.
Further experience should clarify this point as well as explain the
origin of the lead. The Memling and David drawings are as claimed,
i.e., silverpoint. The Legros drawing on the other hand shows no
metalpoint and may be simply a graphite pencil. The prepared paper

may have been used to provide a desired texture. The Raphael drawing
is very puzzling, Ag, Fe and Pb are present in the drawn lines.
Under UV illumination a wing appears which is not visible in normal
light. This shows Fe and Pb. It may be that there is an underlying drawing executed in iron-gall ink with a ground layer and
silverpoint on top.

The Durer ink and wash also showed unexpected results. I had
assumed that the ink was a carbon ink which would be undetectable
by EDX. However, Fe, Zn, Pb and As are associated with both the
lines and the wash. Further research may explain this unexpected
combination of elements.

In spite of the small amounts of material to be detected, these
objects are favourable for further research since it will be a
fairly simple matter to produce quantitative results because
interferring matrix effects are at a minimum. Another useful
possibility for these and other paper objects on which we have done
a small amount of work, is to use the Compton/Coherent scatter ratio
to specify papers which have added fillers and ground layers.

II. Watercolours and Pastels

From a preliminary study of watercolours about a year ago, two
points emerged: (a) for many pigments used as a light wash, sensitivity was being strained; (b) the absence of lead as lead white in
many watercolours was a real asset.

Table III summarizes the results obtained on four items from
the National Gallery of Canada War Collection.

A relatively large spot (about 2 cm diameter) was examined. A
new collimator design is expected to produce spot sizes of about
1 mm. This should permit much greater precision in defining the
area under examination. For the Comfort and Milne watercolours, the
elements observed are entirely consistent with the possible pigments
for the colours listed. For the Aldwinkle, it is not clear what
produces the green and for the Wessel, the blue is uncertain but may
be Prussian Blue - a very intense pigment which in a light wash may
be undetectable. Comfort's white pigment is evidently zinc oxide
and that of Wessel, lithopone. One does not, of course, observe any
organic pigment which may be incorporated in a watercolour.

The results are sufficiently encouraging that it is expected
that most problems involving pigment identification for watercolours,
etc., will be solvable by this simple and rapid technique.

TABLE III

Artist	Title	Colour	Elements Found
C. F. Comfort	Demolished Watch Tower	Green	Cu, Zn
		Brown	Mn, Fe, Zn
		Blue	Co, Zn
		Grey-Blue	Fe, Co, Zn
D. B. Milne (dry brush)	Ripon from Cathedral Tower	Green	Cr, Zn
		light Brown	Fe
		Orange-red	Hg
		Blue	Co (Mn)
		Reddish-brown	Fe (Hg
E. Aldwinkle	Mustangs in Readiness	Red	Ca, Fe
		Blue	Fe, Ca, Mn, (Zn), (As)
		Greenish	Fe, Mn
W. Wessel (pastel)	The Garglione River	Green	Cr, Pb, Zn, Ba, (Sr)
		Blue	Zn, Ba, (Fe), (Pb), (Sr)

III. Photographic Prints

A class of objects of increasing interest is that of photographic prints both old and modern. The print process is frequently
uncertain, as is the nature of any toning procedure. Toning is and
has been used to achieve specific effects and/or stabilize the print
against fading. Our curator of photography wished to compare certain prints in the collection versus documentary material regarding
processes used by various photographers.

The results are summarized in Table IV. The date given for many
of the older prints is only approximate. In all cases it proved
possible to determine the nature of the image forming material iron,
silver or platinum and the toning process if any, e.g., gold,
selenium. Modern glossy prints are on paper loaded with barium sulfate as evidenced by the barium and strontium seen in these cases.
For several prints, separate determinations showed the elements in
the print paper and in the mounting board. An interesting and
possibly valuable sidelight, is the occurrence of many elements in
trace quantities. It may prove possible to "fingerprint" prints
from different workshops. For these as for other works of art on
paper, the signal to background ratio is improved when the work
can be studied free of its mounting board since backscatter is
greatly reduced.

IV. Paintings

The following is taken nearly verbatim from reports written
following brief surveys of two paintings undertaken at the request
of gallery personnel involved in conservation and examination.
These data are given here to illustrate the utility of rapid
examination by energy dispersive X-ray spectroscopy.

<u>Stomer, "The Betrayal of Christ", Dutch, 17th century.</u>
Certain passages were thought to be later restorations and we wished
to check these for possible pigment anomalies. Twelve areas were
analyzed (requiring approximately 3 minutes for each) with the
following general results:
  1) the red areas show Hg, presumably vermilion
  2) yellow and browns show Fe – earth colours
  3) white shows Pb – white lead
  4) Pb & Sb are observed throughout – evidently lead stibnite
     (Naples Yellow) was used to impart a yellowish warm tone.
  5) Ca & Sr in lightly pigmented areas – a calcium ground is
     indicated.
  6) the rear of the canvas shows Cu and Zn – possibly a prior
     preservative treatment.

The evidence for Naples Yellow which was confirmed by X-ray

TABLE IV

Photographer and Subject	Date	Image	Print	Mount	Trace Elements
Evans, "Durham Cathedral"	1900	Pt (Hg?)	nil	Fe, Zn	Mn, Cu, As
Evans, "Ely Cathedral"	1900	Fe		Sr	
Abbot, "Eugene Atget"	1930	Ag, Fe		Fe	
Starbuck, untitled	1970	Ag	Ba, Sr	--	
Notman (JWB print)	1970	Ag, Au	Ba, Sr	--	
Notman, "Parliament"	1866	Ag		--	Fe, Zn, As, Pb, Sr
Du Camp, "Temple Hypestre"	1852	Ag, Au			
Negre, "Palais des papes"	1852	Ag, Au			
Lyons, "Angelica"	1959	Ag, Au	Ba, Sr		
Jones, "Homage à Atget"	1970	Ag, Se	Ba, Sr		
Barrow, "Barbie"	1970	Ag, Au	Ba, Sr		
Piot, "Acropolis"	1852	Ag	Ca, Sr	--	Fe, Cu, Zn, As
Hill, "Rev. Cairns"	1843	Ag, Fe, Zn	--	--	Cu, As, Rb, Sr
Hill, "Two Fishwives"	1843	Ag	--	Fe, Ca, Sr	Cu, As, Rb, Sr
Hill, "Mrs. Bonnar"	1843	Ag (Pt?)	Fe	Fe	As, Sr
Hill, "#6"	1843	Ag (Pt?)	Fe	Fe	As, Sr
Hill, "Byrne"	1843	Ag	Fe	Fe	Cu, Zn, As, Rb, Sr

diffration was of some interest since its period of greatest use was
in the subsequent century. However, all of the pigments indicated
are consistent with period of the painting.

   Anon., "Portrait of an Ecclesiastic", Flemish School, 16th
century. The study was undertaken to determine (a) whether the left
hand strip of board was a later addition; (b) the elements associated with the chevron design revealed by X-ray radiography. The
results from 19 areas, while inconclusive, are indicative.
   1) Ca & Sr indicate the use of a calcium ground on the panel
   2) Pb, i.e., white lead, occurs throughout
   3) green areas show Cu, indicating probably malachite
   4) red areas show Hg, i.e., vermilion
   5) black areas show strong Cu lines possibly indicating either
      a dark overpaint or a colour change, green to black, of
      copper resinate
   6) Zn shows at the extreme edges on both the right and left
      sides
   7) Fe occurs sporadically indicating earth colours

   The chevron design was studied extensively. Ca, Fe, Cu, Zn
and Pb were observed. Variations in the ratio of Fe to Cu were
observed from band to band in the chevron.

   The occurrence of Zn, a 19th century introduction, in the chevron and in the margins as well as that of Cu in the dark areas
suggests substantial alteration from the original.

   Since paintings are multiple layer structures and, since EDX
will detect the elements in underlying layers (10), interpretation
of the results has to be done with great caution and quantitative
in situ analysis is probably impossible by this technique. However,
even with these limitations, extremely valuable results can be
obtained.

                              V.   Jade

   A small suite of jade objects belonging to the National Gallery
of Canada was brought to the attention of the laboratory. If applicable EDX would provide an excellent method of analysis for this
type of object since sampling of small, very valuable pieces is to be
avoided if possible and the sheer number of objects in existence
would preclude lengthy traditional techniques. The term "jade" is
applied to the amphibole, nephrite (a calcium magnesium silicate) and
the pyroxene, jadeite (a sodium aluminum silicate). It is also applied to just about any rock or glass object of vaguely oriental
appearance. It would therefore be useful to be able to differentiate
true jade from other materials and possibly identify the origin.
Since the major elements (except calcium) in both materials would

not be detected using I^{125} excitation, we were interested in the minor and trace constituents. The results for the suite of National Gallery objects are given in Table V. Counting times were typically 6-8 min. The last three objects listed are not considered to be jade.

TABLE   V

Object	Elements[1]
Light green dish	Ca, Fe
Green and brown birds	Fe, Zr
Green vase	Fe, Cu, Cr, (Ni)
Green and brown tray	Ca, Fe, (Zn), (Pb)
Red, brown and green dragon	Fe, Zr, Sr, (Mo)
Black cup	Fe, Ni
Green belt clasps	Fe, Ca, Ni, Sr
Dark green box	Fe
White gilded goddess	Ca, Fe, Rb
Blue seated figure	Zr
Pale purple vase	(Fe)
White and blue vase	Zr

1 - listed in order of decreasing concentration (estimated)

( ) - indicates very minor constituent.

Two encouraging conclusions were reached from this very cursory study. Firstly, there was considerable variation in composition between different objects. Secondly, the composition of a particular object was relatively uniform from one area to another, even for variegated pieces.

In order to extend our experience, a set of mineralogical samples were borrowed from the National Museum of Natural History and from the Geological Survey of Canada. The results are given in Table VI. The basic amphibole mineral is Tremolite; nephrite is a compact variety and actinolite is a variety with greater than 2% Fe. Only one jadeite sample was studied.

TABLE VI

Sample	Location	Elements Observed[1]
1. Tremolite	Canaan, Conn.	Ca,Fe,(Ni)
2. Tremolite	Balmat, N.Y.	Ca,Sr,Mn,Fe
3. Nephrite(wht.xtaline)	?	Ca,Sr
4. Tremolite	Hermon, Ont.	Ca,Fe,Sr
5. Tremolite	Bruceton, Ont.	Ca,Fe,Sr
6. "Chrome" Tremolite	Cardiff Twp., Ont.	Ca,Fe,Cr,Cu,Zn,Sr
7. Actinolite	Banner Lake, Ont.	Fe,Ca,Zr
8. Actinolite in schist	King Mtn., N.C.	Fe,Ni,Ca,Cr
9. Tremolite(var.Hexagonite)	Balmat, N.Y.	Ca,Mn,Fe,Sr
10. Nephrite(var.Actinolite)	B.C.	Ca,Fe,Ni
11. Jade	Wyoming	Ca,Fe,Ni
12. Nephrite	Quiet Lake, Yukon	Ca,Fe,Ni,Cr,(Zn)
13. Nephrite	Lewis River, Yukon	Fe,Sr,Cu,Rb,Y,Mo
14. Nephrite (black)	Lytton Lake, B.C.	Fe,Ni,Cr
15. Nephrite(translucent)	B.C.	Fe,Ca,Cr,Ni,Sr
16. Nephrite	B.C.	Fe,Cu,Ni,Sr
17. Nephrite	Salukwe, S.Rhod.	Fe,Ca,Ni
18. Nephrite	New Zealand	Fe,Ca,Ni
19. Jadeite	Kazahkistan	Fe,Zr,Sr,(Cu)

[1]Samples 4, 6, 13 may contain Zr (interference Sr K_β)
Co may be of frequent occurrence (interference Fe K_β)
- quantitative work could clarify these points.
Elements are listed in decreasing concentration as estimated.
( ) indicate very low signal over background.

    Following this work, our session chairman, Maurice Salmon, brought to my attention a report of a detailed quantitative study of a set of nephrite and jadeite samples (15) and we plan to continue our own studies in a quantitative way, using the extensive jade collection of the Royal Ontario Museum.

CONCLUSIONS AND FUTURE WORK

The technique of EDX offers a valuable addition to the difficult discipline of analysis of museum objects. Clearly, interpretation must be carefully done with due regard for the nature and provenance of each object. Certain classes of materials, e.g., drawings, watercolours, stone and metal objects, are amenable to quantitative analysis. For other, the technique should be regarded as a survey or documentation procedure.

A very promising area in which we have done work, to be reported elsewhere, is that of analysis of ceramics. Both authentication and scholarship of these objects have, until now, been severely handicapped. Utilization of this type of instrumentation in the field of analysis and sorting can be expected to increase, particularly, given improvement in excitation and cryostat design.

ACKNOWLEDGEMENTS

The author would like to express his gratitude to the Director and staff of the National Conservation Research Laboratory for their advice and encouragement and to the staff of the National Gallery for the continuing interest shown by them.

REFERENCES

1. Salmon, M.E., "An X-ray Fluorescence Method for Micro Samples Applied to Chinese Bronze" in Application of Scientific Methods in the Analysis of Works of Art, Boston Museum of Fine Arts, June 1970 (in press).

2. Lechtman, H., "The Gilding of Metals in Pre-Columbian Peru", loc. cit.

3. Stolow, N., Hanlan, J.F., and Boyer, R., "Element Distribution in Cross-Sections of Paintings Studied by the X-ray Macroprobe", Studies in Conservation 14, 139-151 (1969).

4. Gordus, A.A., "Non-Destructive Analysis of Metals by Neutron Activation" in Application of Scientific Methods to the Analysis of Works of Art, Boston Museum of Fine Arts, June 1970 (in press)

5. Cairns, T., "Recent Interfaces between Mass Spectrometry and Art" this symposium.

6. Young, W.J., "The Application of the Laser, etc.", this symposium.

7. Frankel, R., "Detection of Art Forgeries by X-ray Fluorescence Spectroscopy", <u>Isotopes and Radiation Technology</u>, <u>8</u>, 65-88 (1970).

8. Hanson, V.F., "The Curator's Dream Instrument", <u>Application of Science in the Examination of Works of Art</u>, Museum of Fine Arts, Boston, June 1970 (in press).

9. Kennedy, P.L., Tolmie, R.W., Hanlan, J.F., and Grant, J.MacG., "X-Ray Fluorescence Spectra of Artists Pigments", <u>AECL Report</u>, <u>CPSR-266</u> (March 1970).

10. Hanlan, J.F., Stolow, N., Grant, J.MacG., and Tolmie, R.W., "Application of Non-Dispersive X-Ray Fluorescence Analysis to the Study of Works of Art", <u>Bull. IIC-AG</u>, <u>10</u>, No. 2, 25-40 (1970).

11. Hanlan, J.F., "The EDX Spectrometer in Museum Use", <u>Bull. IIC-AG</u>, <u>11</u>, No. 2, 85-90 (1971).

12. Frierman, J.D., Bowman, H.R., Perlman, I., and York, C.M., "X-Ray Fluorescence Spectrography: Use in Field Archaeology", <u>Science</u>, <u>164</u>, 588 (1969).

13. ASTM, "Energy Dispersion X-Ray Analysis", American Society for Testing Materials, Philadelphia, 1971.

14. Giaque, R.D., "A Radioisotope Source-Target Assembly for X-Ray Spectrometry", <u>Anal. Chem.</u>, <u>40</u>, 2075-7 (1968).

15. Ronzio, R.D. and Salmon, Merlyn, "Trace Elements in Jade", <u>Gems and Minerals</u>, (Feb., 1970) 25-26, 45.

CHARACTERIZATION OF MEDIEVAL WINDOW GLASS BY NEUTRON ACTIVATION ANALYSIS

J. S. Olin, B. A. Thompson and E. V. Sayre

Smithsonian Institution and National Bureau of Standards,

Wash., D.C., Brookhaven National Laboratory, Upton, N. Y.

Recent research by a growing number of workers is demonstrating that neutron activation analysis can provide valuable assistance in establishing the provenance of potsherds; coins and other metal artifacts; obsidian; amber; and paintings.(1) One of the authors of this paper has been involved for some years in analytical studies of ancient glass which have included the earliest application of neutron activation with high resolution gamma ray spectroscopy, i.e., Ge(Li) detector counting, to the investigation of archeological artifacts. The study described in this paper was undertaken to determine whether this method could be of similar usefulness in the characterization of medieval window glass. Areas of practical interest include questions of which panes are original and which replacements and the precise provenance of specific windows.

Earlier studies of ancient and medieval glass compositions are too numerous to list here in their entirety. The publications of Wilhelm Geilmann, et al (3) and the more recent progress report by Robert H. Brill (4) are important references to the analysis of medieval window glass. Results of analyses are given in both. At present very little specific information is known of the exact sources of the materials used in the manufacture of medieval window glass and of the exact recipes used for its formulation. One would hope that extensive studies of the kind discussed here would help in determining this information.

Forty-five samples of glass were analyzed in a sequence determined by the information gained as the investigation proceeded and by the availability of samples. Two fragments were sampled at the outset to study the homogeneity of the glass.

33

Thirty-two fragments of scattered provenance and eleven fragments from one panel were next sampled and analyzed.

ANALYTICAL PROCEDURE

    The first experiments were carried out to gain information
about the composition and homogeneity of medieval glass and also
to find out whether useful data could be obtained from the very
small samples (10–100 mg) which could be expected to be made
available from actual windows. The samples used were fragments
of medieval glass from The Cloisters of the Metropolitan Museum of
Art on loan to the Smithsonian Instutition. These fragments will
become part of the collection of the Corning Museum of Glass and
we are grateful to both museums for allowing us to sample the fragments. The Cloisters accession numbers of the fragments are given
when data for one of these fragments is presented.

    Two sets of analyses were carried out, one at the National
Bureau of Standards and one at Brookhaven National Laboratory.
The general approach used at the National Bureau of Standards
for the first analyses was similar to that of Coleman and Wood (5)
in their work with plate glass fragments. First, a small sample
(10 mg) of each glass was irradiated for 2 seconds in the National
Bureau of Standards Reactor at a thermal neutron flux of
6×10^{13} n cm^{-2} sec^{-1}. Analysis of the spectra of gamma-rays
emitted from the activated samples, recorded at several times
after activation, permitted determination of the relatively major
constituents sodium, potassium, calcium, aluminum and manganese.
Vanadium and indium were also observed if present in sufficient
concentrations. A larger sample (50–100 mg) was then irradiated
for 15 min–1 hour and allowed to decay for 1–2 weeks after which
concentrations of minor and trace constituents were determined
from gamma-ray spectra.

    The samples to be irradiated were heat-sealed in medical-grade
polyethylene tubing and a small (1–2 mg) copper foil flux monitor
was taped to the outside of the polyethylene to permit correction
for small differences in neutron flux or irradiation time from
sample to sample. After irradiation the samples were rinsed with
acetone and with 1:1 HNO_3 before counting to remove surface contamination. A 47-cc Ge(Li) detector and a 2048-channel analyzer were
used for all gamma-ray spectroscopy. In all cases concentration
levels were determined by comparison with standards which were
irradiated under the same conditions as the samples and made up to
the same geometry as the samples for counting. Pure Na_2CO_3 was
used for a sodium standard and both $K_2Cr_2O_7$ and K_2CO_3 were used
as potassium standards. For most of the other elements determined
the standards were either the pure metal or oxide. Because of the
high manganese levels in most of the glasses, the analyzer dead
time during the measurement of aluminum was often as high as
30–40%. Although no gain shift or loss of resolution could be

observed, the high dead time created a problem in the choice of a
suitable standard for this element. The problem was solved by
the use of an NBS Standard Reference Material steel, which has
a high manganese concentration, as a standard. SRM 101e with
1.77% manganese was used as the manganese standard. SRM 344,
which has 0.57% manganese and 1.16% aluminum, was used as an
aluminum standard. Using samples of this, it was fairly easy
to get a standard with about the same ^{56}Mn activity and thus
about the same dead time as the sample. This eliminated the need
for decay corrections on these short-lived activities, since all
were counted for the same period and at the same clock time after
irradiation. To remove the possibility of any small differences
in the size and shape of the glass fragments, these were dissolved
in hot $HF-HC_1O_4$ and made up to a standard 40-cc volume in 4-oz
polyethylene bottles for counting.

    Precision in the analysis of a number of replicate samples of
medieval glasses is shown in Table I and was obtained using the
above procedure. For those elements, such as sodium and manganese,
having good sensitivity and no interferences, the precision of
measurement was often as good as $\pm$ 2%, while for elements such as
potassium and iron with poor sensitivity and/or interfering
gamma-rays, the precision was usually much poorer. None of the
samples analyzed showed any evidence of inhomogeneity within the
limits noted above, nor could any differences be detected with
different methods of cleaning the samples. Table I shows the
results obtained with glass of medieval origin and a glass
manufactured in the eighteenth century. The standard deviations
listed in this table are not the more usually reported standard
deviations of the calculated values but are standard deviations
of the groups of replicate measurements. The significance of
these results in that one can therefore draw conclusions about an
entire fragment of glass when only a 10 mg sample of that glass
is used for analysis.

    In the next phase of the investigation, thirty-two frag-
ments of the medieval glass from The Cloisters were activated
for both short-lived and long-lived nuclides using the reactor
facilities at Brookhaven National Laboratory. The samples
for activation were crushed using an agate mortar and pestle;
dried for 48 hours at 200° C; and weighed. Samples of approximately 20 milligrams were weighed and heat-sealed in polyethylene
tubing for a twenty-second activation at a flux of 1.5×10^{14} n
cm^{-2} sec^{-1} in the VII facility of the High Flux Beam Reactor.
The ^{56}Mn was counted after about four hours and the ^{24}Na and ^{42}K
were counted after twenty-four hours. A second group of samples
of 50 milligrams was activated for 16 hours at a flux of
1.5×10^{14} n cm^{-2} sec^{-1}.* After approximately two weeks the

*An activation of 7 hours at a flux of 2.8×10^{14} n cm^{-2} sec^{-1} was
 sometimes used.

Table I

Homogeneity of Medieval and 18th Century Glasses

(Average concentrations in two representative specimens with
the standard deviations of the groups of replicate measurements)

Element	Concentration, %	
	Accession Number 30.73.22	Accession Number 23.229.3-7
	Blue Glass, XIII-XIV Cent. (?)	Clear "Bullseye" Paris, 1775
Na	0.95 ± 0.04[a]	4.08 ± 0.05[c]
K	18.9 ± 2.7[a]	3.51 ± 0.62[c]
La	0.0040 ± 0.00008[b]	0.0015 ± 0.0001[c]
Co	0.114 ± 0.001[b]	0.00066 ± 0.00006[c]
Fe	0.93 ± 0.25[b]	0.57 ± 0.06[c]
Cu	0.38 ± 0.03[a]	
Sb	0.0058 ± 0.0003[b]	
Mn	1.08	0.40
Al	0.82	1.66

[a]Std. deviation; n=4

[b]n=5

[c]n=8

Also observed in the blue glass: In, Ca, Sc, Zn, Rb

Also observed in the "bullseye" glass: Ca, Sc, Zn, Sm

samples were poured out of the aluminum vials into glass vials for counting. The aluminum vials were washed out into glass vials with ethyl alcohol for quantitative transfer. These samples were counted for 80 to 100 minutes using an automatic sample changer. Along with each group of glass samples, four different standard rock samples of similar weight were activated and counted. Since the rock standards contained in measurable amounts all of the elements determined in the glasses, they served as element by element monitors of the neutron flux densities and counting geometries encountered by the samples.

A 35 cc Ge(Li) detector and two 1600-channel SCIPP analyzers, which could be connected in series to produce a 3200 channel spectra, were used for the gamma ray spectroscopy. The data were collected on magnetic tape and fed into a CDC 6600 computer, where a curve fitting program BRUTAL (6) was applied. The output of BRUTAL gave numerical values for the position and integrated intensity of each peak, correct for background.

A list of the properties of the radioisotopes measured at both laboratories are presented in Table II.

The computed value for each peak was corrected for decay of the radioisotope concerned and converted to specific activity. The specific activities were compared with those of the U. S. Geological Survey rocks G-2, GSP-1, AGV-1 and BCR-1 and the concentrations of the oxides of the elements present were calculated from these ratios.

DATA

The general information known about the procedures and techniques used in making medieval glass provided a basis for the selection of important elements and useful comparisons and correlations of the various experimental results in light of the original objectives. Particularly valuable sources of such information are the excellent series of papers on ancient glass and glass making by Turner (7) and the medieval treatise of Theophilus, "On Divers Arts" (8). In the Middle Ages glass was made by mixing an alkali and sand with additives to produce a desired color. The alkali, which might have been a plant ash or a naturally occuring inorganic alkaline salt, can be characterized by being rich in either sodium or potassium. It was not until the 18th century that the differences between potassium and sodium alkalis began to be recognized. Medieval glass is most often characterized by a high potassium content which results from the use of vegetable ash, particularly beechwood ash for glass making. Turner points out that the alkali content of such ash varies not only with the type of ash but also from place to place and from time to time depending

Table II

Characteristics of the Radioisotopes
Used for Measurement

Constituent	Radioisotopes measured	Half-live	Gamma-rays measured (ke V)
Aluminum	^{28}Al	2.3m	1780
Antimony	^{124}Sb	60d	1692
Barium	^{131}Ba	12d	496
Cerium	^{141}Ce	32.5d	145.4
Cesium	^{134}Cs	2.04y	801.8
Chromium	^{51}Cr	27.8d	320.1
Cobalt	^{60}Co	5.26y	1173.2 + 1332.5 (a)
Copper	^{64}Cu	0.533d	511
Europium	^{152}Eu	12y	1408
Hafnium	^{181}Hf	42.5d	482
Iron	^{59}Fe	45.6	1099.3 + 1291.5 (a)
Manganese	^{56}Mn	2.576h	846.8
Potassium	^{42}K	12.45h	1524
Rubidium	^{86}Rb	18.66d	1078
Scandium	^{46}Sc	83.9d	889.4 + 1120.6 (a)
Sodium	^{24}Na	14.96h	1368.53
Tantalum	^{182}Ta	115.1d	1189 + 1222 + 1231 (b)
Thorium	^{233}Pa (c)	27d	312

Table II
(Cont.)

Characteristics of radioisotopes

(a) The integrated areas of the two peaks were summed to improve
     the statistics.

(b) The integrated areas of the three peaks were summed to improve
     the statistics.

(c) Formed by beta decay of the short-lived ^{233}Th, which is
     generated by thermal neutron capture.

on soil content, etc. He also points out that at various times
ash from seacoast plants with high sodium was popular. The
highest concentration of K_2O in any of the samples we have
analyzed was 31.3%. The lowest concentration of K_2O we measured
was 2.4%. The Na_2O concentration ranged from 0.2 to 19.4%;
however, only four of the forty-three fragments had a concentration of Na_2O greater than 5.7%.

The data for the thirty-two samples of glass of scattered
provenance are given in Tables III through V. Eleven samples
from one panel have also been obtained and the data for these
is given in Tables VII and VIII. Hence, forty-three samples
from museum collections in the United States have been analyzed
to date. One of the samples considered for analysis contained
1.37% antimony and could not be analyzed for minor and trace
constituents without prior extraction of the antimony and this
sample was set aside for future work. Each sample is identified
by the museum and accession number of the panel or fragment from
which the sample was taken. It is evident from the data in
these tables that the composition of medieval window glass varied
significantly. Various procedures were used to select samples
with related compositions from among those analyzed.

The data for all the samples were plotted using a program,
Glasplot, written by one of the authors, E. V. S., for the
CDC 6600 at Brookhaven's Computer Center. The logarithms of the
concentrations were plotted in a systematic fashion so that the
complete composition pattern for each sample was displayed.
These plots were overlaid over a light box and plots having
similar patterns were selected. The groups which resulted were
tested using a second computer program, Glastat. This program
generates the geometric mean; the group and mean standard
deviation; the mean plus and minus one and two standard devia-

Fig. 1. Fragments of French glass from The Cloisters, The
Metropolitan Museum of Art. (Data is shown on Table III.)

Fig. 2. Fragments of French and/or English glass from The Cloisters,
The Metropolitan Museum of Art. (Data is shown on Table IV.)

tions of the group; the mean plus and minus one and two standard
deviations of the mean; and the percents the group and mean
standard deviations are of the mean. (6)

From the thirty-two fragments of scattered provenance, two
compositional groups emerged. These are the data in Tables III
and IV and the fragments are shown in Figures 1 and 2 respectively. It is apparent that the grisaille fragments of the
first matching group, shown in Figure 1, are closely related in
style as they are in composition. To assure independent
evaluation on the basis of composition alone, however, compositional comparison was made prior to and without reference
to stylistic comparison. Hence, the fragments in Figure 1 were
not identified as being stylistically similar until after they
were matched together on the basis of their composition. At
the present time it is still too early to say what correlation
in composition can be expected within a group of fragments from
different pieces of glass made in the same workshop. But, this
might well be what the data in Tables III and IV represent.

Table V lists those fragments of the thirty-two analyzed
for sixteen elements which did not form groups of more than two
using the procedures described earlier. The ranges of concentrations for the oxides of the elements analyzed on the samples
studied to date are as follows: Na_2O, 0.30 - 19.4%; K_2O, 2.41 -
31.3%; BaO, 0.04 - 2.6%; MnO, 0.36 - 1.9%; Fe_2O_3, 0.34 - 1.2%;
Rb_2O, 12.3 - 834 ppm; Cs_2O, 0.10 - 27 ppm; Sc_2O_3, 0.92 - 4.2
ppm; CeO_2, 13.2 - 84.8 ppm; Eu_2O_3, 0.16 - 0.9 ppm; HfO_2, 0.97 -
5.5 ppm; ThO_2, 0.93 - 4.3 ppm; Ta_2O_5, not detected - 0.6 ppm;
Cr_2O_3, 8.9 - 34.6 ppm*; CoO, 5.7 - 166 ppm; and Sb_2O_3, 0.84 -
20.8 ppm. These wide ranges demonstrate clearly that diversity
in composition is to be expected in unrelated specimens of
medieval glass. Therefore, it would seem unlikely that compositional correlations involving a large number of constituent
elements, such as exist in the two compositional groups previously described, would arise fortuitously. One is justified in
seeking related origins for such matching specimens.

The mean concentration values and group standard deviation
ranges for components of the two groups of glass with distinct
compositions (Tables III and IV) have been compared in Figures
3 and 4. The logarithms of concentrations have been plotted
because it has been found that a Gaussian curve is obtained
when, for a large number of related archeological samples, the

*On Table V, the sample with accession number 23.229.5-10 will
  be reanalyzed for Cr_2O_3 in view of the value of 649 ppm obtained
  and our experience with fragment number 2 in the grisaille
  panel.

Table III

First Matching Group of Grisaille Fragments

Identified Using Accession Numbers of the Cloisters Museum

(See Figure 1)

Accession Number	Na_2O	K_2O	BaO	MnO	Fe_2O_3	Rb_2O	Cs_2O	Sc_2O_3	CeO_2	Eu_2O_3	HfO_2	ThO_2	Ta_2O_5	Cr_2O_3	CoO	Sb_2O_3
	Percent					Parts Per Million										
23.229.4-8	0.33	18.2	2.54	1.21	0.44	250	2.0	2.29	37	0.179	2.8	3.4	0.36	16.5	9.5	1.35
23.229.4-9	0.30	18.2	2.19	1.16	0.31	260	2.2	1.73	26	0.214	2.3	3.4	0.36	13.5	5.7	1.61
23.229.4-11	0.30	19.4	1.42	1.22	0.41	260	2.6	2.13	37	0.207	2.7	4.2	0.38	16.7	8.7	1.56
30.73.208	0.30	17.9	2.69	1.32	0.43	290	2.4	2.28	33	0.450	2.9	3.7	0.38	22.7	11.0	0.92
Mean	0.31	18.4	2.15	1.23	0.39	265	2.3	2.09	33	0.244	2.7	3.7	0.37	17.1	8.5	1.33
No. Averaged	4	4	4	4	4	4	4	4	4	4	4	4	4	4	4	4
Group Std.Dev.%	4.9	3.6	33.5	5.5	17.6	6.6	12.0	14.1	18.1	51.3	10.8	10.5	3.2	23.9	33.0	29.3

Table IV

Second Matching Group of Grisaille Fragments

Identified Using Accession Numbers of the Cloisters Museum

(See Figure 2)

Accession Number	Na_2O	K_2O	BaO	MnO	Fe_2O_3	Rb_2O	Cs_2O	Sc_2O_3	CeO_2	Eu_2O_3	HfO_2	ThO_2	Ta_2O_5	Cr_2O_3	CoO	Sb_2O_3
	Percent					Parts Per Million										
23.229.5-4	2.5	14.1	0.135	1.01	0.51	220	0.62	1.74	61	0.76	3.2	2.1	0.31	14.5	61	5.72
30.73.206	2.7	14.5	0.147	1.23	0.52	240	0.68	1.70	82	0.45	2.8	2.0	0.30	16.4	8	2.08
30.73.207	3.1	14.9	0.127	1.08	0.51	230	0.95	1.75	79	0.51	3.1	2.2	0.31	16.8	33	1.08
30.73.209	2.6	15.3	0.141	1.05	0.51	240	0.90	1.62	71	0.44	2.7	2.4	0.31	13.5	48	3.08
30.73.213	2.6	13.6	0.146	1.03	0.50	250	0.86	1.67	66	0.40	2.7	2.3	0.25	16.4	48	3.28
30.73.215	4.0	12.5	0.138	1.00	0.53	250	0.77	2.00	58	0.52	2.3	2.2	0.33	18.5	40	4.84
Mean	2.9	14.1	0.139	1.06	0.51	238	0.79	1.74	69	0.50	2.8	2.2	0.30	15.9	42	2.93
No. Averaged	6	6	6	6	6	6	6	6	6	6	6	6	6	6	6	6
Group Std. Dev. %	19.5	7.5	5.6	7.9	2.0	5.1	18.2	7.6	14.9	25.3	12.5	6.7	10.0	11.9	32.7	83.1

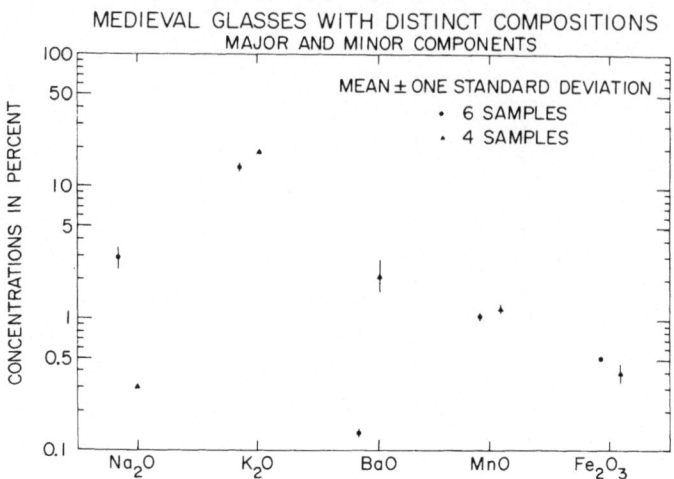

Fig. 3. Plot of data for four fragments shown in Fig. 1 and six
fragments shown in Fig. 2. (Major and minor components.)

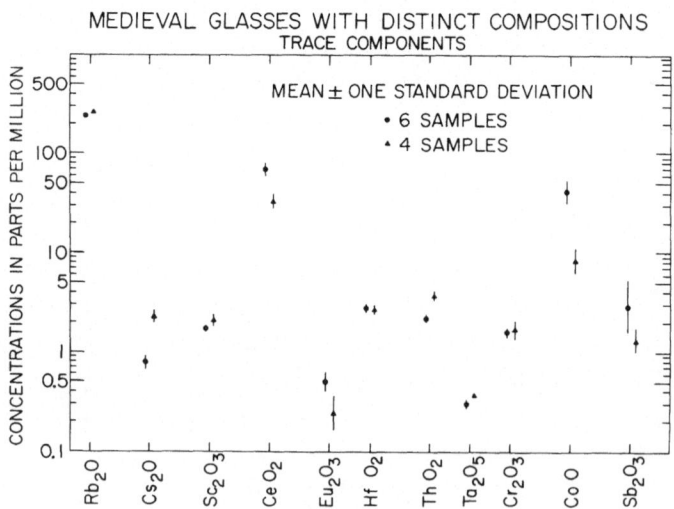

Fig. 4. Plot of data for four fragments shown in Fig. 1 and
six fragments shown in Fig. 2. (Trace components.)

frequency of occurence within ranges of concentrations of
individual chemical elements is plotted against the logarithms
of these concentrations. It is evident from these figures that
for these two groups the differences in concentrations are much
greater for some elements than for others. Na_2O and BaO have
distinctly different ranges; Cs_2O, CeO_2 and CoO have smaller
but definitely different ranges. It is clear, however, from the
ranges of concentrations given earlier for all the specimens
analyzed, that with other groups different sets of elements may
show the distinctions. Comparisons between the concentrations
of Na_2O, K_2O, Cs_2O and Rb_2O on the forty-three samples analyzed
to date for these four elements show a high degree of positive
correlation between the concentrations of Rb_2O and K_2O and Rb_2O
and Cs_2O. There is a less positive correlation between K_2O and
Cs_2O and negative correlation between Na_2O and K_2O. It is
possible that these correlations may become useful in interpreting
the compositions of the glass when a larger number of samples have
been analyzed. The patterns of relationships of concentrations of
potassium and rubidium in geological materials have been studied.
(9) In the case of medieval glass the ratio of potassium to
rubidium will be determined by the concentration of each in the
ash and the sand and the proportions of ash and sand used to
prepare the glass. It is interesting to note in Tables VII and

Fig. 5. Stained glass panel with grisaille interlace; French,
circa 1280 from The Art Museum, Princeton University. Accession
number 43-65. (Data is shown on Tables VII and VIII.)

Table V

Miscellaneous Non-matching Group of Grisaille Fragments

Identified Using Accession Numbers of the Cloisters Museum

Accession Number	Na_2O	K_2O	BaO	MnO	Fe_2O_3	Rb_2O	Cs_2O	Sc_2O_3	CeO_2	Eu_2O_3	HfO_2	ThO_2	Ta_2O_5	Cr_2O_3	CoO	Sb_2O_3
	Percent					Parts Per Million										
23.229.2-2	0.76	12.4	0.111	0.79	0.53	140	0.58	1.81	76	0.93	3.65	3.60	0.43	19.0	17.3	7.02
23.229.2-3	0.30	22.3	0.492	1.83	0.35	460	2.45	1.22	16	0.24	1.87	1.81	0.29	8.9	13.4	1.87
23.229.2-10	1.29	16.7	0.181	1.01	0.81	250	0.17	3.34	64	0.55	5.52	3.35	0.44	27.0	14.9	2.15
23.229.2-14	0.78	16.0	0.082	0.36	0.91	110	0.48	1.55	41	0.26	2.73	1.85	0.30	25.2	6.4	14.50
23.229.4-1	0.21	20.8	0.182	1.01	0.49	530	3.40	1.66	13	0.23	1.98	1.96	0.24	9.1	23.2	4.10
23.229.4-6	0.61	12.1	0.090	0.61	0.69	100	0.69	2.89	55	0.28	4.16	3.09	0.57	31.2	12.7	2.08
23.229.4-7	1.73	12.0	0.230	1.09	0.59	360	0.97	1.69	59	0.57	2.88	2.93	0.39	16.1	17.7	4.32
23.229.4-10	0.68	12.0	0.052	1.26	0.41	200	1.09	0.92	27	0.47	1.41	0.93	0.25	15.7	37.7	3.59
23.229.5-2	5.66	4.5	0.186	1.10	0.83	40	0.45	2.72	27	0.17	2.59	2.74	0.46	20.3	9.3	1.30
23.229.5-5	19.40	2.4	0.042	0.44	1.19	12	0.10	4.22	29	0.38	3.28	4.01	0.53	23.1	10.2	3.80
23.229.5-6	1.95	16.6	0.082	0.74	0.58	360	1.41	1.50	54	0.15	1.36	1.53	0.16	12.6	11.7	1.14

Table V
(CONT.)

Accession Number	Percent					Parts Per Million										
	Na_2O	K_2O	BaO	MnO	Fe_2O_3	Rb_2O	Cs_2O	Sc_2O_3	CeO_2	Eu_2O_3	HfO_2	ThO_2	Ta_2O_5	Cr_2O_3	CoO	Sb_2O_3
23.229.5-7	2.02	18.1	0.220	1.93	0.48	210	2.24	1.09	61	0.31	1.08	1.08	*N.D.	9.9	93.1	0.99
23.229.5-8	3.96	10.6	0.184	1.77	0.52	290	2.34	1.63	50	0.24	2.88	2.01	0.29	15.0	41.8	1.82
23.229.5-9	0.88	11.7	0.036	0.74	0.63	81	0.34	1.44	22	0.16	1.19	1.01	0.15	12.0	5.8	3.07
23.229.5-10	13.20	2.4	0.241	0.65	0.72	39	0.17	1.09	14	0.19	0.97	1.06	*N.D.	649.0	166.0	20.80
30.73.210	2.83	11.4	0.123	0.37	0.82	187	1.10	3.11	61	0.62	4.69	3.89	0.52	34.6	64.8	3.49
30.73.211	15.00	4.8	0.046	0.51	1.00	42	0.39	3.89	47	0.45	4.07	4.22	0.47	25.7	8.8	1.24
30.73.212	3.04	12.6	0.179	1.25	0.46	450	2.35	1.34	85	0.44	2.38	2.11	0.52	16.6	69.5	5.56
30.73.214	1.25	12.7	0.167	0.97	0.54	320	1.73	2.14	63	0.41	3.78	3.62	0.49	24.1	53.4	3.11
30.73.216	0.63	10.7	0.060	0.40	1.09	182	1.30	3.65	27	0.40	3.18	3.54	0.50	30.4	34.1	1.73
30.73.217	2.44	11.0	0.124	0.65	0.84	187	1.20	3.28	62	0.57	4.71	3.49	0.55	35.1	53.3	4.70
30.73.218	1.15	12.7	0.190	1.13	0.69	158	0.49	2.18	72	0.47	3.14	2.86	0.46	22.5	60.3	4.14

*Not determined

VIII that K_2O and Rb_2O show an inverse relationship in glass from the same panel; however Rb_2O and Cs_2O are both greater in one group than the other.

CASE STUDY

The panel selected for study is shown in Figure 5. This panel is thirteenth-century French grisaille and it is in the Art Museum of Princeton University (Accession Number 43-65). Eleven grisaille fragments were selected for sampling out of forty-four which are present.

The publications in which the panel is discussed give no specific information of its provenance. The panel was originally in the collection of a former curator of the Metropolitan Museum of Art, Bashford Dean. Mr. Dean referred to the panel in one of his publications (10). The panel was also discussed by Paul Frankl (11). The latter presents an interesting stylistic interpretation of the provenance of the panel.

The fragments were removed from the leading prior to sampling (12). A corner of each fragment was cleaned with a grinding tool with a tungsten carbide burr and washed with water. The glass was then scratched with a stylus and the sample was broken off. A sample of about 100 mg was taken.

These specimens were analyzed for elements with short-lived activities in part at the National Bureau of Standards and in part at Brookhaven National Laboratory using the methods previously described. Two of the specimens were analyzed at both institutions. The agreement between the duplicate results from the two sets of analyses, which is well within anticipated experimental error, is shown in Table VI. Additional elements with long-lived activities were determined at Brookhaven on all specimens. The accumulated data on all specimens are given in Tables VII and VIII in which average values are given for duplicate runs.

The eleven fragments quite clearly can be divided into two compositional groups. Eight of the fragments formed the group shown on Table VII and the other three are shown on Table VIII. The fragment numbers on these two tables correspond to the fragment numbers on Figure 5 and it is interesting to observe that fragments 2 and 8 correspond to each other in shape and appear symmetrically placed in the panel. The concentration of Cr_2O_3 in fragment 2 is not shown on Table VII. Two samples of fragment 2 were analyzed for Cr_2O_3 and the results were 773 ppm and 256 ppm. Microscopic examination of the two samples of fragment 2 which were analyzed revealed that they contained some particulate impurities which had apparently affected the chromium

Table VI

Separate Analyses from Two Laboratories

Grisaille Glass from Thirteenth-Century French Window
Art Museum, Princeton University

Fragment	Laboratory	Na$_2$O	K$_2$O	MnO	CaO
MGP 3	(Brookhaven National Laboratory)	0.49	23.6	1.21	N.D.
MGP 3	(National Bureau of Standards)	0.53	26.7	1.33	10.7
MGP 8	(Brookhaven National Laboratory)	0.63	22.9	0.79	N.D.
MGP 8	(National Bureau of Standards)	0.61	21.0	0.92	13.4

Table VII

Grisaille Panel

Collection of the Princeton University Art Museum, Accession Number 43-65

Fragment Number	Na$_2$O	K$_2$O	CaO	BaO	MnO	Fe$_2$O$_3$	Rb$_2$O	Cs$_2$O	Sc$_2$O$_3$	CeO$_2$	Eu$_2$O$_3$	HfO$_2$	ThO$_2$	Cr$_2$O$_3$	CoO	Sb$_2$O$_3$
	Percent						Parts Per Million									
1	0.57	31	N.D.	0.42	1.30	0.42	590	10.0	1.38	15.2	0.28	2.2	2.7	9.6	35	1.27
3	0.51	25	10.7	0.45	1.27	0.45	580	9.8	1.38	16.4	0.28	2.1	2.0	13.8	40	1.23
4	0.55	29	N.D.	0.46	1.23	0.45	630	10.2	1.50	18.3	0.30	2.1	2.2	10.3	37	1.29
5	0.56	28	10.6	0.47	1.43	0.44	640	11.2	1.48	18.7	0.22	2.2	2.3	11.4	41	1.26
6	0.53	28	N.D.	0.41	1.25	0.40	610	10.6	1.33	17.9	0.20	2.3	2.0	17.6	36	0.90
7	0.54	28	N.D.	0.44	1.23	0.42	600	10.0	1.41	22.4	0.25	2.4	2.2	8.7	23	1.06
9	0.49	27	N.D.	0.42	1.37	0.42	620	10.2	1.37	16.7	0.23	2.2	2.0	9.1	27	1.23
10	0.54	29	11.3	0.45	1.38	0.43	600	9.8	1.39	18.5	0.25	2.4	2.1	10.3	38	1.31
Mean	0.54	28	–	0.44	1.31	0.43	608	10.2	1.40	17.9	0.25	2.2	2.2	11.1	34	1.19
No. Averaged	8	8	–	8	8	8	8	8	8	8	8	8	8	8	8	8
Group Std. Dev. %	5.1	6.4	–	5.0	6.0	4.1	3.5	4.7	4.1	12.2	14.7	5.4	10.7	26.6	24.2	13.8

Table VIII

Grisaille Panel (continued from Table VII)

Fragment Number	Na$_2$O	K$_2$O	CaO	BaO	MnO	Fe$_2$O$_3$	Rb$_2$O	Cs$_2$O	Sc$_2$O$_3$	CeO$_2$	Eu$_2$O$_3$	HfO$_2$	ThO$_2$	Cr$_2$O$_3$	CoO	Sb$_2$O$_3$
	Percent						Parts Per Million									
2	0.61	21	15.0	0.36	0.94	0.58	830	23	1.60	30.3	0.28	1.86	3.4	---	21.5	1.97
8	0.62	22	13.4	0.33	0.86	0.53	830	26	1.70	27.7	0.30	3.58	2.6	12.2	18.9	1.20
11	0.61	21	13.6	0.36	0.93	0.52	801	27	1.68	17.2	0.28	2.92	2.6	13.3	21.1	1.51
Mean	0.61	21	14.0	0.35	0.91	0.54	820	25	1.66	24.3	0.29	2.69	2.8	12.7	20.5	1.53
No. Averaged	3	3	3	3	3	3	3	3	3	3	3	3	3	2	3	3
Group St.Dev.%	0.9	2.8	6.3	5.2	5.0	6.0	2.3	8.8	3.3	35.6	4.1	39.8	16.8	6.3	7.2	28.2

determination. The chromium values were therefore discarded as
unreliable. However, there was agreement between the duplicate
runs for the determined values of the other elements. Hence, it
seemed reasonable to accept them as representing the true glass
composition for this specimen.

The homogeneity of glass within the two compositional groups
is quite striking, and indeed the two groups themselves are
basically similar in composition. A graphical comparison of the
statistical constants for these groups is given in Figures 6 and
7. Within either one of the groups the individual specimens are
so similar in composition to each other as to raise the question
as to whether they had all been products of a single batch of
glass preparation. Also the second group would seem to be
sufficiently similar to the first that it might have represented
a second batch prepared in the same workshop. Of course, other
equally reasonable explanations for the degree of similarity can
be proposed.

It is to be noted that specimens belonging to the two compositional groups differ slightly but definitely in their outer
surface appearance. Assuming that the two groups of glass are
contemporaneous, one could ascribe this difference to a mechanical
difference in the method of formation of the glass groups or to a
difference in the susceptibility of the two groups to weathering.
The fairly significant difference in total alkali content of the
two glasses makes this latter a distinct possibility. It is also
possible, however, that one of the two groups of fragments
corresponds to repair additions inserted later and hence subject
to less overall weathering attack. This latter possibility does
not seem highly probable in light of the near compositional
similarity of all the glass and the stylistic similarity of the
grisaille design in all of the analyzed pieces. Mr. Rowan
Le Compte, who very kindly examined the panel with us after the
analyses had been completed and gave us his expert opinion on
its structure and style, felt he could not discern significant
differences in the artistic treatment of the various sections
that had been analyzed.

Of course, as with all glass panels of this age there were
some obvious regions of repair. It has sometimes been argued
that one should expect to find a glass panel of this age to be
largely an accumulation of repairs containing little original
glass. The compositional homogeneity of this panel, however,
would argue against extensive and varied alterations having
occurred within it.

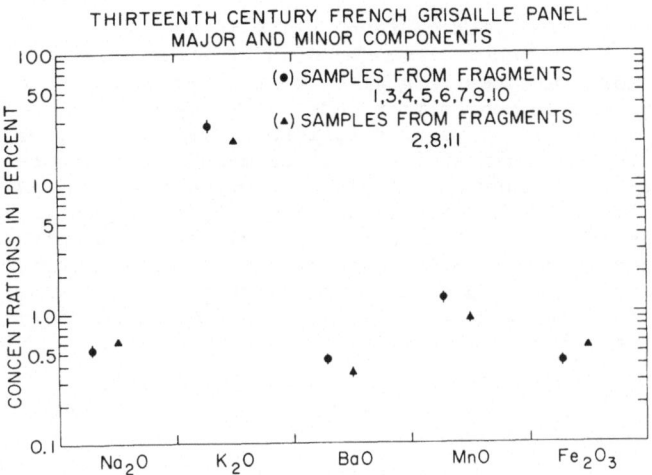

Fig. 6. Plot of data for samples from French panel shown in Fig. 5. (Major and minor components.)

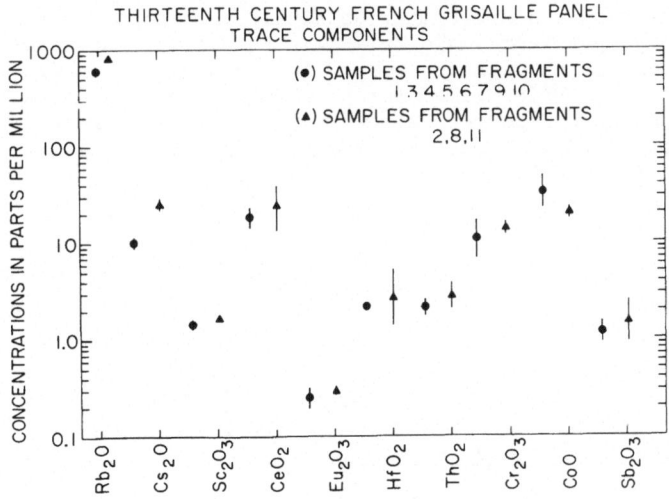

Fig. 7. Plot of data for samples from French panel shown in Fig. 5. (Trace components.)

CONCLUSIONS

    The heterogeneity of medieval glass in general and the homogeneity of the panel discussed argue favorably for the possibility of characterizing glasses of a given provenance on the basis of the composition data gained by neutron activation analysis. Several questions do remain, however. These involve the organization of glass manufacture in the Middle Ages and whether or not glass from one workshop was ·distributed to more than one monument. A further question is what changes in composition occurred in glasses manufactured over a period of years for a given monument. This information, as it develops, may be useful in establishing chronological sequences of glass production as a consequence of particular methods and sources of materials having been used at a given time.

ACKNOWLEDGEMENTS

The authors extend their thanks to Miss Jane Hayward of The Cloisters for bringing to their attention the collection of medieval window glass fragments with which this study began. We thank Miss Frances Jones of the Art Museum, Princeton University for generously allowing us to sample a panel in that collection and Mr. Dieter Goldkuhle for cooperating in the sampling procedure. To Mr. Rowan Le Compte, who later examined the panel with us, we are most grateful. We are indebted to Mr. Eric Miller for performing a number of the analyses as an NSF Summer Fellow at the National Bureau of Standards. This work was supported in part by the National Endowment for the Humanities.

References

(1) Sayre, E. V., and Meyers, P., "Applications of Nuclear Activation Methods to the Study of Art and Archeological Materials; An Annotated Bibliography", Supplement to Vol. 8 Number 4, Art and Archeology Technical Abstracts, 1972.

(2) Sayre, E. V., "Refinement in Methods of Neutron Activation Analysis of Ancient Glass Objects through the Use of Lithium Drifted Germanium Diode Counters", Proceedings of the VIIth International Congress on Glass. Paper 220, Brussels, 1965.

(3) Geilmann, W. and Bruckbauer, T., "Beitrage zur Kenntnis alter
    Glaser II. Der Mangangehalt alter Glaser", Glastechnische
    Berichte 27 (1954) pp 456-459, and Geilmann, W., "Beitrage
    zur Kenntnis alter Glaser III, "Die Chemische Zusammensetzung
    einiger alter Glaser, insbesondere deutscher Glaser des 10.
    bis 18. Jahrhunderts", ibid. 28 (1955) pp 146-156.

(4) Brill, Robert H., "Scientific Studies of Stained Glass: A
    Progress Report", Journal of Glass Studies, XII (1970),
    pp 185-192.

(5) Coleman, R. F. and Wood, G. A., AWRE Report No. 03/68
    (April 1968).

(6) Listings of computer programs are available upon request.

(7) Turner, W. E. S., J. Soc. Glass Technology 40, 39 (1956);
    ibid. 40, 162 (1965); ibid. 40, 277 (1966).

(8) Theophilus, "De diversis artibus", manuscript treatise, ca.
    A. D. 1123; Latin text and translation by C. R. Dodwell
    (London, 1961); translation with technical notes by J. G.
    Hawthorne and C. S. Smith (Chicago, 1963).

(9) Shaw, D. M., "A review of K-Rb fractionation trends by
    covariance analysis", Geochimica et Cosmochimica Acta,
    32, 573 (1968).

(10) Dean, Bashford, "The Exploration of a Crusader's Fortress
     (Montfort) in Palestine", The Bulletin of the Metropolitan
     Museum of Art, New York, XXII, 1927, part II.

(11) Frankl, P., Records of Museum of Historic Art, Princeton
     University, 3, No. 1, 8 (1944).

(12) Mr. Dieter Goldkuhle performed the restoration of this panel.
     It was during the process of removing the glass from the old
     leading that one of the authors, JSO, sampled the eleven
     grisaille fragments.

MOLECULAR SPECTROSCOPY

INFRARED SPECTRA OF SOME DIMETHYLCYCLOHEXANES AT VERY HIGH PRESSURES

J. L. Lauer[*] and M. E. Peterkin

Sun Research and Development Company
P. O. Box 426
Marcus Hook, Pa.   19061

Absorption spectra in the 500-1100 cm^{-1} region were obtained for cyclohexane, methylcyclohexane, and all the dimethylcyclohexanes except 1,1-dimethylcyclohexane, at pressures estimated to range between ambient and 40 kb. A diamond anvil cell was used to contain the samples. Simple modifications to an RIIC Model FS-720 Far Infrared Interferometer allowed it to operate in this spectral region. All the samples are liquids at ambient pressure and become very viscous before crystallization except trans-1,3-dimethylcyclohexane, which could not be crystallized. It is shown theoretically that high pressure can change the distribution of conformers of some of the isomers, and spectral changes can be interpreted to support this deduction. Just as cyclohexane undergoes a phase change in the solid state, so do both cis- and trans-1,2-dimethylcyclohexane, for their spectra change with increasing pressure. The first crystalline structure formed under increasing pressure from the latter material appears to be rather unstable and to correspond to the biaxial conformer.

* to whom correspondence should be addressed

INTRODUCTION

The conformational preferences in the substituted cyclohexanes have been studied since Pitzer and coworkers[1] have
been able to account for most of the differences between
calorimetric and spectroscopic entropies by rearrangements
between chair forms. Moreover, these investigators were
able to show that correlations between enthalpy, entropy,
density, index of refraction and boiling point of the
dimethylcyclohexanes depended more on the axial or equatorial
location of the methyl groups than on their cis or trans
location with respect to the plane of the cyclohexane ring.
Nmr methods later fully substantiated these assignments and
by these and other procedures, equilibrium compositions and
most of the thermodynamically significant parameters were
determined over a wide range of temperatures.

For a study of conformational preferences and crystal
habits under high pressures of materials representative of
petroleum-based lubricants, the diamond anvil cell coupled
to high-resolution infrared spectrophotometer appeared to
be the simplest and most direct procedure. Alkyl cyclohexanes, and particularly the methyl cyclohexanes, are
representatives of such fluids, but they are not easily
analyzed in this manner, for identifiable group frequencies
are lacking and most of the vibrational modes are very complex.
Halogens and various other polar substituents are preferable
to alkyl groups in this regard, so that cyclohexyl halides,
for example, have been used as models for conformational
analysis. The recent work by Klaeboe[2] and coworkers on
cyclohexyl halides has correlated the frequencies of a
number of absorption bands in the skeletal region of the
infrared spectrum with particular conformations. By extra-
polations from their data and some reasonable deductions from
frequency displacements and spectral intensity shifts, the
many pressure-induced changes in the infrared spectra of some
of the dimethylcyclohexanes appeared susceptible to interpretation in terms of changes of conformational composition.
Furthermore, infrared spectral analysis of the pressure-
solidified 1,2-dimethylcyclohexanes was expected to provide
evidence for the existence of transitions between crystalline
states for each of the two isomers, one state being much less
stable than the other, similar to the case of the two modifications of solid cyclohexane. Cryogenic evidence for two
crystalline modifications in trans-1,2-dimethylcyclohexane
is very tenuous, since a difference in melting point of only
0.006°C was postulated by Huffman and coworkers[3] because of
discrepancies between series of heat capacity measurements.

This paper is concerned with experimental procedures, a description of the results obtained, and a discussion of their significance. A more detailed analysis including a comparison of low temperature and high pressure spectra will be the subject of another communication.

EXPERIMENTAL

Most of the spectra were obtained with an RIIC FS-720 far infrared interferometer modified so as to extend its upper frequency limit from 400 to 1100 cm^{-1}. The basic operation of this instrument and the procedures used in processing the data were discussed in an earlier paper[4]. Primarily for comparison, a few spectra were also run on a Block Engineering Model FT-14 interferometer in Professor E. R. Lippincott's laboratory at the University of Maryland. A diamond anvil cell purchased from High Pressure Diamond Optics, Inc., McLean, Virginia, was the sole device for obtaining high pressure[5]. In accordance with advice from Dr. J. W. Brasch, the cell was approximately calibrated by a determination (with a tensile strength tester) of the elastic constants of the spring opposing the motion of the main screw of the device; this quantity and the geometry of the cell holder and cell related the fluid pressure in the cell to the number of turns of the screw. For the work reported here, one turn of the screw corresponded to about 2.5 kb of pressure between the faces of the two diamonds. The spectra were taken through these two diamonds, the more recent ones with the high pressure cell located in the holder of a 4X reflecting beam condenser made by the Perkin Elmer Corporation.

The principal modifications of the interferometer are now stated. The beam splitter was a 2.5 μm thick (8 gauge) film of Mylar replacing the thinnest (6 μm, 25 gauge) film normally supplied. The thinner film made the alignment somewhat more difficult than normal, but not seriously. The triggering circuitry had to be changed so that detector readings were recorded for every 4 um of optical pathlength difference between the two interferometer beams (2 μm of mirror motion). The mirror travel between readings was, therefore, half the smallest distance for which the instrument was designed. This halving was done by an electronic modification supplied by the manufacturer. With the smaller interval, the theoretical range is, therefore, 0-1250 cm^{-1}, provided (a) the beam splitter efficiency is adequate and (b) an optical filter is available that eliminates all the source radiation of frequencies exceeding 1250 cm^{-1} but transmits all the frequencies below this value. Even if such a filter existed, the dynamic range of the interferometer would be excessive,

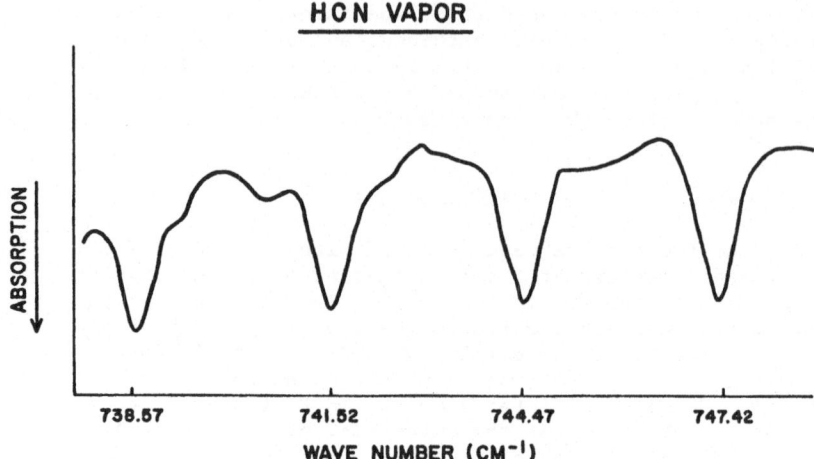

Figure 1 Part of the Rotational Spectrum
(R8 to R11) of the $0^1 0$–$00^0 0$ (ν_2) Band of HCN

because of the very wide frequency range. An interference
filter transmitting about 80 percent of the radiation between
600 and 1200 cm^{-1} and about 20 percent between 400 and 600
cm^{-1} and essentially nothing beyond these limits (available
from Optical Coatings Laboratory, Santa Rosa, California) was
found to work very well in the spectral region of interest.
To reduce the dynamic range still further, work is now
underway to use more filtering so as to narrow the spectral
range covered in any given run. It will then be possible
to increase the spacing between mirror positions, at which
readings are taken, back to 8 µm to obtain greater resolution
with an equal number of points read (but over a narrower
frequency range). For the work reported here the theoretical
resolution was 1.22 cm^{-1} for 1024 single-sided interferometer
mirror positions (readings) per scan. Our instrument's
capability with the 4 um spacing arrangement, the 2.5 um beam
splitter and the above-mentioned filter is illustrated in
Figure 1. It contains a portion of the rotational spectrum
of HCN vapor (300 torr pressure, 10 cm pathlength) run with
5000 actually read points upped to 8192 points by "zero-filling"
(Ref. 4) to give a theoretical resolution of 0.15 cm^{-1}.
This spectrum, which was also used for calibration of the wavenumber scale, compares very well in structure and position
with the reference spectrum given in the book by Rao et al.[6]
No special requirements were needed to process the data, but
the cost of digital computation was nearly proportional to
the inverse of the resolution (expressed in terms of the
minimum wavenumber interval resolved). The computational cost

of the other spectra shown in this paper averaged $6 per spectrum (for our G. E. Co. 635 Computer System).

Gaskets for the diamond cell were made of tempered Inconel sheet, 0.010-0.018 inches thick. We are grateful to the International Nickel Company for supplying us with necessary amounts of these materials. The procedures for making these gaskets were explained to us by Dr. J. W. Brasch of the Department of Chemistry, University of Maryland, College Park, Maryland, and we are hereby expressing our gratitude to him.

HIGH PRESSURE SPECTRA

All the spectra were obtained on pure materials with reasonable care and the ambient pressure spectra were compared with API-44 published spectra. In all cases excellent agreement was found in all details, in spite of our need for time-averaging and some smoothing because of the exceedingly small sample volumes provided by the diamond cell and the very low energies falling onto the detector. No attempt was made to grow single crystals, but at every pressure the samples were examined under the polarizing microscope. Except for trans-1,3-dimethylcyclohexane, all the samples were visibly microcrystalline and anisotropic at the highest pressure. They were also crystalline at most of the intermediate pressures studied, where their state had to be deduced from spectral changes, since phase boundaries between viscous superpressed liquids and cubic or orthorhombic crystals, in particular, are difficult to see.

Cyclohexane (Figure 2) was run for comparison of our technique with that of Obremski[7]. This compound has been known for a long time to solidify on cooling in one of two forms, (a) a plastic phase (I) exhibiting no birefringence, thus indicating a cubic structure just below the freezing point (6.6°C) and (b) an anisotropic phase (II) at temperatures below -126.6°C. Obremski was able to obtain single crystals of the two modifications with the high pressure cell, but he reported spectra for both single crystals and polycrystalline materials. In the frequency range of Figure 2 a difference between polycrystalline Phase II and Phase I or the liquid is the peak of the absorption band near 900 cm^{-1} falling below 901 cm^{-1} (i.e. 896-901 cm^{-1}) for the former and above 901 cm^{-1} (i.e. 902-909 cm^{-1}) for the latter. Another difference can be seen in the relative intensities of absorption of the components of the 1014/1040 cm^{-1} doublet; the former is more intense than the

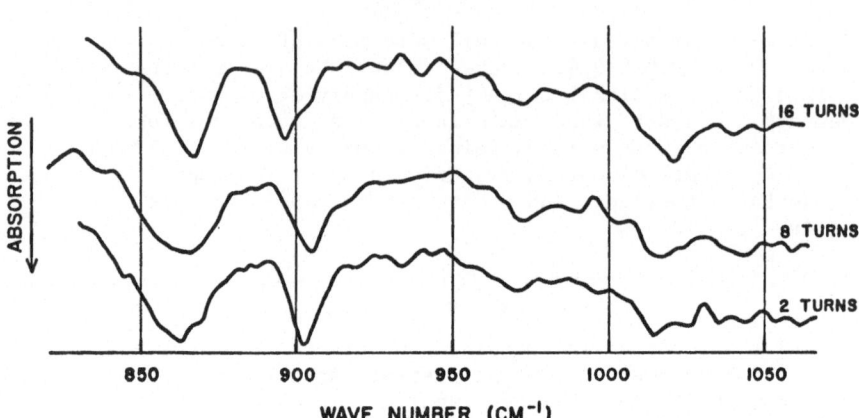

Figure 2   High Pressure Infrared Spectra of Cyclohexane
(One turn is approximately equivalent to 2.5 kb)

latter for Phase II and somewhat more so for the liquid, but
less intense for Phase I, the intermediate phase. Our spectra
of Figure 2 are consistent with those of Obremski for the three
phases. The broadness of the bands of the spectrum
corresponding to the intermediate pressure is a characteristic
of plastic crystals; the molecules are undergoing some sort of
restricted rotation in the unit cell. The absorption bands
in our spectra are broader than those of Obremski, however,
because our samples were not annealed and the bands in these
and all the other spectra appear to be even broader because
of the scales used for both coordinates.

Methylcyclohexane has the lowest freezing point (Table I)
of any of the materials examined; it was polycrystalline only
at the very highest pressure and certainly liquid at the
intermediate pressure for which spectra are shown in Figure 3.
It will be noted that the lowest and highest pressure spectra
resemble each other more closely than either resembles the
intermediate pressure spectrum. This is because a number of
absorption bands are well resolved in the latter spectrum but
are only shoulders in either of the others. Since
methylcyclohexane at ordinary temperature and pressure consists
of an equilibrium mixture of conformational isomers in which
the methyl group is located equatorially and axially with
respect to the ring in the ratio of 95:5 (Ref. 8) and since an
increase of pressure favors the latter isomer because of its
smaller volume, it is reasonable to assume that the bands that
are enhanced in the intermediate pressure spectrum are largely
derived from the axial conformation. In the crystalline solid

TABLE I

PROPERTIES OF ALKYL CYCLOHEXANES

Compound*	Molecular Volume 20°C ml/mole	Freezing Point °C	Conformations	Conformational Enthalpy[13] Difference kcal/mole
1	108	6.6	–	–
2	128	-127	e,a	1.7
3	141	– 50	ae	–
4	144	– 88	ee,aa	2.7
5	146	– 76	ee,aa	5.5
6	143	– 90	ae	–
7	143	– 87	ae	–
8	147	– 37	ee,aa	3.6

* 1 – cyclohexane
  2 – methylcyclohexane
  3 – cis-1,2-dimethylcyclohexane
  4 – trans-1,2-dimethylcyclohexane
  5 – cis-1,3-dimethylcyclohexane
  6 – trans-1,3-dimethylcyclohexane
  7 – cis-1,4-dimethylcyclohexane
  8 – trans-1,4-dimethylcyclohexane

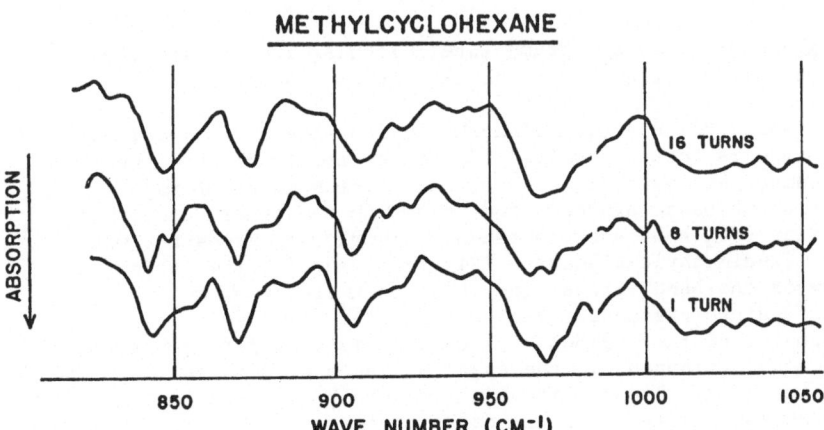

Figure 3 High Pressure Infrared Spectra of Methylcyclohexane
(One turn is approximately equivalent to 2.5 kb)

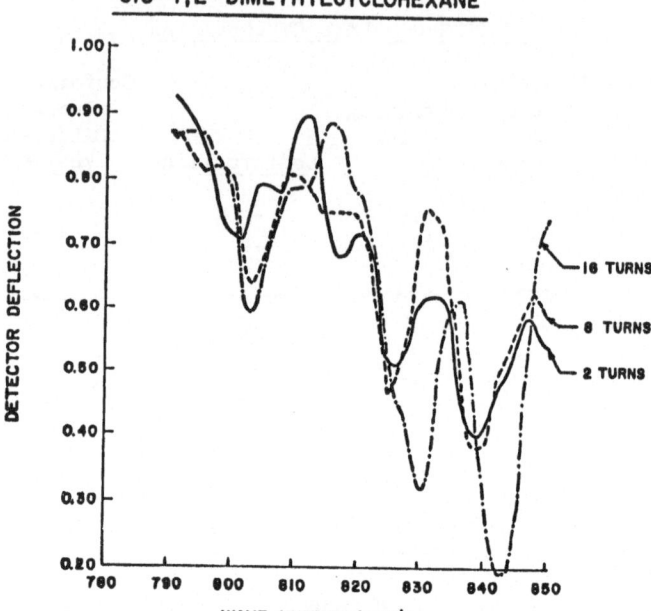

CIS-1,2-DIMETHYLCYCLOHEXANE

Figure 4   Infrared Spectra of Cis-1,2-dimethylcyclohexane
in the 790–850 cm^{-1} Region over a Series of Pressures

(highest pressure spectrum) intermolecular forces play the
major role and they favor the equatorial conformation.

While the spectrum of methylcyclohexane at an intermediate
pressure is better resolved than those at higher and lower
pressure, no reversal in the elative intensities of major
absorption bands has occurred here. On the other hand, in a
similar series of spectra taken at increasing pressures with
cis-1,2-dimethylcyclohexane (Figures 4 and 5), the ratios of
some of the band intensities reach an extremum at an
intermediate pressure. For example, the low frequency
component of the 829/841 cm^{-1} doublet is stronger than the high
frequency component in the spectrum taken at 4 turns (10 kb)
but weaker in the spectra taken at both higher and lower
pressures. There is little doubt that the 4-turn pressure
spectrum refers to a different crystalline modification than
the higher pressure spectra, which probably contain more
than one crystal structure and indeed, cis-1,2-dimethylcyclo-
hexane has been known[9] to exist in two crystalline forms at

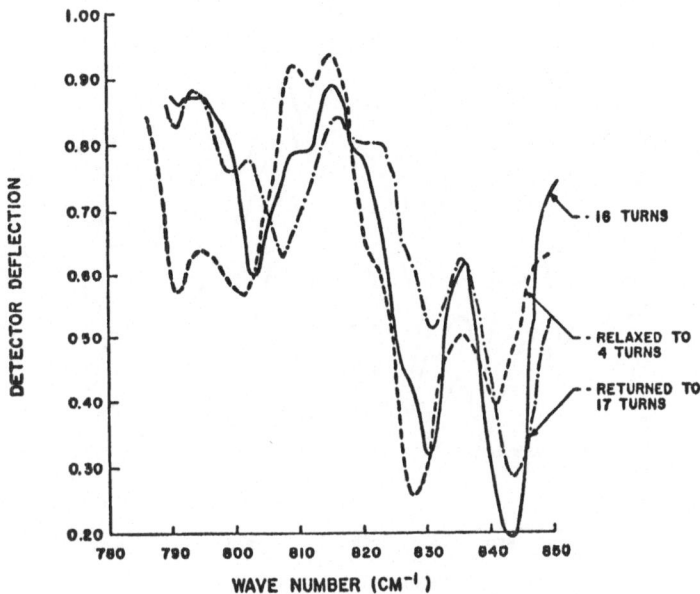

Figure 5   Infrared Spectra of Cis-1,2-dimethylcyclohexane
in the 790-850 cm^{-1} Region over a Series of Pressures

low temperatures, with a transition temperature of -100.7°C.
However, this compound can exist in only one conformation, viz.
with one methyl group in the equatorial and the other in
the axial position (Table I).

Are there also two crystal forms in trans-1,2-dimethyl-
cyclohexane? Huffman and coworkers[3] had some very tenuous
evidence for a second crystalline form from their low
temperature heat capacity measurements, but their postulated
transition temperature between the two solid phases was only
0.006°C below the freezing point. While realizing that the
crystal structures produced by cooling and by pressure are not
necessarily equal[9], it nevertheless appeared to be worth
looking for them under high pressure. Indeed we stumbled on
the "unstable" form because of our apparent inability to
obtain reproducible spectra at 8-9 turns (~22 kb) of pressure.
Accordingly, we went through the pressure cycle shown in
Figure 6. Letting the 9-turn sample stand for one hour at
100°C changed its spectrum--obtained after cooling to room

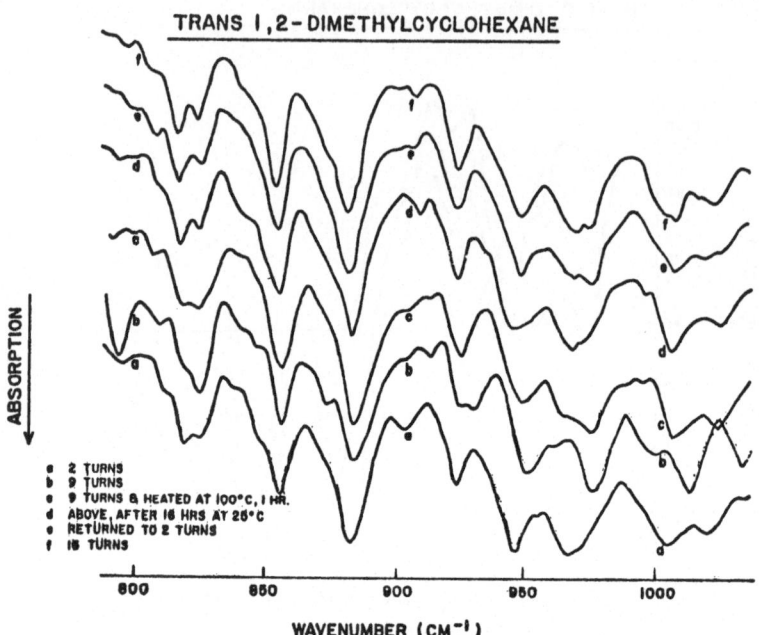

Figure 6   High Pressure Infrared Spectra of Trans-1,2-dimethyl·
cyclohexane (One turn is approximately equivalent to 2.5 kb)

temperature--and letting it stand at room temperature even
longer, changed it further. The spectra of another cycle
(over a more limited spectral region) are shown in Figure 7.
These data show an extremum at 8 or 9 turns in the relative
strength of the 817/825 cm^{-1} doublet (in one experiment, the
first component had vanished). This as well as other major
differences in the spectra are therefore consistent with the
existence of at least two crystal structures for trans-1,2-
dimethylcyclohexane. There appears to be evidence of
reforming the unstable modification, at least in part, when the
pressure is relaxed from 16 turns to 2 turns (Figure 6).
Although no crystals were visible under the polarizing
microscope at any but the very highest (16-17 turns) pressures,
this observation does not exclude the possibility of plastic,
cubic, or even orthorhombic crystals[10]. Furthermore, trans-1,
2-dimethylcyclohexane, as well as cis-1,3- and trans-1,4-
dimethylcyclohexane, can exist in two different conformations
(both methyl groups equatorially or both axially oriented,
with the former arrangement the predominant one) and has the

TRANS -1,2-DIMETHYLCYCLOHEXANE

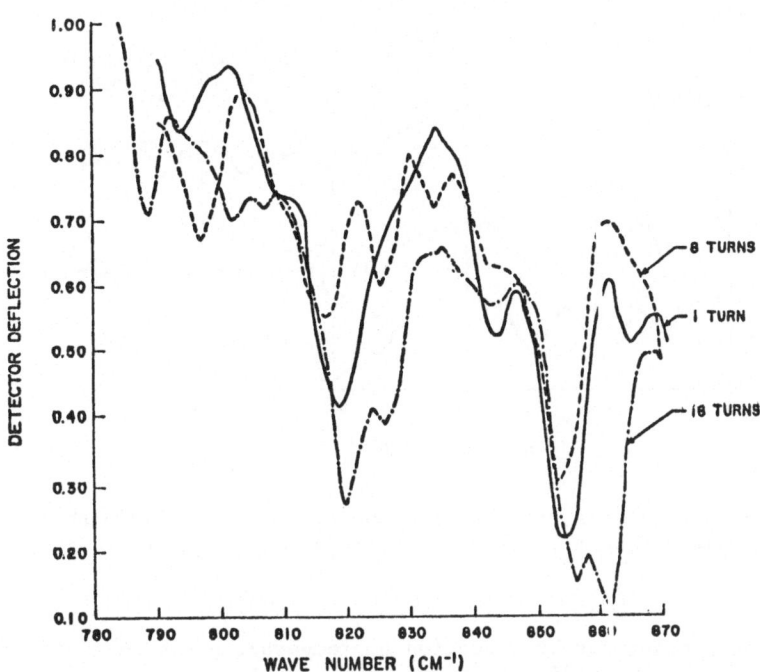

Figure 7   Infrared Spectra of Trans-1,2-dimethylcyclohexane in the 790-850 cm^{-1} Region over a Series of Pressures

smallest enthalpy difference between two conformers of the dimethylcyclohexanes. The "unstable" form very likely contains an appreciable amount of the higher energy conformer.

The spectra of cis-1,3-dimethylcyclohexane (Figures 8 and 9) show only a uni-directional change with pressure. At the highest pressure the spectrum has become much simpler, as for example, the band at   825 cm^{-1} has disappeared. Since the solid is more likely to consist of only one conformation than the liquid, this band could have been largely a contribution of the higher energy conformer. On the other hand, essentially no changes with pressure are apparent in the spectra of trans-1,3-dimethylcyclohexane (Figure 10). As mentioned earlier, this compound was the only one we were unable to crystallize even at the highest pressure. After making many purity checks and still finding ourselves unable to crystallize this sample--

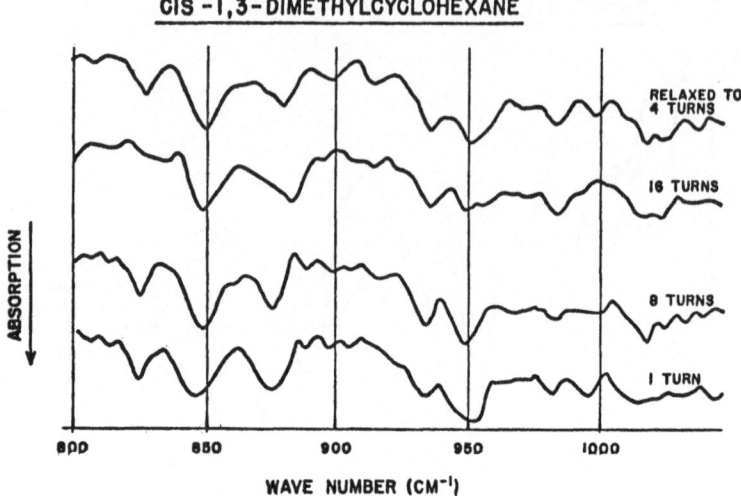

Figure 8   High Pressure Infrared Spectra of Cis-1,3-dimethyl-
cyclohexane (One turn is approximately equivalent to 2.5 kb)

either by pressure or by reduction of temperature--we were at
least consoled by finding that Huffman[3] and his coworkers,
using the purest samples for API Project 44, had similar
experiences.

Cis-1,4-dimethylcyclohexane (Figure 11) showed only minor
spectral changes with pressure and these mostly at the highest
pressure when the sample crystallized. Since this compound
can exist in only one conformational form (Table I), this
result came out as expected.

Since the spectra of trans-1,2- and cis-1,3-
dimethylcyclohexane both show evidence for the existence of
two conformers and the difference in enthalpy of the
conformers of trans-1,4-dimethylcyclohexane falls between the
conformational enthalpy differences of the other two isomers
mentioned, one would expect that the spectra of the last
isomer should also show evidence for the presence of two
conformations. Spectrum c of Figure 12 presumably refers
to a partially liquid material because of its simple structure,
which is similar to Spectrum a. However, Spectrum c has a
strong band at 845 cm^{-1}, which is nearly absent in Spectrum a.
This band is very likely caused by the axial conformation
because it occurs at high pressure where the axial confor-

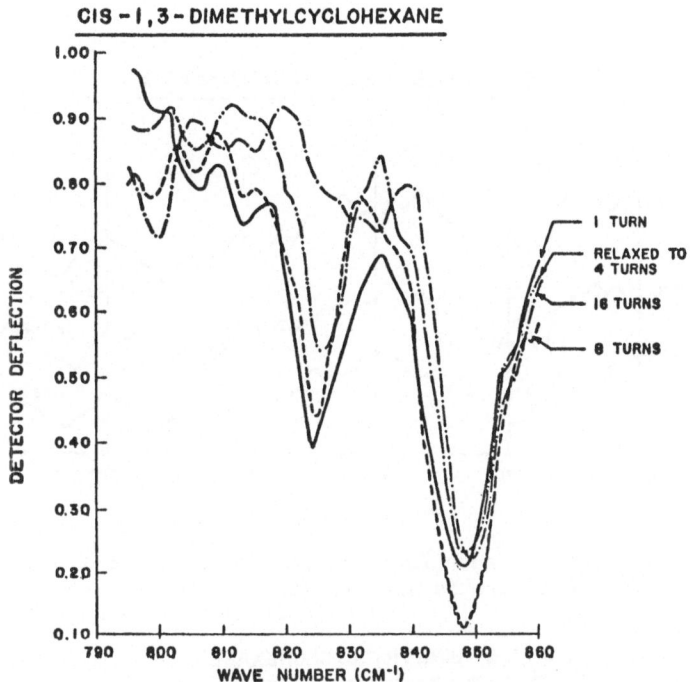

CIS - 1,3 - DIMETHYLCYCLOHEXANE

I TURN

RELAXED TO
4 TURNS

16 TURNS

8 TURNS

Figure 9 Infrared Spectra of Cis-1,3-dimethylcyclohexane in the 790-850 cm^{-1} Region over a Series of Pressures

mation is favored. Spectrum f, which also shows this band, therefore indicates that the crystalline form at the extreme pressure might be in the axial conformation. Future work at low temperatures will help confirm this interpretation.

DISCUSSION

The methyl derivatives of cyclohexane do not possess specific infrared group frequencies from which the conformational makeup of a particular compound under given conditions can be immediately inferred. For, as has been pointed out by Kaznetsova and coworkers[11] and more analytically by Schachtschneider[12], the normal modes of these materials mostly contain a mix of linear and angular deformations involving a great number of valence bonds and angles simultaneously. It is known from nmr, dipole moments, and other physical properties

TRANS - 1,3 - DIMETHYLCYCLOHEXANE

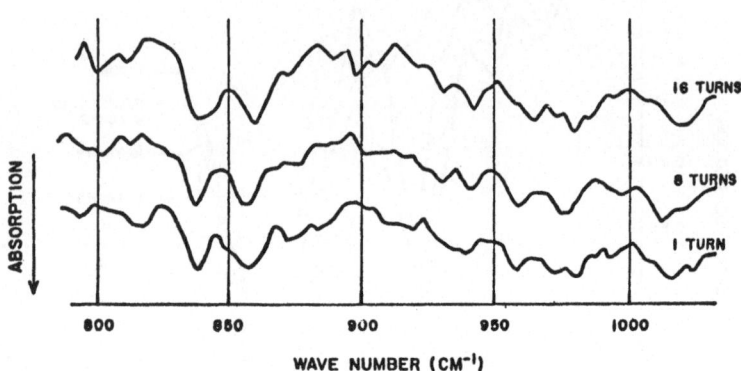

Figure 10   High Pressure Infrared Spectra of Trans-1,3-dimethyl-
cyclohexane (One turn is approximately equivalent to 2.5 kb)

CIS - 1,4 - DIMETHYLCYCLOHEXANE

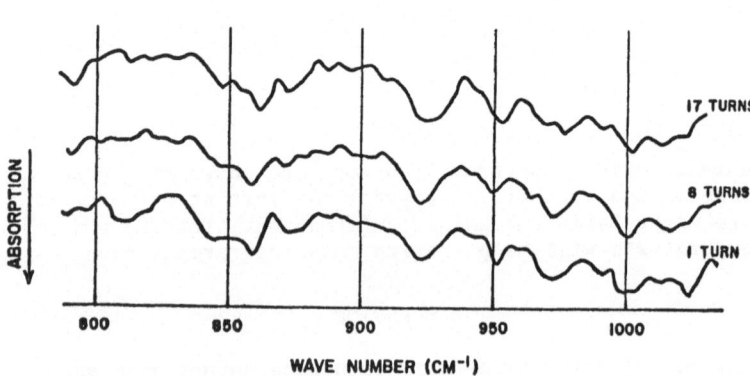

Figure 11   High Pressure Infrared Spectra of Cis-1,4-dimethyl-
cyclohexane (One turn is approximately equivalent to 2.5 kb)

TRANS 1,4 - DIMETHYLCYCLOHEXANE

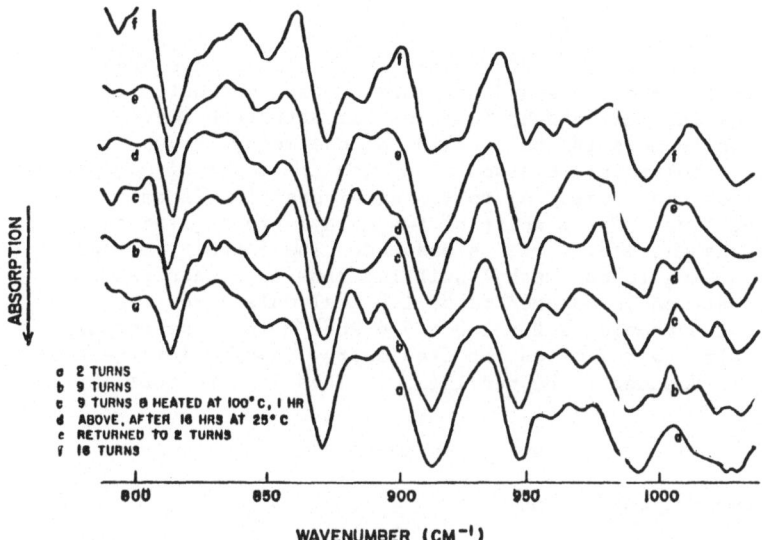

a 2 TURNS
b 9 TURNS
c 9 TURNS & HEATED AT 100°C, 1 HR
d ABOVE, AFTER 16 HRS AT 25°C
e RETURNED TO 2 TURNS
f 16 TURNS

WAVENUMBER (CM⁻¹)

Figure 12 High Pressure Infrared Spectra of Trans-1,4-dimethyl-
cyclohexane (One turn is approximately equivalent to 2.5 kb)

that, except under very unusual conditions, only the mode
of substitution of the cyclohexane ring--axial or equatorial--
can be shifted by pressure and temperature[13], but that the ring
itself remains in the chair form. Since the molar volume of a
methyl-substituted cyclohexane derivative is smaller when the
methyl group is in the axial position than when it is in
the equatorial position, the application of pressure is likely
to increase the proportion of the axially substituted conformer
in an equilibrium composition. The axially substituted
conformer also happens to be the one with the greater internal
energy and, for this reason, is present only in very small
amount--if at all--at ambient temperature and pressure. The
situation is different for polar substituents, e.g. for the
halocyclohexanes, for then the arrangement of the dipole
moments, and not just the geometry, helps to determine the
conformational composition for minimum energy.

To estimate whether the pressures accessible to the
diamond cell can cause an appreciable change in the conformation
of the cis- and trans-dimethylcyclohexanes the following
reasoning was applied:

The molar volumes of cis and trans dimethylcyclohexanes
at ambient temperature and pressure differ by about 3 cc.
Since the only difference in molecular structure between cis
and trans isomers is the location of one methyl group, the
change in molar volume associated with the shift of one methyl
group from the equatorial to the axial position is assumed to
be about 3 cc. For a shift of two methyl groups, twice the
volume change, i.e. 6 cc, is assumed. The enthalpy change
corresponding to one methyl shifting from an equatorial to an
axial position is assumed to be 1.8 kcal/mole ; for two methyl
groups, the change is 2 x 1.8 = 3.6 kcal/mole. The enthalpy
change also approximates the free energy change, for the entropy
change is assumed to be negligible. Thus for the equilibrium

$$ee \rightleftharpoons aa$$

under ambient conditions

$$\Delta G = 3.6 \ kcal/mole = -RT \ln K$$

so that

$$K = 0.0025$$

and the equilibrium composition is essentially all ee.
Since

$$(\partial \Delta G / \partial p)_T = \Delta V$$

then on the assumption that the volume change ΔV is independent
of pressure,

$$\ln[K(p_2)/K(p_1)] = -\Delta V(p_2 - p_1)/RT$$

and with

$$\Delta V = -6cc \quad and \quad p_2 - p_1 = 25kb$$

one obtains

$$K(p_2)/K(p_1) \approx 400$$

which would make

$$K > 1.0$$

at ambient temperature. Accordingly a shift from a diequatorial to a diaxial configuration under high pressure appears to be thermodynamically feasible.

It is also likely that pressure would help accelerate an e $\rightarrow$ a shift, for an equilibrium constant can be decomposed into a forward and backward first order rate constant toward an activated complex, which has been assumed to be the twistboat form [13] whose volume is smaller than that of the diequatorially substituted chair form, viz.

$$\left(\frac{\partial \ln K}{\partial p}\right)_T = \left(\frac{\partial \ln k_2}{\partial p}\right)_T - \left(\frac{\partial \ln k_1}{\partial p}\right)_T$$

$$= \frac{v_c - v_1}{RT} - \frac{v_c - v_2}{RT} = -\frac{\Delta v_c}{RT}$$

Here Δv_c is the difference in volume between the activated complex and the reagent or product. Assuming the back reaction to be negligible,

$$\left(\partial \ln k / \partial p\right)_T = -\left(\Delta v_c / RT\right)$$

so that the effect of pressure on the rate constant of reactions in the liquid phase is determined by the volume change in the course of formation of the activated complex.

These considerations, showing feasibility of an e → a shift by pressure, apply only to a constant liquid phase; the formation of crystals involves other forces and energies, which can be very large. Because the crystal bonding energies are usually quite different for different conformational isomers, a crystalline phase is generally composed of one conformational isomer only. However, Brasch[9] has shown that this statement is more applicable to solids formed by freezing than by compression and that, in any case, as mentioned earlier, the crystal structure formed by these two means can be different. Furthermore, the effectively lower symmetry at a crystal site can remove degeneracies in the vibrational spectrum of a molecule and therefore give rise to absorption bands not present in the liquid. When an absorption band is present in the liquid and not present in the solid--and splitting has apparently not occurred--the band is usually assigned to a conformer not present in the liquid, but when an absorption band is present in the solid and not in the liquid, it is likely to be associated with a

TABLE II

COMPARISON OF SKELETAL FREQUENCIES

Substituents

(a) Mono-substi-tuted Cyclohexane	Cl	Br	I	CH_3	Klaeboe's Conformational Assignment
	868	864	862	866sh	a
	852	–	–	849	a
	1014	1010	1006	1020	a
	993	988	988	1010	e

Substituents

(b) Disubsti-tuted Cyclo-hexane	Cl,Cl	Br,Br	Br,Cl	CH_3,CH_3	
trans-1,2	826	812	819	823	aa
	865	861	862	872	aa
	1005	999	1000	1010	aa
	1032	1032	1032	1032	aa
	820	808	815	818	ee
	845	840	842	852	ee
	909	906	940	921	ee
	980	972	976	1002	ee
trans-1,4	1003	1004	–	1009	aa
	988	989	–	992	ee

molecular vibrational mode, which is degenerate in the
liquid state (it can also be a lattice mode but these are
generally of a frequency lower than the range covered here).

Klaeboe's skeletal frequency assignments for the
monosubstituted and disubstituted[2] halocyclohexanes have
helped us confirm some of our conclusions from the previous
considerations. Thus there seems to be little doubt (Table
II) that Klaeboe's "axial" frequency of 868 cm^{-1} for cyclohexyl chloride, 864 cm^{-1} for the bromide, and 862 cm^{-1} for the
iodide, corresponds to the shoulder at 866 cm^{-1} seen only in
our 8-turn spectrum of (still liquid) methylcyclohexane.   The
other comparisons of Table II are also consistent.
Methylcyclohexane, therefore, consists of axial and equatorial
conformers.   Increase of pressure favors the former at the
expense of the latter as long as the sample remains liquid.
However, crystalline methylcyclohexane, though produced by

pressurization, is mostly the equatorial conformer. Similar comparisons for the dihalo and dimethylcyclohexanes are also given in Table II.

CONCLUSIONS

Analyses of the infrared spectra at various pressures have led to the following conclusions:

(a) Cyclohexane, as already shown by other authors, can be solidified under pressure to form either one of two crystal structures, which can be transformed into one another. The molecular structure, however, is the same in either crystal form. Cis-1,2-dimethylcyclohexane seems to behave quite similarly.

(b) Trans-1,2-dimethylcyclohexane can also be solidified under pressure into one or two crystal forms, the lower pressure form being rather unstable and readily transformed into the other. The former solid appears to contain a significant proportion of the aa-conformer, the latter is apparently exclusively ee.

(c) There appears to be evidence for the presence of some aa-conformer in the compressed liquid of cis-1,3-dimethylcyclohexane, but the solid formed under very high pressure contains primarily the ee-conformer.

(d) Of all the dimethylcyclohexanes, only trans-1,3-dimethylcyclohexane could be crystallized neither under pressure nor under reduced temperature, as others had found out before. No appreciable changes in the infrared spectrum with pressure were observed for this isomer.

(e) Methylcyclohexane contains a significant proportion of the a-conformer in the compressed liquid phase, but crystallizes under pressure mostly as the e-conformer.

ACKNOWLEDGEMENT

We would like to thank Dr. J. W. Brasch of the University of Maryland for introducing us to the diamond anvil cell. Our machinist, Mr. Thomas Smith built us a holder and housing for it and thereby contributed much to make this work possible. Above all, we are grateful to the Sun Oil Company for sponsoring this research and for their permission to publish this paper.

REFERENCES

1. C. W. Beckett, K. S. Pitzer, and R. Spitzer, J. Am. Chem. Soc., 69, 2488 (1947)
2. P. Klaeboe, Acta Chem. Scandinavica, 23, 2642-2552 (1969); 25, 695-711 (1971)
3. H. M. Huffman, S. S. Todd, and G. D. Oliver, J. Am. Chem. Soc., 71, 584 (1949)
4. J. L. Lauer and M. E. Peterkin, Developments in Applied Spectroscopy (Plenum Press, New York, 1971) Vol. 9, p. 73-107
5. (a) C. E. Weir, A. Van Valkenberg, and E. R. Lippincott, J. Res. Nat. Bur. Stds. 63A, 55 (1959); (b) E. R. Lippincott, F. E. Welsh, and C. E. Weir, Anal. Chem. 33, 137 (1961); (c) E. R. Lippincott, C. E. Weir, A. Van Valkenberg, and E. N. Bunting, Spectrochim. Acta 16, 58, (1960); (d) J. W. Brasch and R. J. Jakobsen, Spectrochim. Acta 21, 1183 (1965)
6. K. N. Rao, C. J. Humphreys, and D. H. Rank, Wavelength Standards in the Infrared (Academic Press, New York, 1966), p. 139
7. R. Obremski, Vibrational Spectra of High Pressure Solids (Thesis, Univ. of Maryland, 1968)
8. M. Hanack, Conformation Theory (Academic Press, New York, 1965), p. 93.
9. J. W. Brasch, Private Communication, November 12, 1971
10. J. W. Brasch, J. Chem. Phys. 43, 3473 (1965)
11. T. I. Kuznetsova and M. M. Sushchinskii, Optics and Spectroscopy, Supplement 2, Molecular Spectroscopy, English Edition, p. 74-77, 1966
12. R. G. Snyder and J. H. Schachtschneider, Spectrochimica Acta, 21, 169-195 (1965)
13. E. L. Eliel, N. L. Allinger, S. J. Angyal, and G. A. Morrison, Conformational Analysis (Interscience Publishers, New York, 1965), p. 50f.

RAMAN SPECTROSCOPY OF HIGH PRESSURE PHASES OF SOLIDS[*]

Malcolm Nicol, Jane R. Kessler, Yukiko Ebisuzaki,
William D. Ellenson, Mei Fong, and C. Sherman Gratch
Department of Chemistry[+]
University of California
Los Angeles, California

This report concerns one role of Raman spectroscopy for observing the behavior of solids under high pressures--that is, Raman spectroscopy is an experimental technique for characterizing in situ phases of solids that occur only at high pressures. This task has several aspects, three of which receive particular attention in the work described here. These are:

a. the detection of phase transformations in solids under pressure including determination of phase boundaries and other characteristics of the transition;

b. assignment of the Raman spectrum in terms of plausible structures and other properties of the high-pressure phases; and

c. interpretation of pressure dependences of vibrational spectra in terms of interatomic interactions that may explain the occurrence of known transitions, with the goal of generalizing the analysis to describe properties of related materials under conditions that may not be accessible for experimentation.

Results obtained from experimental studies of three systems currently in progress at UCLA will be used to demonstrate how these objectives can be met.

The report is divided into three major parts. In the first part, experimental details are briefly described. Many satisfactory discussions of the fundamentals of Raman spectroscopy are available elsewhere[1]; only experimental techniques unique to these studies are

discussed here. Studies of fluorite and related divalent fluoride
systems are described in the second part to illustrate the experimental and theoretical possibilities of this application of Raman
spectroscopy. The report concludes with a discussion of aspects
of rutile, calcite, ammonium chloride and ammonium bromide
that demonstrate some specific advantages of this experimental
technique and unusual types of behavior.

EXPERIMENTAL TECHNIQUES

Except for the high-pressure cell, the spectroscopic techniques
used for these studies[2,3] are typical of laser Raman work. A
Spectra-Physics Model 165 argon laser is now being used for
excitation, and Raman spectra are analyzed with a Spex double mono-
chromator. For Raman frequencies below about 100 cm^{-1}, the laser
is operated on a single longitudinal mode, and the intensity of the
Rayleigh component of the scattered radiation is reduced with an
iodine filter.

The design of the high-pressure cell used for the work described
here is illustrated schematically in Fig. 1, a composite cross-
sectional view of the cell that illustrates most of its important
features. The cell is a modification of one designed by Drickamer.[4]

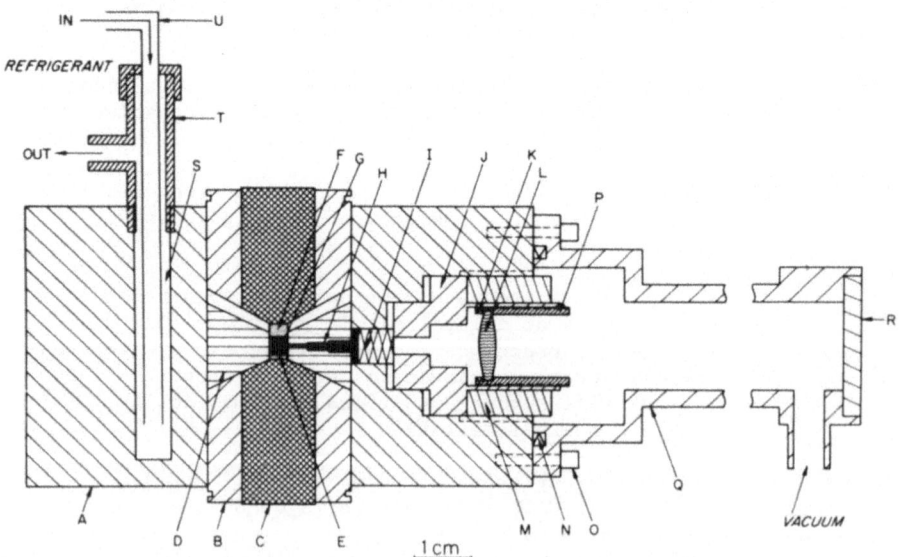

Fig. 1. Cross-sectional side view of the variable-temperature
high-pressure optical cell depicting one of three optical windows
and one of four refrigerant channels.

The sample is confined within a cylindrical chamber (G) by a pair of pistons (E and F) through which force is applied. Within the cylinder, sodium chloride is used to transmit pressure to the sample as well as to orient the sample during assembly of the cell. The right side of the illustration depicts one of the three optical windows. It consists of a compacted salt inner window (H) that is kept under compression by a sapphire (I) and steel piston assembly (J,M). A small, collimating lens (L) is mounted in the piston assembly to increase the optical efficiency of the cell. An outer evacuable jacket (Q) is used during low-temperature operation of the cell to prevent condensation in the optical path. Four refrigerant ports (S) have been drilled in the main cell jacket (A) through which coolants like liquid nitrogen can be circulated to control the temperature of the cell. By a combination of circulating coolant and resistance heating, the temperature of the cell can be closely controlled throughout the range 77 to 500 kelvins. The temperature of the sample is measured with a copper-constantan thermocouple located adjacent to the upper piston (F).

The pressure within the sample chamber (G) cannot be measured directly and must be calibrated in terms of the force (load) applied to the anvils (B). The primary calibration of this cell is based upon observation by Rayleigh and Raman scattering of the applied loads at which several known phase transitions occur. The transitions studied are listed in Table I, and the linear dependence of

TABLE I

High Pressure Phase Transitions Used to Calibrate the Cell

	Phase Transition	Reported Pressure	Precision of Measured Pressure
1.	KNO_3 II-III	2.4 kbar $(54°C)$[a]	± 0.5 kbar
2.	KNO_3 III-IV	3.9 kbar $(54°C)$[a]	± 0.5 kbar
3.	RbCl	7.3[b], 7.5[c] kbar	± 0.3 kbar
4.	$AgNO_3$ I-II	9.78 kbar[d]	± 0.6 kbar
5.	KBr	18.1 kbar[e]	± 0.3 kbar
6.	KCl	19.7 kbar[e]	± 0.3 kbar
7.	$AgNO_3$ II-?	21.2 kbar[d]	± 0.6 kbar

a. P.W. Bridgman, Proc. Am. Acad. Arts Sci. 51, 60 (1916); ibid 72, 46 (1937).
b. G.A. Samara, Solid State Comm. 4, 279 (1966).
c. D.B. Fitchen, Phys. Rev. 134A, 1599 (1964).
d. G.W. Kennedy, in D.E. Gray, ed., Amer. Inst. of Phys. Handbook (McGraw-Hill, New York, 1963).
e. P.W. Bridgman, Proc. Am. Acad. Arts Sci. 76, 55 (1948).

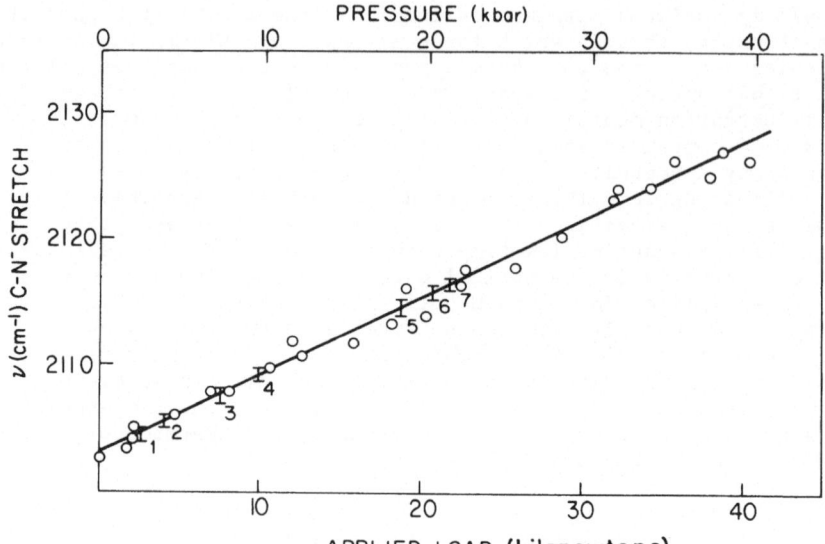

Fig. 2. A plot of ν_{CN} (open circles) vs. load (in kilonewtons) applied to the anvils of the cell and the pressure (in kbar) within the cell as determined by the loads at which seven known phase transitions are observed. The transitions are identified in Table I. The solid line represents the linear dependence of ν_{CN} on applied load as determined by least-squares analysis.

the transition pressure on the applied load is represented in Fig. 2. At other pressures, the calibration is related to these fixed points by observation at room temperature (23 ± 3°C) of the frequency, ν_{CN}, of the Raman spectrum due to the C-N stretch of the cyanide ion substitutionally dissolved in NaCl. This frequency (in cm^{-1}) is related to the pressure by the equation:

$$\nu_{CN} = 2103 + 0.62(\pm 0.03)P$$

where P is the pressure in kbar.[3] As illustrated in Fig. 2, ν_{CN} varies essentially linearly with applied load up to loads corresponding to pressures greater than 40 kbar when the pressure cylinder (D) is relatively new.

Pressures determined by this calibration are less precisely known than pressures measured in hydrostatic optical cells. For measurements of frequency shifts and mode Gruneisen constants, this lower precision is a disadvantage which is compensated by the higher pressure capability of this cell. A limited amount of calibration work done at other temperatures suggests that the room-temperature calibration is at least a good approximate calibration over the entire range of temperatures for which this cell is used, although more work is required to confirm this at very low temperatures.

RESULTS

Fluorite and Related AF_2 Compounds. The Raman spectrum of fluorite is among the simplest of any solid--it consists of a single, narrow line. Thus, a discussion of the spectra of this and related compounds is a good introduction to effects of pressure on Raman spectra. The simplicity of the fluorite spectrum is related to the simplicity of the fluorite structure, depicted in Fig. 3 in terms of three interpenetrating fcc lattices--one of calcium ions and two of fluoride ions. The basis of this structure is the simple, linear F-Ca-F unit whose symmetry modes of vibration are depicted at the bottom of the figure. The cubic symmetry of the lattice requires that each mode is triply degenerate except for a to-lo splitting of the infrared-active mode due to long-range electrostatic effects. Because the calcium ion is located at a site of inversion symmetry, the mutual exclusion rule is obeyed; and only the one symmetric mode is Raman active. For this mode, the two fluoride sublattices move against each other, while the calcium sublattice remains stationary. This yields the single-line Raman spectrum which for CaF_2 at room temperature and atmospheric pressure occurs at $323 \, cm^{-1}$.

Application of pressure to CaF_2 and the resulting uniform compression of the lattice shifts the frequency of this spectrum to higher frequencies as depicted in Fig. 4. The rate of shift, $+0.74 \, cm^{-1} \, kbar^{-1}$, is in close agreement with the value, $+0.72 \, cm^{-1} \, kbar^{-1}$, obtained in work with a hydrostatic cell by Mitra.[5] This corresponds to a mode Gruneisen constant γ_R of the value:

$$\gamma_R = \left. \frac{-\delta \ln \nu}{\delta \ln V} \right)_T = \left. \frac{B_T}{\nu_o} \frac{\delta \nu}{\delta P} \right)_T = 1.8_5 \, (\pm 0.1_0)$$

where Wong and Schuele's value[6] has been used for the isothermal bulk modulus, B_T, and ν_o is the value of the Raman frequency at 1 atm.

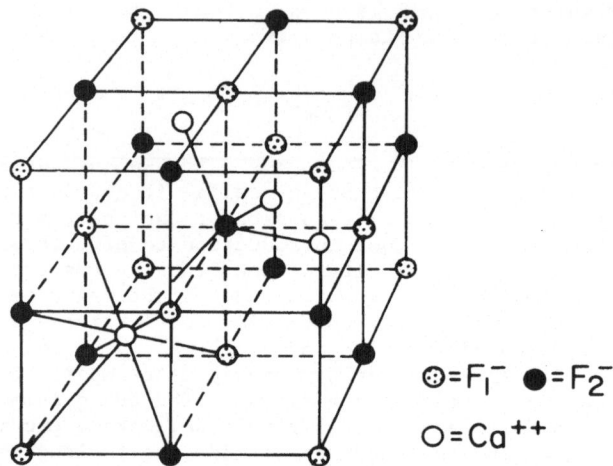

$$\odot = F_1^- \quad \bullet = F_2^-$$
$$\bigcirc = Ca^{++}$$

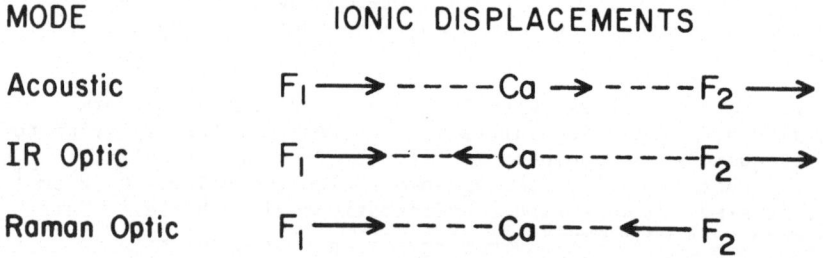

Fig. 3. Upper: The fluorite structure. Lower: Symmetry modes of the F-Ca-F basis of fluorite.

Several other divalent fluorides have the fluorite structure including SrF_2, BaF_2, CdF_2, and PbF_2; and the pressure dependences of their Raman frequencies are quite similar to that of CaF_2. Fig. 5, for example, depicts the apparently linear pressure dependence of the Raman frequency of BaF_2 which shifts at the rate $+0.85$ cm^{-1} kbar^{-1}, corresponding to a γ_R of approximately 1.9 or 2.0, depending upon the value selected for B_T. These and similar data for SrF_2 and CdF_2 are summarized in Table II. The positive frequency shifts are a general phenomenon and can be qualitatively attributed to the increasing importance of the short-range repulsive interactions between neighboring ions as they are brought closer together. For the fluorites, these numbers also can be interpreted in terms of quantitative lattice dynamical models like that due to Axe[7], and quite reasonable agreement between calculated and observed pressure dependences can be obtained.

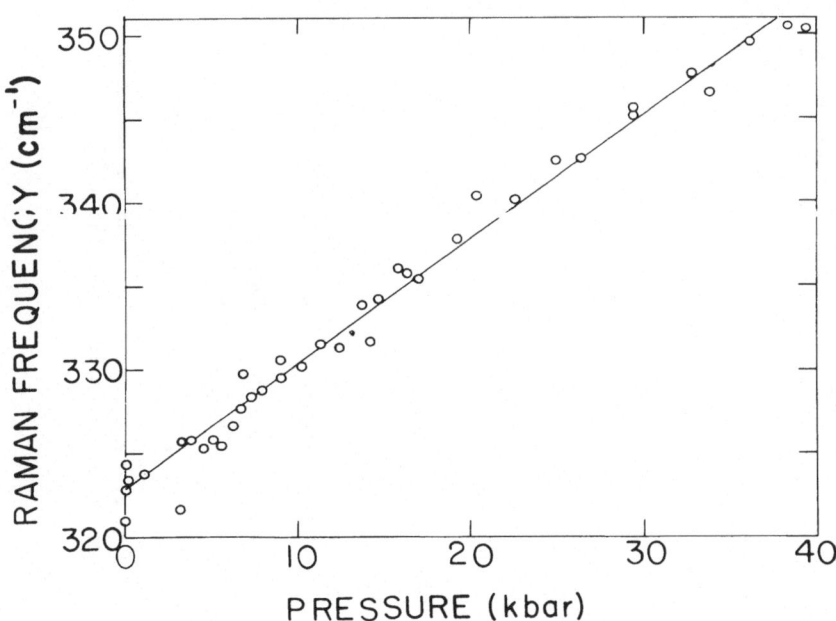

Fig. 4. A plot of the frequency of the Raman-active mode of CaF_2 vs. pressure. The equation of the solid line is ν(cm^{-1}) = 323 + 0.74 P(kbar).

TABLE II.

Pressure Dependences of the Raman Frequencies
of Fluorite Phases of Some AF_2 Compounds

Compound	ν_0 (cm^{-1})	$\dfrac{d\nu}{\partial p}$ (cm^{-1}kbar^{-1})	γ_R
CaF_2	323	0.74 (±0.04)	1.8_5[a] (±0.1_0)
		0.72[b]	1.9[b]
SrF_2	285[c]	0.70[b]	1.6[d]
BaF_2	241	0.85 (±0.06)	1.8_7 (±0.1_0)[d]
			2.0_0 (±0.1_0)[a]
		0.80[b]	1.8[b]
CdF_2	318	0.47 (±0.04)	1.3_6[d] (±0.1_0)
		(See text)	

a. Calculated from ν, $\frac{d\nu}{\partial p}$, and the value of B_T given by C. Wong and D. Schuele,
    J. Phys. Chem. Solids 29, 1309 (1968).

b. S.S. Mitra (unpublished data) cited by J.F. Ferraro, H. Horan, and A. Quattrochi
    J. Chem. Phys. 55, 664 (1971).

c. J.D. Axe, Phys. Rev. 139, A1215 (1965).

d. Calculated from ν_0, $\frac{d\nu}{\partial p}$, and the value of B_T given by P.W. Bridgman, Proc.
    Am. Acad. Arts Sci. 66, 185 (1931).

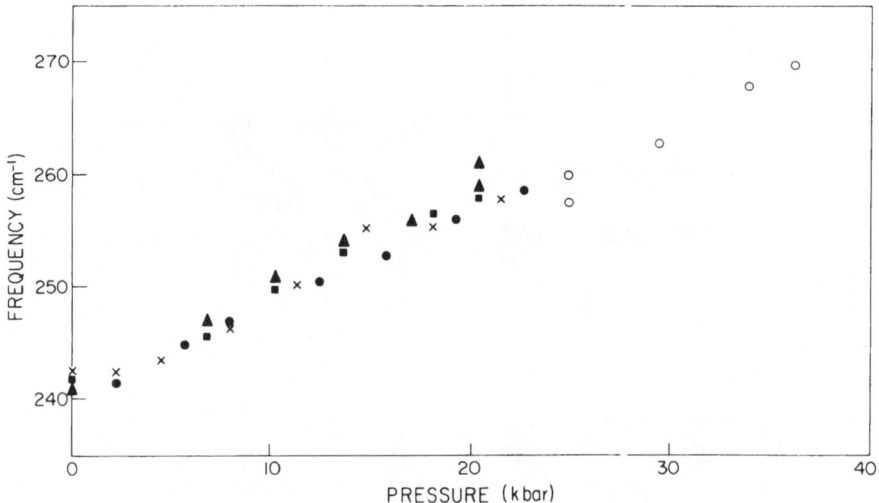

Fig. 5. A plot of the frequency of the Raman-active mode of
$BaF_2(O_h^5)$--solid points--vs. pressure. The open circles represent
data for BaF_2 II.

The pressure dependence of the Raman frequency of BaF_2 depicted
in Fig. 5 can be somewhat misleading, however, if it is not interpreted with care. Although the apparently linear shift of this
frequency with pressure can be represented up to 40 kbar and higher,
BaF_2 undergoes a phase transformation at about 23 kbar with a volume
discontinuity of about 3%. The data points for the higher pressures
are for a closely related but different mode of the high-pressure
orthorhombic phase. The nature of this transformation is depicted
in Fig. 6. In the fluorite structure on the left, [111] planes of
cations are sandwiched between pairs of [111] planes of one or the
other anion sublattices. Each cation is surrounded by eight
equivalent anions. The transformation involves moving half of the
cations in a [111] plane up along [111] into the immediately adjacent
anion plane while the other cations move down along [111] into the
other anion plane. The effects of this transformation are to
increase the number of atoms in the primitive cell from three to
twelve and to increase the cation coordination to nine, although
these nine neighbors are not all equivalent. The increase of the
size of the primitive cell, together with the reduction of the space
group symmetry from O_h^5 to V_h^{16}, increases the possible number of
bands in the first-order Raman spectrum from one to eighteen and
yields the far more complex spectrum shown in Fig. 7. This figure

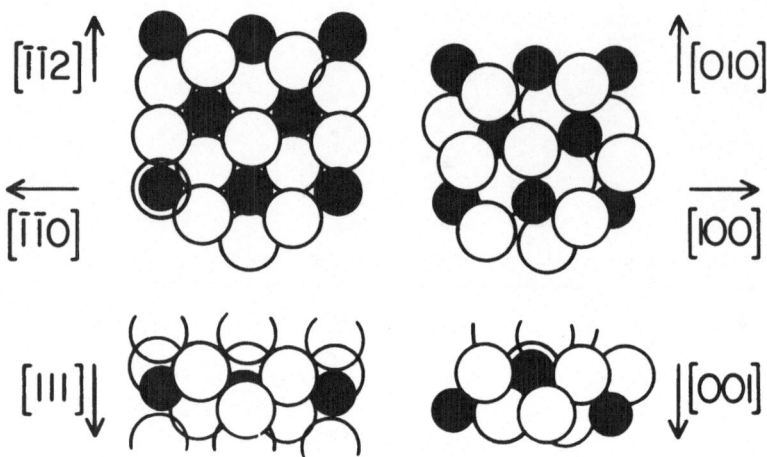

Fig. 6. Left: Two projections of the structure of the O_h^5 phase of BaF$_2$. Right: The corresponding projections of the V_h^{16} structure of BaF$_2$.

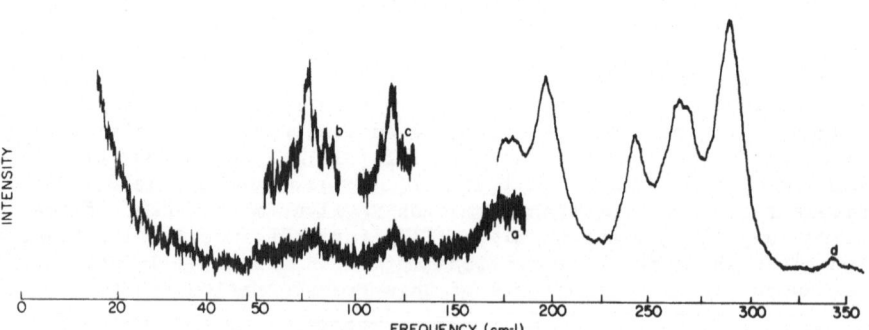

Fig. 7. A composite of Raman spectra of BaF$_2$II at about 25 kbar. An iodine filter was used to reduce the Rayleigh intensity for a,b, and c. Note change of abscissa at 50 cm^{-1}.

is a composite of several BaF$_2$II spectra, taken with various filters and slit widths to show most of the ten features that have been identified and to give an impression of their relative intensities.

This BaF$_2$II spectrum can be analyzed both qualitatively and quantitatively in terms of the corresponding features of the phonon spectrum of the fluorite phase.[8] The two features of the low-frequency region correspond closely in frequency to critical points of the ta bands of the fluorite phase, while the 150-225 cm^{-1} region may be associated with the la and to phonon branches. The three bands between 225 and 300 cm^{-1} are related to the Raman branch of the fluorite phase; thus, it is not entirely coincidental that the frequency of the 264 cm^{-1} band appears where the Raman band of the fluorite phase would be expected at the same pressure. It is also interesting that the peak intensity of this band is about one-fourth of that of the Raman line of the fluorite phase. The weak band near 340 cm^{-1} derives from the lo branch of the fluorite phase.

The quantitative basis of these relationships is more clearly demonstrated in Table III which compares the Raman frequencies of BaF$_2$II at about 25 kbar with the to, Raman, and lo frequencies of the fluorite phase at 1 bar and with two sets of calculated frequencies for the BaF$_2$ II spectrum based upon reasonable estimates of the structure and interionic potentials for BaF$_2$II. The structure assumed for both calculations is based upon lattice constants of a sample of BaF$_2$ II recovered at 1 bar determined by Dandekar and Jamieson[9] and positional parameters typical of the V_h^{16} structures of the other barium halides.[10] A rigid-ion approximation of Axe's shell model of the fluorite phase of BaF$_2$, as modified by Wong and Schuele,[11] was used for the interionic potential in Model I. In this approximation, the ions are assumed to carry their full formal charge, and an exponential repulsive potential is used for the anion-cation short-range potential. The anion-anion short-range potential is taken to be the sum of an exponential repulsive term and an attractive r^{-6} term that is used to make this interaction attractive for the fluoride-fluoride separation of BaF$_2$ at 1 bar. The short-range potential between cations is neglected due to the large separations involved.

In Model II, the magnitude of the attractive r^{-6} term of Model I is reduced by about half, which significantly improves the fit. This reduction may be considered as a compensation for use of lattice constants appropriate to 1 bar to calculate a 25-kbar spectrum. Although the approximations involved are crude, the fit obtained for Model II is surprisingly good and suggests that the model should be studied further to examine the nature of the instabilities that lead to the transition.

TABLE III. Comparison of Observed Raman Frequencies of BaF_2 II with IR and Raman Frequencies of BaF_2 I and with Frequencies Calculated by Two Models

Observed		Calculated	
O_h^5	V_h^{16}	Model I	Model II
		$57, A_g^1$	
		$64, B_{2g}^1$	$67, B_{2g}^1$
	79		$80, A_g^1$
		$87, B_{3g}^1$	$83, B_{3g}^1$
	120	$113, A_g^2$	$118, A_g^2$
		$133, A_g^3$	$132, A_g^3$
			$160, B_{1g}^1$
	163	$164, B_{3g}^2$	$168, B_{1g}^2$
		$167, B_{1g}^1$	$174, B_{3g}^2$
	175	$170, B_{1g}^2 + B_{2g}^2$	$177, B_{1g}^3$
	181		
184, to			
	195		$196, B_{2g}^2$
241, R	241	$255, A_g^4$	$236, A_g^4$

264		$271,\ B_{2g}^{3}$
286		$297,\ B_{3g}^{3}$
	$303,\ A_{g}^{5}$	$301,\ B_{1g}^{4}$
	$305,\ B_{1g}^{4}$	
	$318,\ B_{2g}^{3}$	$319,\ A_{g}^{5}$
	$328,\ B_{3g}^{3}$	
326, 1o	$330,\ B_{1g}^{5}$	$331,\ B_{1g}^{5}$
	$335,\ A_{g}^{6}$	
340	$356,\ B_{1g}^{6}$	$369,\ A_{g}^{6}$
		$395,\ B_{1g}^{6}$

a. Model I. Lattice constants for BaF_II at 1 bar from D.P. Dandekar and J.C. Jamieson, Trans. Amer. Cryst. Assn. 5, 19 (1969). Positional parameters from BaCl_2; E.B. Brackett, T.E. Brackett, and R.L. Suss, J. Phys. Chem. 67, 2132 (1963). Potential parameters from C. Wong and D. Schuele, J. Phys. Chem. Solids 29, 1309 (1968).

b. Model II: Same as Model I except B constant of attractive r^{-6} term of F-F short range interaction reduced from 0.435 to 0.225.

Lead fluoride undergoes a similar transition from an O_h^5 to a
V_h^{16}, $\alpha PbCl_2$, structure at about 4 kbar at room temperature. The
Raman spectrum of the high-pressure phase has recently been
obtained, and the 30-kbar spectrum is depicted in Fig. 8. It
resembles the spectrum of BaF_2 II in many respects, although any
weak bands that might be related to the lo branch of the fluorite
phase are apparently hidden by the tail of the 266 cm^{-1} band. In
Table IV, this spectrum is compared with spectra of the V_h^{16} phase
of $PbCl_2$[12] which suggests the possibility of additional bands at
about 18 and 35 cm^{-1} that have not yet been detected. The frequencies of the high-frequency modes of PbF_2 II are similar to
those computed from $\alpha PbCl_2$ and $PbBr_2$ frequencies[12] simply by cor-
recting for the different anion masses, and very preliminary calculations suggest that it will not be difficult to fit this spectrum
with models similar to those used to fit the BaF_2 II spectrum.
Because the transition pressure is so low, the pressure dependence
of the phonon frequencies of the PbF_2 II phase can be accurately
determined which may prove useful in resolving some of mode
assignments that are not easily discriminated by other criteria.
For this as well as the BaF_2 system, it now seems to be practical
to examine theoretical models that hopefully can adequately explain

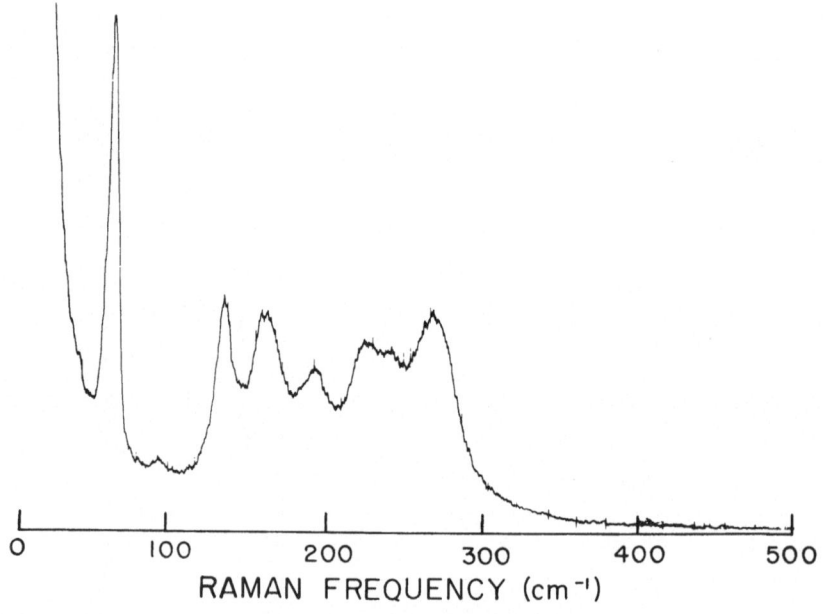

RAMAN FREQUENCY (cm⁻¹)

Fig. 8. The Raman spectrum of PbF_2 II at 30 kbar.

TABLE IV. Comparison of the Raman Frequencies of PbF_2 II
with Phonon Frequencies of Related Compounds

Compound:	PbF_2	PbF_2	$PbCl_2$ [b]
Phase:	O_h^5	V_h^{16}	V_h^{16}
Pressure:	1 bar	30 kbar	1 bar
			18, A_g
			25, B_{1g}
		61	35, $A_g + (B_{1g}?)$
			58, B_{1g}
		91	86, $B_{2g} - B_{3g}$
	102^a to		
		133	100, $(B_{1g}?)$ [137]
		158	126, B_{1g} [173]
		190	134, B_{2g} [184]
		223	
		236	156, $A_g + B_{2g}$
	257 r	266	178, B_{1g} [243]
	337 lo		

---

a. J.D. Axe, J.W. Gaglianello, and J.E. Scardefield, Phys. Rev. 139, 1211 (1965).

b. G.A. Ozin, Can J. Chem. 48, 2931 (1970). Frequencies enclosed in brackets are obtained by multiplying the α $PbCl_2$ frequencies by 1.37, the square root of the ratio of the atomic weights of chlorine and fluorine, to estimate the approximate effect of the anion's mass on these high frequency modes that predominately involve anion motions.

both the effects of pressure on the lattice dynamics of several
phases of each system as well as the atomic bases of transformations
between them. Furthermore, only modest increases of the pressure
range of these experiments will permit examination of similar
transitions in other fluorite analogues including CdF_2, SrF_2 and
CaF_2. Such studies could be an important basis for interpretation
of high-pressure Raman spectra of more complicated systems.

Systems with Unusual Pressure Dependences. **Rutile.** The
remainder of the report describes effects of pressure on Raman
spectra of four systems that have unusual characteristics,
represent unsolved problems, and demonstrate some of the unique
capabilities of the variable-temperature cell. The first of these
systems is the rutile phase of TiO_2 whose structure is depicted in
Fig. 9. The primitive cell is tetragonal and contains two O-Ti-O
formula units. Each Ti is approximately octahedrally coordinated.
One unusual aspect of this structure is the short separation of
nearest-neighbor oxygen ions along [110]; this distance is about
0.5 Å less than a typical diameter of an oxygen ion.

Even at atmospheric pressure, the Raman spectrum of rutile,
shown in Fig. 10d, is unusual because several of the intense
features--e.g., that at 233 cm^{-1}--are due to second- or higher-
order Raman processes.[13] However, the effects of pressure on the
parts of the spectrum due to these high-order processes are typical
of effects on other phonons. The pressure dependences of the
frequency of the first-order component due to the B_{1g} phonon at
about 145 cm^{-1} is unusual because this frequency decreases nearly
linearly with increasing pressure at about -0.3 cm^{-1} kbar^{-1} as
shown in Fig. 11. This rate is independent of the orientation of
the rutile crystal within the high pressure cell; thus, this effect
cannot be attributed to any anisotropy of the stress distribution

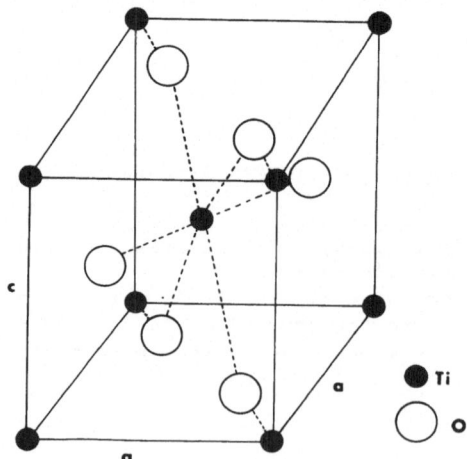

Fig. 9. The tetragonal
structure of rutile.

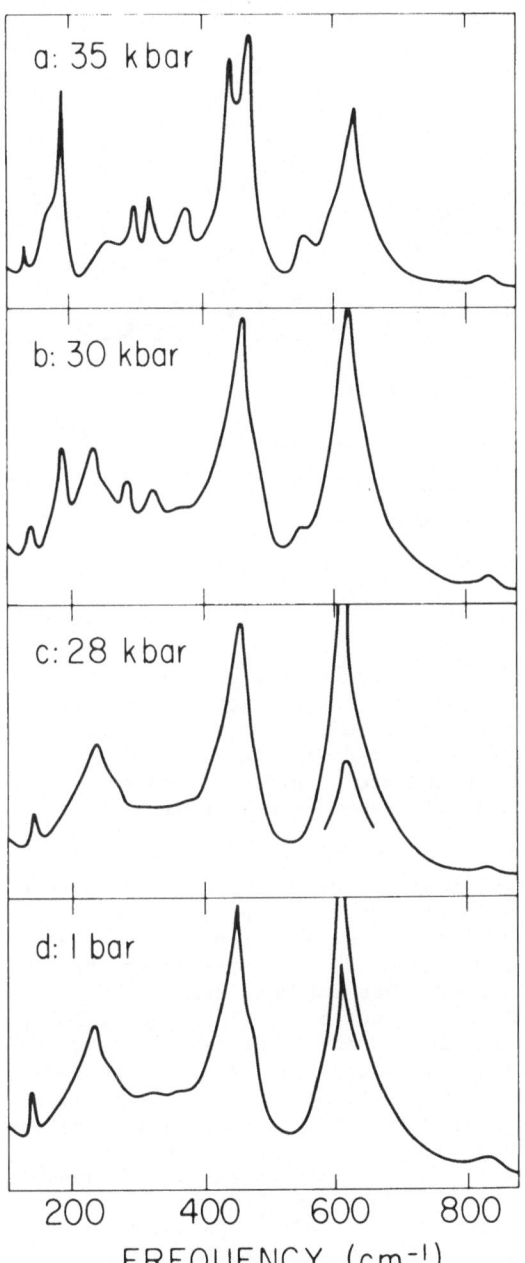

Fig. 10. Raman spectra obtained on compression of a rutile crystal at several pressures. The crystal was oriented with its tetragonal axis normal to the cylindrical axis of the pressure cell.

a: 35 kbar

b: 30 kbar

c: 28 kbar

d: 1 bar

FREQUENCY (cm⁻¹)

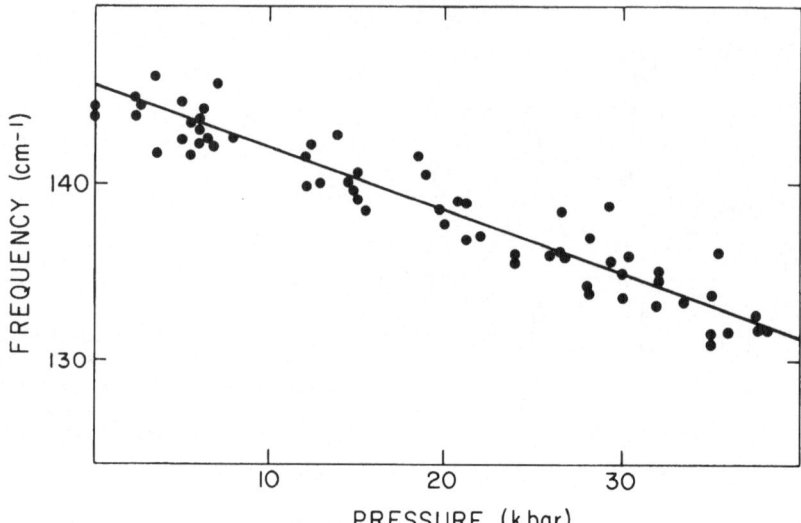

Fig. 11. A plot of the frequency of the B_{1g} mode of rutile vs. pressure.

within the cell. The negative frequency dependence can be ration-
alized in terms of the motion associated with this mode, which ro-
tates the O-Ti-O groups around the c axis and, thereby, tends to
reduce the repulsive interactions associated with the short O-O
distance. Extrapolation of this dependence to higher pressures
suggests that the rutile structure will become unstable with respect
to some lower symmetry phase.

     Indications of a phase transformation from rutile to a structure
of lower symmetry have been detected for rutile samples mounted
with their tetragonal c axes normal to the axis of the high pressure
cylinder, as shown in Fig. 10--compare the 30- and 35-kbar spectra
with those at lower pressures. No evidence for this transition
was observed for samples mounted with c axes parallel to the cylinder
axis which implies that some stress anisotropy is necessary to drive
the transformation. This transition was discussed by Nicol and Fong[14]
in terms of the V_h^{14} or αPbO_2 structure of TiO_2, known as TiO_2 II,
that has been recovered from rutile samples subjected to high-pressure
shocks, although available Raman evidence by itself is adequate
to define the new structure. An alternative assignment of the new
structure has recently been proposed by Nagel and O'Keeffe[15] who
identify it as a V_h^{12} structure like that of $CaCl_2$. This structure
is related to rutile by a combination of the B_{1g} rotation and an
orthorhombic distortion of the tetragonal lattice that could be

induced by a uniaxial stress normal to the tetragonal c axis. For any other material, the number of components of the Raman spectrum of the high-pressure phase--at least 10 and possibly 14 have been identified--would be inconsistent with the $V_h{}^{12}$ structure. However, as in the rutile spectrum, several of these components may arise from higher-order Raman processes. Further studies of related systems and, if possible, of recovered TiO_2 II are required to resolve this problem.

Calcium Carbonate. Compression of the calcite phase of $CaCO_3$ at room temperature results in two phase transformations to phases of unknown structures.[16] At 14 kbar, calcite transforms to $CaCO_3$ II which then transforms to $CaCO_3$ III at 18 kbar. Raman spectra of these two high-pressure phases, prepared by this room-temperature process, have been reported by Fong and Nicol[17] and are shown in Fig. 12. Unfortunately, the II-III transformation appears to be incomplete at room temperature; and Raman spectra of $CaCO_3$ III prepared in this manner often appear to include features characteristic

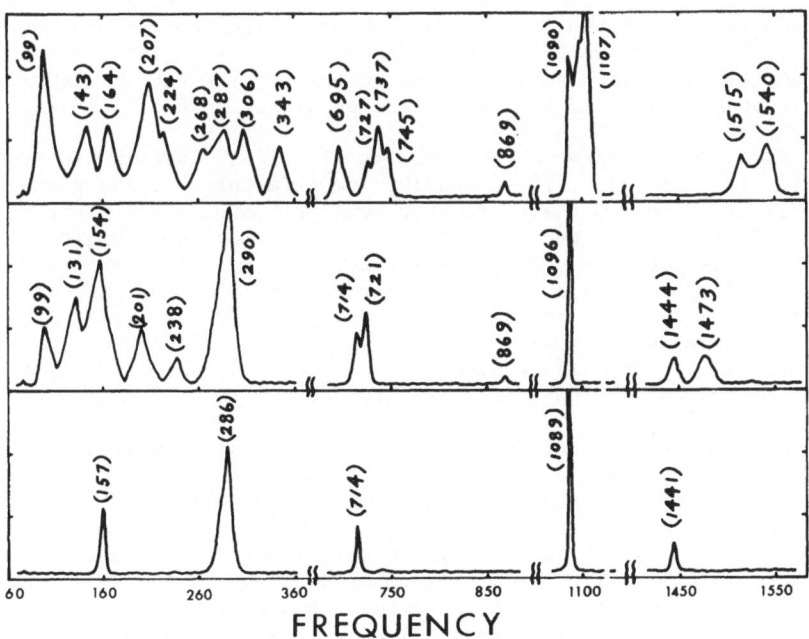

FREQUENCY

Fig. 12. Raman spectra of calcite (lower), $CaCO_3$ II and $CaCO_3$ III (upper) obtained upon compression of calcite at room temperature.

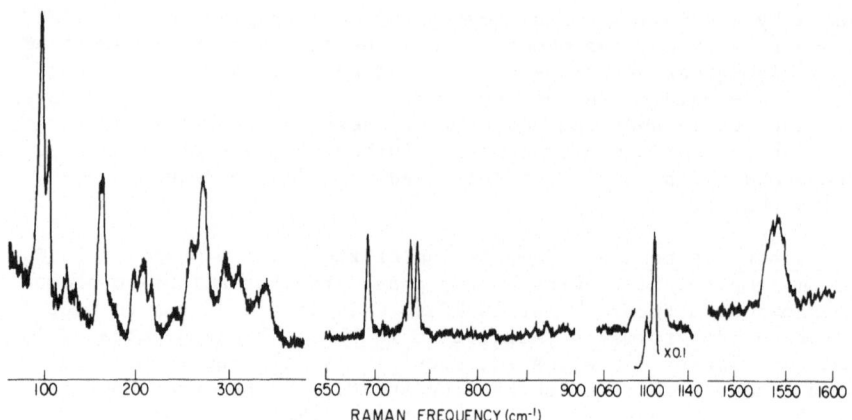

RAMAN FREQUENCY (cm⁻¹)

Fig. 13. Raman spectrum of $CaCO_3III$ at 18 kbar and 77K, prepared
directly from calcite by compression at 77K.

of $CaCO_3II$ which interfere with the identification of the spectrum
and structure of $CaCO_3III$. In order to obtain Raman spectra of
$CaCO_3III$ free from $CaCO_3II$ contamination, an alternative preparation
of $CaCO_3III$ was attempted in which the low-temperature capability of
this high-pressure cell was exploited to transform calcite directly
to $CaCO_3III$ at 77K which is well below the calcite-$CaCO_3II$-$CaCO_3III$
triple point. At 77K, the transition pressure is about 11±3 kbar.[18]

The $CaCO_3III$ spectrum obtained at 77K and 18 kbar is reproduced
in Fig. 13. The 715-, 727-, and 1099-cm⁻¹ features of the room-
temperature spectrum that were attributed to $CaCO_3II$ are clearly
absent from this spectrum, and the line-narrowing that results at
the lower temperature permits identification of three previously
unresolved features: a weak line near 243 cm⁻¹, a line of medium
intensity at 260 cm⁻¹ that appears as an unresolved shoulder of the
269-cm⁻¹ line, and a weak unresolved shoulder of the 170-cm⁻¹ line
near 179 cm⁻¹. This illustrates one of the important advantages of
the low-temperature capability of this cell.

Although the structure of $CaCO_3III$ cannot be unambiguously
identified from these spectra--and other data are of limited value,
the Raman spectra provide some clues about possible structures. The
two lines in the ν_1 region imply that the primitive cell contains at
least two carbonate ions; but, if there are only two formula units
in the primitive cell, the total number of lines in the spectrum
require that the symmetry of the structure must be sufficiently low
that all phonons would be Raman-allowed. This possibility is not
inconsistent with other available data.[19-22] Structures with larger
primitive cells cannot be excluded; however, both known $CaCO_3$

structures with cells containing four formula units can be excluded for a combination of reasons.[17,19]

Ammonium Chloride and Ammonium Bromide. The room-temperature structures of both ammonium chloride and ammonium bromide are interpreted in terms of the simple model depicted in Fig. 14.[23] The ammonium ion occupies the center of the primitive cell with the chloride at the vertices; the hydrogens are located along four of the eight [111] axes. In any cell, therefore, two equivalent orientations of the ammonium ion are possible; and, at room temperature and 1 bar, the ions are randomly distributed between these orientations. Both at lower temperatures and at higher pressures, however, ammonium chloride and ammonium bromide order with respect to ammonium ion orientations. At room temperature, the transition pressures of NH_4Cl and NH_4Br are about 10 kbar and 19 kbar, respectively. (The behavior of NH_4Br is complicated by the occurrence of phases that are said to represent intermediate degrees of ordering.) The order-disorder transitions and other unusual properties of these ammonium halides are responsible for some interesting effects of pressure on their Raman spectra, a few of which are described here.

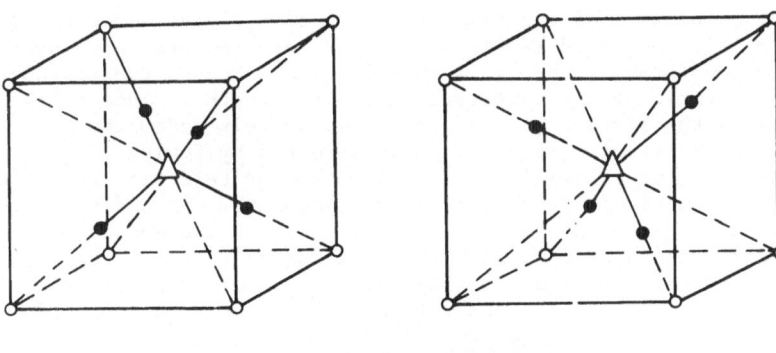

○ **HALOGEN**
△ **NITROGEN**
● **HYDROGEN**

Fig. 14. The structure of the room-temperature phase of NH_4Cl showing the two ammonium ion orientations.

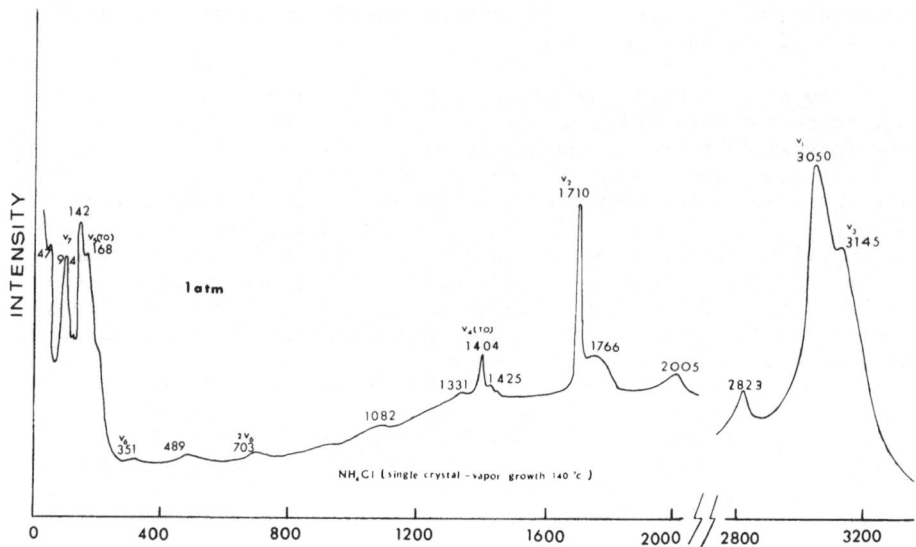

Fig. 15. Raman spectrum of a NH$_4$Cl crystal grown from the vapor phase at 1 bar and room temperature.

The complexity of the Raman spectrum of the disordered phase of NH$_4$Cl is illustrated in Fig. 15. Many more features are observed than are predicted by the usual selection rules based upon trans- lational invarience and conservation of wave vector. Indeed, the spectrum in the lattice mode region qualitatively resembles the calculated density of states recently reported by Cowley[24] which suggests that the relaxation of the wave vector selection rule due to the orientational disorder is nearly complete. In addition to other intense bands due to the internal modes of the ammonium ion, the spectrum includes many weak features due to the lo lattice mode, the librational mode and its overtone, and a variety of combination and overtones, some of which have not been assigned.

The effect of compression and the resulting ordering on the lattice mode spectrum is shown in Fig. 16 which compares the spectrum of the disordered phase at 1 bar with spectra obtained at 8.3 kbar, near the ordering transition, and at 26 kbar where the crystal is ordered. The effect of ordering is to reinstate the wave vector conservation rule that permits scattering only by phonons at the center of the zone. Similar intensity effects related to ordering are observed for the librational fundamental, ν_6, which is Raman-forbidden in the ordered phase, and for the to component of ν_4.

The pressure dependences of the Raman spectra of the internal modes of the ammonium ion can be seen in Fig. 17 which compares the room temperature spectrum of the disordered phase of NH$_4$Br at

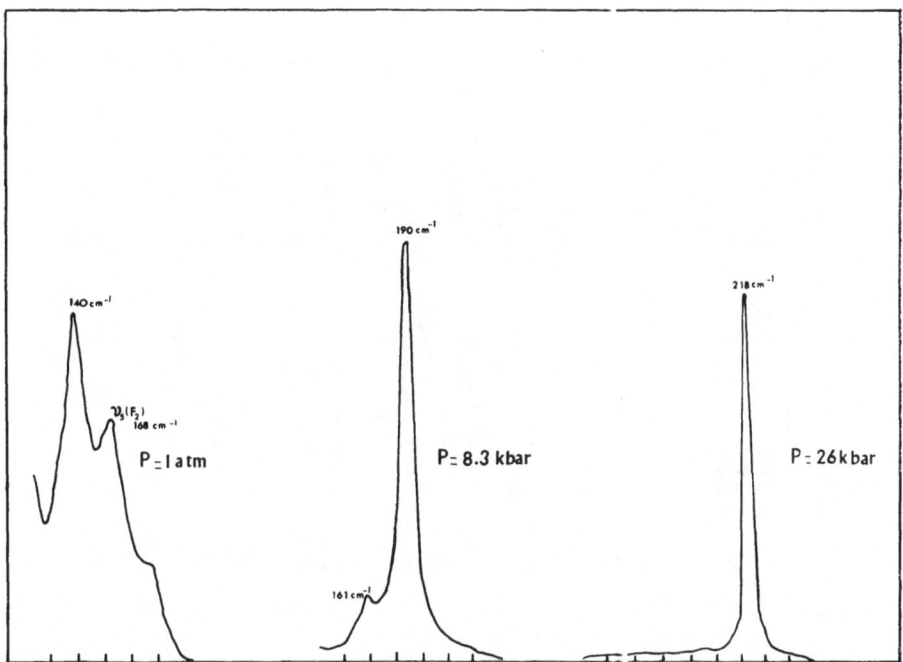

Fig. 16.   Room-temperature lattice-phonon Raman spectrum of NH₄Cl
at three pressures. Different intensity scales have been used for
each spectra; thus, the relative intensities of the different
spectra are not accurately represented.

1 bar with the spectrum of an ordered phase of NH₄Br at 27 kbar.
Except for some aspects of the lattice phonon region of NH₄Br that
are not completely understood, the pressure dependences of the
NH₄Cl and NH₄Br spectra are very similar. The frequencies of the
ν_1, ν_3, and ν_4 modes decrease with increasing pressure, although
the decrease of ν_1 for NH₄Br is less than experimental error, while
the frequency of the ν_2 mode increases. These effects can be
attributed to the role of hydrogen bonding on the internal modes.

The assignments of the combination bands can be checked by
comparing the pressure dependences of their frequencies with the
pressure dependences of the corresponding fundamentals. The pressure
dependence, $\frac{\partial \nu}{\partial P}$, of the combination should be the sum of the pressure
dependence of the fundamentals which it combines if the dispersions
of the relevant bands are small or the scattering originates from the
same point of the Brillouin zone. On this basis, the assignment as
$\nu_2 + \nu_6$ of the 2005-cm^{-1} band of NH₄Cl and the 1958-cm^{-1} band of
NH₄Br can be challenged--the combination bands shift by only about
half as much as the sum of the fundamentals; and preliminary data
suggest other assignments require reexamination.

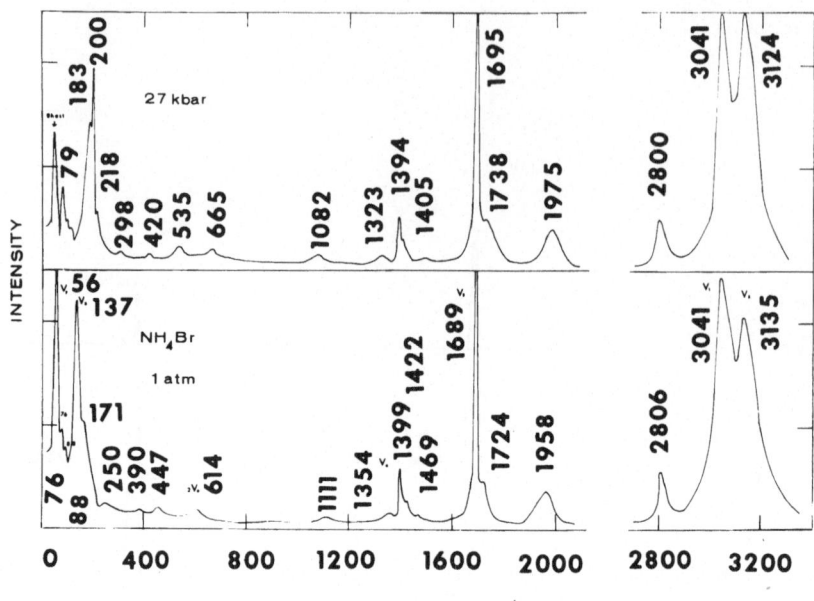

WAVENUMBER (cm⁻¹)

Fig. 17. Raman spectra of NH₄Br at 296 K and 1 atm (lower trace) and 27 kbar (upper trace). A grating ghost at approximately 50 cm⁻¹ is not resolved from the 56 cm⁻¹ line of the 1 bar spectrum which corresponds to the 79 cm⁻¹ line of the 27 kbar spectrum.

The pressure dependence of the librational phonon, ν_6, is of interest because it provides information about the interaction responsible for the ordering. The overtone band, $2\nu_6$, can be studied in both phases, although the fundamental has been resolved only in the disordered phase of NH₄Cl. The pressure dependences of the overtone frequency are plotted in Fig. 18. Each set of data shows two distinctive regions, characteristic of the low-pressure disordered and high-pressure ordered phases. The density dependences of $2\nu_6$ for the disordered phases are significantly larger than for the ordered phases. The value for the disordered phase of NH₄Cl is too large—and the value for the ordered phase is too small—to be consistent with the librational potential proposed by Gutowsky, Pake, and Bersohn[25]. These data suggest that repulsion between hydrogens on adjacent ammonium ions[23,26]—which may be significant in the disordered phases but should be negligible in the ordered phases— have an important role in determining the pressure dependences of the librational frequencies in the disordered phase.

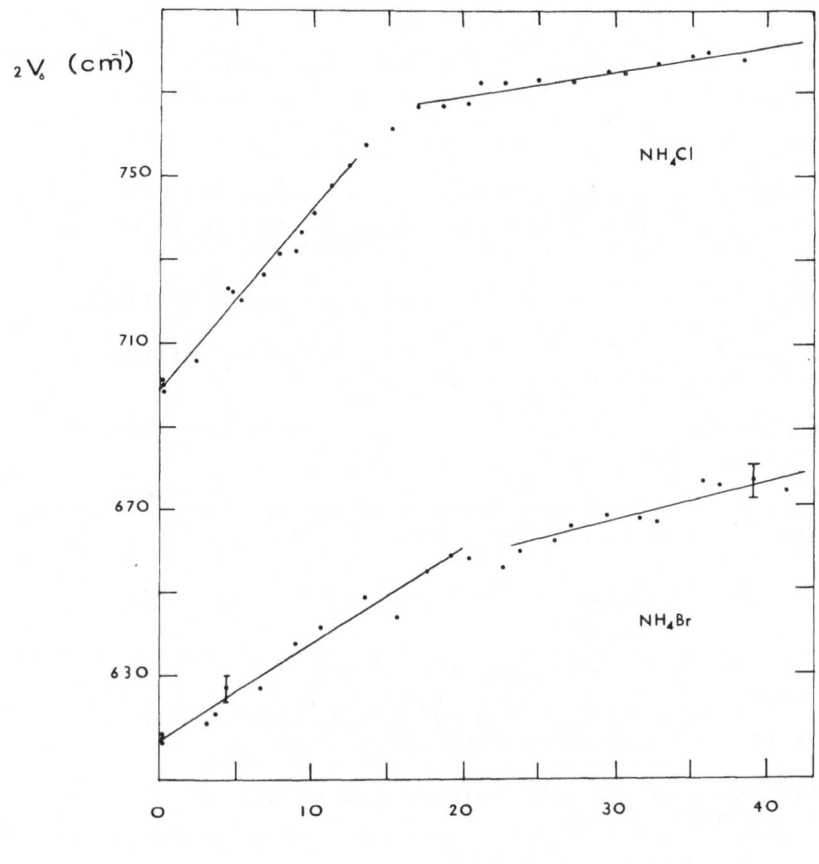

Fig. 18. A plot of the frequencies of the librational overtones, $2\nu_6$, of NH_4Br (lower) and NH_4Cl (upper) vs. pressure at 296 K. The solid lines represent least squares fits of the data for the various phases.

ACKNOWLEDGMENT

     The authors are very grateful for the valuable technical assistance of Mr. Ahzez Karim and the programming support kindly provided by Miss A. Kay Wilkerson.

REFERENCES

*   The support provided by U.S. Atomic Energy Commission Contract
    AT(04-3)34, P.A. 88 is gratefully acknowledged.

+   Contribution No. 2931

1. See, for example, S.P.S. Porto "Light Scattering with Laser
    Sources", in G.B. Wright, ed., Light Scattering Spectra of
    Solids (Springer-Verlag, New York, 1969), p.1.

2. J.F. Asell and M. Nicol, J. Chem. Phys. 49, 5395 (1968).

3. M. Nicol, Y. Ebisuzaki, W.D. Ellenson, and A. Karim, Rev.
    Sci. Instr. (to be published).

4. R.A. Fitch, T.E. Slykhouse and H.G. Drickamer, J. Opt. Soc.
    Amer. 47, 1015 (1957).

5. S.S. Mitra (unpublished data) cited by J.R. Ferraro, H. Horan
    and A. Quattrochi, J. Chem. Phys. 55, 664 (1971).

6. C. Wong and D.E. Schulele, J. Phys. Chem. Solids 28, 1225
    (1967).

7. J.D. Axe, Phys. Rev. 139, A1215 (1965).

8. J.P. Hurrell and V.J. Minkiewicz, Sol. State. Comm. 8, 463
    (1970).

9. D.P. Dandekar and J.C. Jamieson, Trans. Amer. Cryst. Assn.
    5, 19 (1969).

10. E.B. Brackett, T.E. Brackett and R.L. Suss, J. Phys. Chem. 67,
    2132 (1963).

11. C. Wong and D.E. Schuele, J. Phys. Chem. Solids 29, 1309 (1968).

12. G.A. Ozin, Can. J. Chem. 48, 2931 (1970).

13. S.P.S. Porto, P. Fleury, and T.C. Damen, Phys. Rev. 154, 522
    (1967).

14. M. Nicol and M.Y. Fong, J. Chem. Phys. 54, 3167 (1971).

15. L. Nagel and M. O'Keeffe, Mat. Res. Bull. (to be published).

16. P.W. Bridgman, Am. J. Sci. 237, 7 (1939).

17. M.Y. Fong and M. Nicol, J. Chem. Phys. 54, 579 (1971).

18. M. Nicol and W.D. Ellenson, J. Chem. Phys. 56, 000 (1972).

19. J.C. Jamieson, J. Geol. 65, 334 (1957).

20. R.N. Schock and S. Katz, Amer. Mineral. 53, 1910 (1963).

21. C.E. Weir, E.R. Lippencott, A. Van Valkerburg and E.N. Bunting, J. Res. Natl. Bur. Stds. 63A, 55 (1959)

22. D. Cifrulak, Amer. Mineral. 55, 815 (1970).

23. See, for example, C.W. Garland and J.S. Jones, J. Chem. Phys. 41, 1165 (1964) and references therein.

24. E.R. Cowley, Phys. Rev. B3, 2743 (1971).

25. H.S. Gutowsky, G.E. Pake, and R. Bersohr, J. Chem. Phys. 22, 643 (1954).

26. T. Nagamiya, Soc. Chim. Phys., Compt. Rend. Reunion Ann. Comm. Thermodynam. Union Intern. Phys. (Paris), 251 (1952).

THE USE OF PRESSURE IN INFRARED SPECTROSCOPIC STUDIES OF

HYDROGEN BONDING*

R. J. Jakobsen and J. E. Katon

Battelle, Columbus Laboratories, Columbus, Ohio   43201

Chemistry Dept., Miami University, Oxford, Ohio   45056

INTRODUCTION

Until 1965, the use of pressure as a variable in hydrogen
bond spectral studies was extremely limited. There were two
reasons for this: (1) the high cost and cumbersome nature of high
pressure equipment and (2) the early pressure experiments[1] did
not show that there was a large frequency shift with increasing
pressure for OH vibrations. In fact, Pimentel and McClellan[1]
felt that the effect of pressure on OH stretching vibrations was
far less important than the effect of other variables such as
temperature, concentration, or solvents.

The advent of the diamond-window high pressure cell made in
situ spectral measurements at ultra-high pressures both relatively
simple and inexpensive to obtain. In addition, the ultra-high
pressures (up to 150 kbars) that could be obtained with this cell
helped workers to demonstrate that increases in pressure indeed
did produce large frequency shifts for OH vibrations. In fact, it
was possible to show that not only was pressure as important a
variable in hydrogen bond spectral studies as temperature, but in
many instances the use of pressure produced information not available by any other means.

* Presented at the Symposium on Spectroscopy of Materials Under
  High Pressure, Tenth National Meeting, Society for Applied
  Spectroscopy, St. Louis, Missouri, October, 1971.

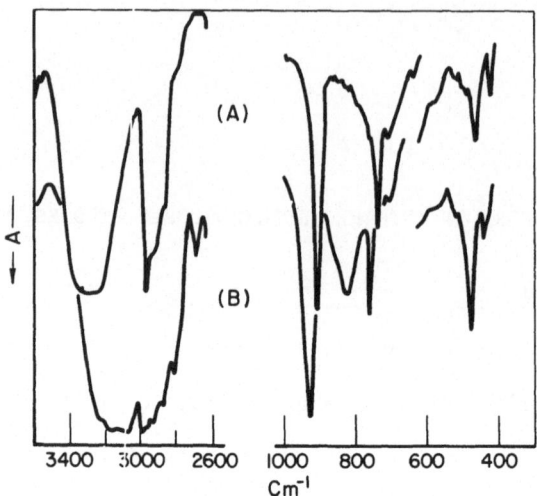

Figure 1. Infrared Spectra of Single Crystals of t-Butanol (A) at Low Pressures and (B) at High Pressures

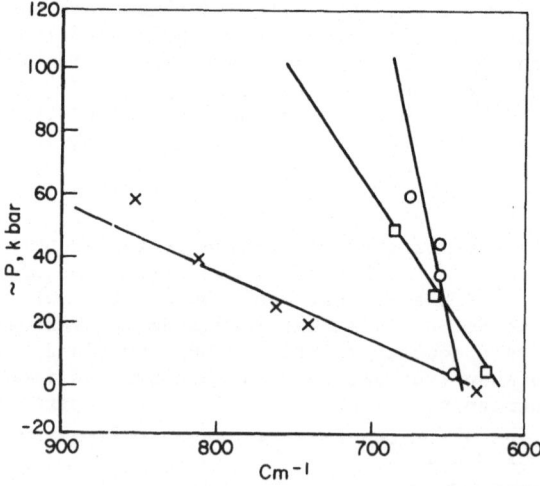

Figure 2. Plot of Pressure vs OH Deformation Frequency of (O) n-Butanol, ( ) 3-Pentanol, and (X) 3-Methyl-3-Pentanol

It is this diamond-window high pressure cell hydrogen bond research that is the subject of this paper. First the published
research since 1965 is briefly reviewed along with some work that
has only been covered in oral presentations at various meetings.
Lastly, the current and as yet unpublished work is presented.

PAST WORK

High pressure techniques have been used in two ways for spectroscopic studies of hydrogen bonding. In one way, studying frequency shifts with variations in pressures gives information on
hydrogen bond structures and configurations and also aids in the
assignment of hydrogen bond vibrations. Variations in pressure
also allows polymorphism in hydrogen bonded crystals to be studied.
The second way high pressure techniques have been useful centers
about the relative ease of growing single crystals in the diamond-
window high pressure cell[2]. The polarized infrared spectra of
these single crystals then can be used to either deduce the crystal
structure of the hydrogen bonded molecule or as another aid in making vibrational assignments.

One of the first examples of pressure techniques in hydrogen
bond studies involved the use of pressure induced frequency shifts
as an aid in the assignment of OH out-of-plane deformation frequencies of alcohols[3]. This is illustrated in Figure 1 which
shows single crystal spectral of t-butanol at both low and high
pressures. In the top spectrum, there is no absorption band in
the 600-900 cm^{-1} region which can obviously be assigned as the out-
of-plane OH deformation. However, a series of spectra at gradually
increasing pressures show a band slowly shifting to higher frequencies until finally it reaches 810 cm^{-1} (as shown in the bottom
spectrum of Figure 1). This significant shifting to higher frequencies with increasing pressure leaves little doubt that the
810 cm^{-1} band is the OH out-of-plane deformation and that it is
not clearly seen at low pressures because it is obscured by the
sharp absorption band in the 760 cm^{-1} region.

The variation with pressure of the OH out-of-plane deformation
for several alcohols[3] is shown in Figure 2. Just like the pressure induced shifts of the OH stretching vibration, the shifts of
the OH deformation vary with type of alcohol (largest shifts for
tertiary alcohols and smallest frequency shifts for primary alcohols). In fact, a good qualitative estimate is that the pressure
induced frequency shifts of the out-of-plane OH deformation will
be about half as large as the pressure induced frequency shifts of
the OH stretching vibration.

In 1966, Mikawa, Jakobsen, and Brasch[4] grew a single crystal of formic acid in a diamond cell and while changing the

pressure, observed a solid-solid phase transformation. The combination of visual observation of the phase change and differences in the infrared spectra of the two phases positively established the existence of two polymorphs of formic acid. Structures were proposed for the two polymorphs which were in conformity with the spectral data. The existence of two polymorphic forms of formic acid cleared up inconsistencies between previous infrared and X-ray data by demonstrating that the infrared data were obtained from one polymorph while the X-ray data were obtained from the second polymorph.

Also in 1966, Ferraro, Mitra, and Postmus[5] reported on successfully using high pressure techniques in the far infrared spectral region. While they did not emphasize hydrogen bond studies, they did report that they observed no pressure induced frequency shifts for the intermolecular hydrogen bond vibrations of organophosphorus compounds. McDevitt[6], et al, did, however, observe large pressure induced frequency shifts for the hydrogen bond stretching mode of phenol. The effect of pressure on the low frequency hydrogen bond stretching vibration would be heavily influenced by the hydrogen bond geometry so it is not necessarily surprising that frequency shifts would be observed in one case, but not in another.

Combining far- and mid-infrared pressure data, Brasch, Jakobsen, McDevitt, and Fateley[7] showed that for phenol the frequency of the far infrared hydrogen bond stretching mode is dependent on the strength of the hydrogen bond. Single crystals of two polymorphs of phenol were grown in the pressure cell and mid- and far-infrared spectra recorded for each form. The frequency shifts of the OH stretching vibrations of the two polymorphs was inversely related to the frequency shifts of the OH---O stretching vibrations of the different forms. Since the shift of νOH is accepted as a measure of hydrogen bond strength, it was concluded that the frequency shift of ν (OH---O) was also related to hydrogen bond strength.

Using both pressure techniques and low temperature isotopic dilution studies, Jakobsen, Brasch, and Mikawa[8] investigated the behavior of the OH stretching frequency of solid alcohols. These experiments showed that all of the solid alcohols studied had two OH stretching vibrations which collapsed to a sharp singlet when decoupled by isotopic dilution. The coupled OH doublet generally consisted of one sharp and one broad component which gave different frequency shifts with increasing pressure and which gave different polarization properties. This behavior could only be explained by crystal splitting of the OH stretching vibration with the splitting being due to nearest neighbor or first order coupling of the vibrations of adjacent OH groups along the hydrogen bond chain. Since

the decoupled OH stretching vibration of these alcohols gives a
narrow singlet absorption, it was concluded that νOH of solid alcohols is not inherently broad, but gains breadth as a result
(direct or indirect) of the coupling between OH groups.

Polarized infrared spectra of single crystals grown in a
diamond-window high pressure cell have been useful in two
cases[9,10]. In the first case[9] the polarized infrared spectra
were used along with the crystal structure determined from X-ray
work to make a complete vibrational assignment of propanoic acid.
In the second case[10] a crystal structure was proposed for ethanol
based on the dichroic measurements. All of the polarized infrared
bands of ethanol could be interpreted on the basis of the proposed
structure. It is important to note that the complete vibrational
assignments made for these two molecules formed the basis for the
interpretation of the low frequency hydrogen bond vibrations of a
series of acids[11] and alcohols[12].

By measuring the effect of pressure on the free (unbonded) OH
stretching vibration and the bonded OH stretching vibration of a
series of hindered (with methyl groups) liquid alcohols, Jakobsen,
Mikawa, and Brasch[13] showed that increasing the pressure does not
produce the same effect as lowering the temperature for these alcohols. Upon lowering the temperature all the free OH absorption
disappeared indicating a shift in the polymer equilibrium toward
longer change length. Pressure increases either did not affect the
free OH absorption or increased its intensity indicating no change
in the polymer equilibrium or a change towards shorter chain
lengths. These spectra also demonstrated that frequency shifts of
the bonded OH stretching vibration do not always indicate a change
in polymer equilibrium. For in these experiments the bonded OH
frequency vibration showed a greater frequency shift with pressure
than with temperature in spite of the fact that the intensity of
the free OH vibration showed that pressure did not cause a change
in the polymer equilibrium. Thus the pressure induced frequency
shifts of the bonded OH vibration were attributed to a decrease in
the length of the hydrogen bond caused by compression of the alcohol molecules. This research also indicated that the free OH absorption observed in the spectra of these alcohols arose from the
polymer end group and not from a monomer alcohol molecule. Therefore the chains were fairly short and linear, not cyclic.

In a continuation of this work, Jakobsen and Mikawa[14] attempted to ascertain the hydrogen bond chain length of these
hindered alcohols. In this study the pressure cell was only used
as a sampling device to hold the alcohol at constant volume while
the sample was being heated. Figure 3 shows the free and bonded
OH stretching vibration of 2,2,4-tri-methyl-3-pentanol at various
temperatures. As the sample was heated, the bonded OH stretching

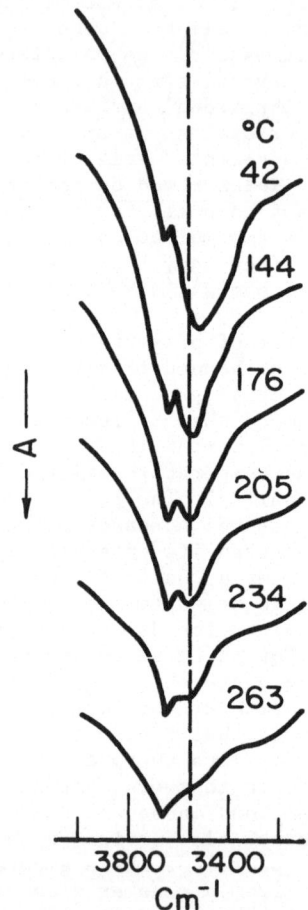

Figure 3.   Infrared Spectra of the Free and Bonded OH Stretching
Frequencies of 2, 2, 4-Trimethyl-3-Pentanol at the Indicated
Temperatures

frequency gradually shifted to higher frequencies and decreased in intensity. The free OH stretching vibration gradually increased in intensity as the temperature was raised. At 176 C, the two bands appeared to be equal in intensity, which was confirmed by the plot of optical densities shown in Figure 4. Also from this temperature on, the frequency of the bonded OH stretching mode remained fairly constant (Figure 5) and any further increase in temperature only resulted in a decrease in the bonded OH mode intensity. All this indicated that at 176 C, the alcohol was in the form of a hydrogen-bonded dimer with the free OH frequency arising from the free OH end group of the dimer. From comparison of the optical density of the free OH bond at 176 C and at room temperature, the number of bonded OH groups was estimated. For this series of hindered alcohols the hydrogen bond chain length ranged between five and ten hydrogen bonds.

CURRENT WORK

In the previous sections it has been shown that crystal structures of various hydrogen bonded solids can be determined using the diamond-window high pressure cell. However, this process is both tedious and time-consuming involving the use of polarized spectra of single crystals of the molecule in question, calculations of transition moments, and trial and error fitting of the observed data to all possible crystal structures. In many instances, the crystal structure is not needed--only the hydrogen bond structure (linear or cyclic, dimer or polymer). Thus a simpler method would be highly desirable, especially a method that would show the differences in hydrogen bond configuration between polymorphs of the same compound.

To this end single crystals of two polymorphs (α and β) of chloroacetic acid were grown in the diamond-window pressure cell. The polarized infrared spectra of these polymorphs are shown in Figures 6 (α-form) and 7 (β-form). In the ν OH region of Figure 6 there are indications of the presence of two broad bands in this region. This has been confirmed by other spectra taken at slightly different polarization angles. The situation also occurs for the carbonyl stretch in the 1700 cm^{-1} region. Although the intensity changes in the rest of the bands appear to be small, careful measurements demonstrate that the peaks (between the two polarized forms) occur at different frequencies. Thus the polarized spectra are serving to resolve overlapping doublets.

In the polarized spectra of the β-polymorph (Figure 7) there are very few frequency shifts which would indicate overlapping bands. Instead the usual intensity changes associated with polarization studies are observed, that is bands go from strong to weak

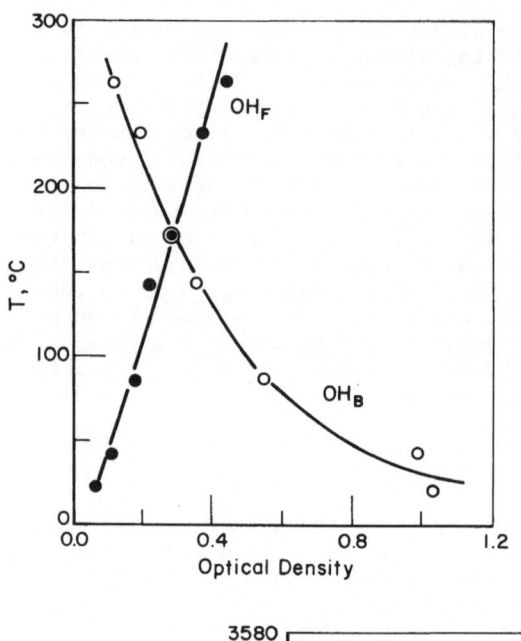

Figure 4. Plot of temperature vs Optical Density of
the OH Stretching Frequency
of 2, 2, 4-Trimethyl-3-Pen-
tanol: (O) Bonded OH and (O)
Free OH

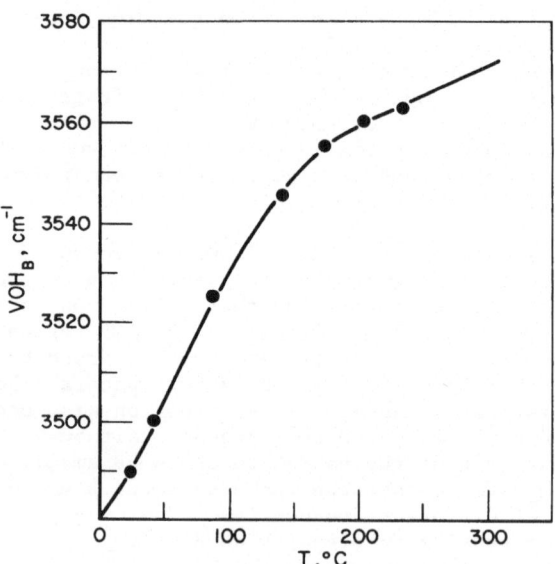

Figure 5. Plot of Bonded OH Stretching Frequency vs Temperature
for 2, 2, 4-Trimethyl-3-Pentanol

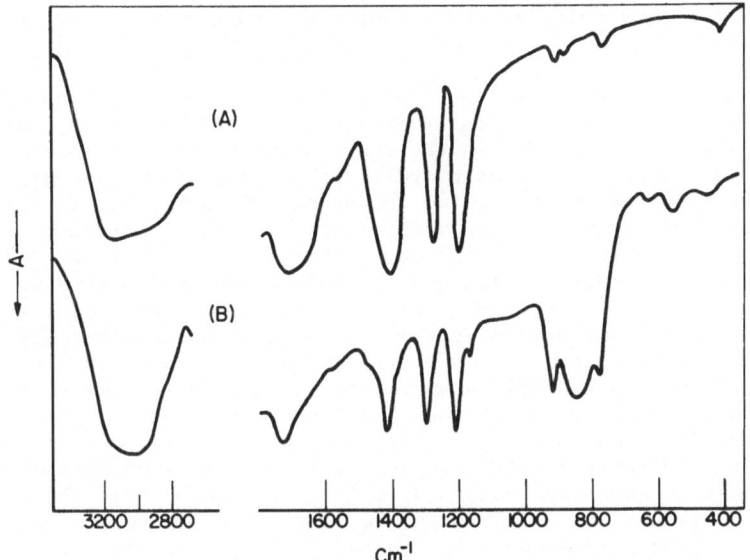

Figure 6. Polarized Infrared Spectra of Single Crystals of α-Chloroacetic Acid: (A) 0°, (B) 90°

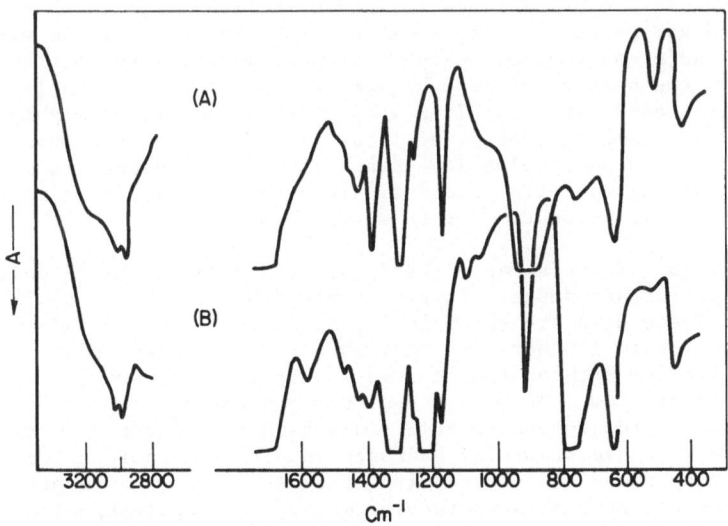

Figure 7. Polarized Infrared Spectra of Single Crystals of β-Chloroacetic Acid: (A) 45°, (B) 135°

intensities as the polarization is changed (and remain at the same frequency).

The different polarization behavior of the two polymorphs of chloroacetic acid are the first indication of differences in hydrogen bond structure. If, as is the case for many acids, chloroacetic acid hydrogen bonds as a cyclic dimer, the normal vibrations will occur in pairs, with the two components of each pair differing only in that the motion of one monomer unit is either in-phase or out-of-phase with respect to the same vibration in the other monomer unit. The in-phase vibrations are only Raman active because they are symmetric to the center of symmetry, but often appear as weak bands in the infrared spectrum. The out-of-phase vibrations are infrared active, but also often appear as weak bands in the Raman spectrum. In general each pair of vibrations involve the same motion and will occur at the same frequency unless coupling with other vibrations occur. In practice such coupling does occur with dimer structures and the components of the pairs of fundamental modes do sometimes absorb at different frequencies. Because of the infrared and Raman activity the infrared polarization behavior for dimer acids should be "normal", i.e., band intensities vary greatly with polarization changes. This is the observed behavior for the β-form of chloroacetic acid indicating a dimer hydrogen bond structure for that polymorph.

Since hydrogen bonded polymers do not have a center of symmetry, there is coincidence of Raman and infrared bands. The polymer structure can also have symmetric-antisymmetric in-phase or out-of-phase pairs of vibrations, but the vibration is now with respect to adjacent monomer units. In other words, a polymer acid should show the same spectral behavior as shown by the alcohol polymers discussed earlier. Thus while these absorptions might occur near the same frequency, at least in some cases, the absorption bonds should be resolvable. Since the polarized spectra of the α-form of chloroacetic acid indicate such behavior, a hydrogen bonded polymer is suspected for this polymorph.

Comparison of the Raman[15] and infrared data for these polymorphs lends further support to the proposed hydrogen bond configurations. These data are shown in Tables 1 and 2. For α-chloroacetic acid (Table 1) there are many infrared and Raman coincidences, while for β-chloroacetic acid (Table 2), the number of coincidences are few. Thus by the easy expedient of growing single crystals in a pressure cell and obtaining polarized infrared spectra of these crystals, hydrogen bond configurations for these polymorphs can be deduced without extensive work attempting to determine the crystal structure or without elaborate spectral interpretation.

TABLE 1. VIBRATIONAL FREQUENCIES (cm^{-1})
FOR α-CHLOROACETIC ACID

Raman	Infrared
1738	1730
1690	
1421	1420
1400	1407
	1300
1208	1215
1170	1173
933	930
914	914
789	780

TABLE 2. VIBRATIONAL FREQUENCIES (cm^{-1})
FOR β-CHLOROACETIC ACID

Raman	Infrared
1746	
	1720
1700	
	1437
1415	
1395	1392
	1308
	1265
1204	
1191	
	1183
1170	
935	930
	918
907	
781	780
658	653
556	
	545
	460
436	
425	

TABLE 3.   VIBRATIONAL FREQUENCIES (cm^{-1})
FOR β-CHLOROACETIC ACID

Low Pressure	High Pressure	$\Delta\nu$	Mode
1720	1705	-15	C = O
1437	1475	+38	OH, C - O
1392	1398	+6	CH_2
1308	1360	+52	OH, C - O
1265	1269	+4	CH_2
1221	1255	+34	OH, C - O
1183	1191	+8	CH_2
918	990	+72	OH
--	945	--	CH_2
918	931	+13	C - C
780	809	+29	C - Cl
653	679	+26	CO_2
545	547	+2	CO_2
460	495	+35	CCO

For the polymorphs of chloroacetic acid, the pressure induced frequency shifts can be extremely helpful in making the vibrational assignment of these different forms. The spectra of one polymorph (β-form) at both low and high pressure are shown in Figure 8 and the frequencies are listed in Table 3. From Figure 8 and Table 3 it is seen that there are large pressure-induced frequency shifts for vibrations involving OH or CO bonds as compared to those involving CH bonds. It should be noted that three frequencies are assigned to the coupled OH, C-O vibration where only two are expected. Possibly this is a case where both components of the symmetric-antisymmetric pair are being observed although the intensities observed would not support this.

While hydrogen bond configurations can often be easily determined from just the polarized infrared spectra of polymorphs of the same molecule, attempts were made to further simplify this procedure. Nonpolarized single crystal spectra of the two polymorphs of formic acid[4] are shown in Figure 9. Although not readily apparent from the figure (because of sample thickness) the three bands in the top spectrum near 1620, 1380, and 1220 cm^{-1} are split into doublets in the lower spectrums. The crystal structures of the two forms are given in Reference 4; here it suffices to point out that Crystal 1 is a hydrogen bonded polymer without a center of symmetry and Crystal 2 is a linear hydrogen bonded polymer with a center of symmetry. In Figure 10, the spectra of the α- and β-polymorphs of chloroacetic acid are shown where the α-form is an asymmetrical hydrogen bonded polymer and the β-form is a cyclic hydrogen bonded dimer. Here the spectra of the two polymorphs do not show the singlet-doublet differences seen in the formic acid spectra, but the differences are mainly differences in bond intensities and number of bands.

For both formic and chloroacetic acids, the hydrogen bond configurations were deduced by means that involved more than just obtaining the spectra of the different polymorphs. Yet when just spectra (Figure 11) of two polymorphs of acetic acid were obtained, it is obvious that the spectral differences between these polymorphs closely resemble those of formic rather than chloroacetic acid. The C = O stretching frequency and the two bands near 1400 cm^{-1} split into doublets on going from one polymorph to the other. From this comparison it can be reasonably assumed that the two polymorphs of acetic acid consist of a symmetrical and an antisymmetrical hydrogen bonded polymer (as in formic acid). Thus hydrogen bond configurations of polymorphs of the same molecule can often be determined by only obtaining infrared spectra of the two polymorphs.

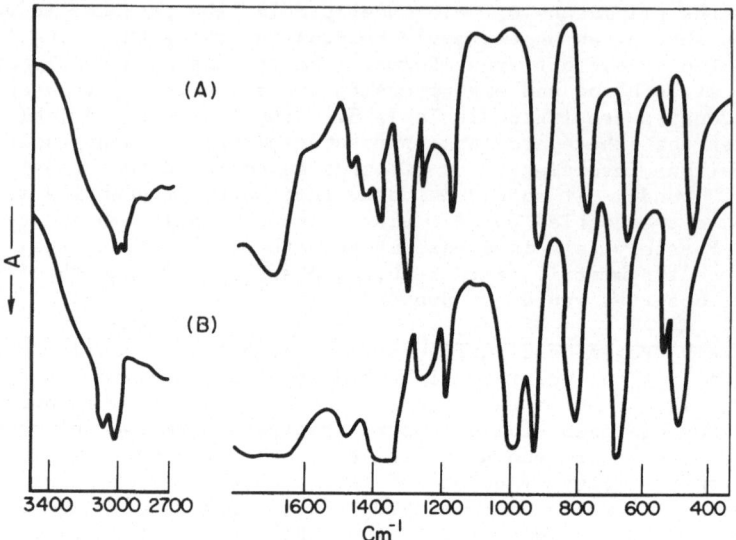

Figure 8.   Infrared Spectra of Single Crystals of β-Chloroacetic
Acid (A) at Low Pressures and (B) at High Pressures

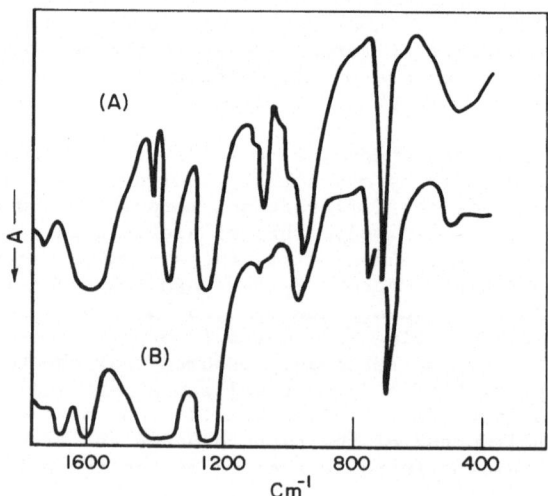

Figure 9.   Infrared Spectra of Single Crystals of Two Different
Polymorphs (A,B) of Formic Acid

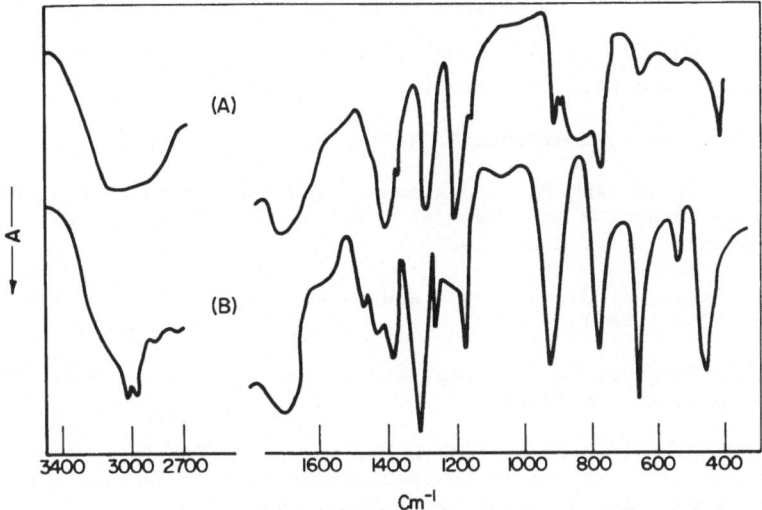

Figure 10.   Infrared Spectra of Single Crystals of (A) α-Chloroacetic Acid and (B) β-Chloroacetic Acid

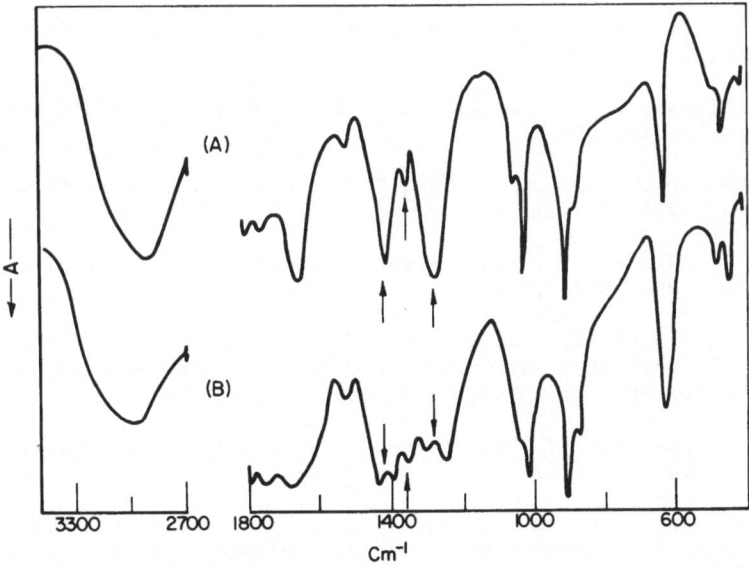

Figure 11.   Infrared Spectra of Single Crystals of Two Different
Polymorphs (A, B) of Acetic Acid

REFERENCES

(1)   G. C. Pimentel and A. L. McClellan, The Hydrogen Bond, W. H. Freeman and Co., San Francisco, 1960, pp 81-82.

(2)   J. W. Brasch, Spectrochim. Acta., 21, 1183 (1965).

(3)   J. W. Brasch and R. J. Jakobsen, Paper presented at the Ohio State Symposium on Molecular Structure and Spectroscopy, Columbus, Ohio, June, 1965.

(4)   Y. Mikawa, R. J. Jakobsen, and J. W. Brasch, J. Chem. Phys., 45, 4750 (1966).

(5)   J. R. Ferraro, S. S. Mitra, and C. Postmus, Inorg. Nucl. Chem. Letters, 2, 269 (1966).

(6)   N. T. McDevitt, R. E. Witkowski, and W. G. Fateley, Paper presented at the Ninth European Congress on Molecular Spectroscopy, Madrid, Spain, September, 1967.

(7)   J. W. Brasch, R. J. Jakobsen, W. G. Fateley, and N. T. McDevitt, Spectrochim. Acta., 24A, 203 (1968).

(8)   R. J. Jakobsen, J. W. Brasch, and Y. Mikawa, J. Mol. Structure, 1, 309 (1967).

(9)   Y. Mikawa, J. W. Brasch, and R. J. Jakobsen, J. Mol. Structure, 3, 103 (1969).

(10) Y. Mikawa, J. W. Brasch, and R. J. Jakobsen, Spectrochim. Acta., 27A, 529 (1971).

(11) R. J. Jakobsen, Y. Mikawa, and J. W. Brasch, Spectrochim. Acta., 25A, 839 (1969).

(12) J. W. Brasch, Y. Mikawa, and R. J. Jakobsen, Paper presented at the Ohio State Symposium on Molecular Structure and Spectroscopy, Columbus, Ohio, September, 1968.

(13) R. J. Jakobsen, Y. Mikawa, and J. W. Brasch, Appl. Spectroscopy, 24, 333 (1970).

(14) R. J. Jakobsen and Y. Mikawa, Paper presented at the Mid-America Symposium on Spectroscopy, Chicago, Illinois, May, 1969.

(15) I. E. Poliakova and S. S. Raskin, Optics and Spectroscopy, 6, 220 (1959).

HIGH-PRESSURE, HIGH-TEMPERATURE SPECTROPHOTOMETER CELL

FOR IN SITU CATALYST IDENTIFICATION

H. Burnham Tinker and Donald E. Morris

Central Research Department, Monsanto Company

St. Louis, Missouri 63166

ABSTRACT

A high-pressure, high-temperature spectrophotometer cell has
been designed and constructed in order to examine reacting
solutions at conditions which are above ambient temperature and
pressure. Large windows, a variable-path length, and a relatively
compact design provide considerable versatility and permit the use
of the cell interchangeably among a variety of commercially
available IR and UV-Visible spectrophotometers without the need
for special instrument modification. The cell has been used to
study solutions at a variety of reaction conditions up to 100
atmospheres of pressure at 200° C without any suggestion of window
or cell failure. The design and fabrication of the cell are
reported. In addition, specific applications of the cell to
studies of the rhodium-catalyzed hydroformylation of olefins are
described.

INTRODUCTION

Most industrial reactions and many reactions of academic
interest must be carried out at elevated temperatures and
pressures in order to achieve desirable reaction rates or
measurable equilibria. When superambient conditions are necessary
for reaction, the structure and even the existence of transient
intermediates are more frequently based on chemical intuition than
on experimental observables. More relevant to the present
discussion regarding homogeneous catalysis by transition-metal
complexes is the fact that the active catalyst may in no way
resemble the metal complex which was originally added to the

reaction solution. Consequently, chemists must base their reaction
mechanisms on the proposed structure of the metal complex at
reaction conditions. To overcome this difficulty, we have designed
and constructed a high-pressure, high-temperature spectrophotometer
cell in order that metal complexes can be characterized during the
natural progress of reactions which are only observable at superambient conditions.

The Spectrophotometer Cell

Several spectrophotometer cells have been reported for UV-
Visible studies at very high pressure and/or temperature and for
high-pressure IR studies of gases [1-8]. A number of reports [9-11]
have implied the use of high pressure spectrophotometer cells for
in situ IR studies of reacting solutions but did not provide
details of the equipment used for the experiments. Infrared
studies of liquids at both high pressures and high temperatures
simultaneously do present some unique problems. For example, very
little is known concerning the effect of temperature on the
elastic limits of window material suitable for IR studies. In
addition, these materials are typically soft with relatively large
coefficients of thermal expansion. Consequently, the window
material and the metal-to-window seal presented the most serious
problems during the design, construction, and use of the cell.

Two cells, similar to the one described here, had been
designed independently and have recently been reported [12, 13].

The cell described here incorporates certain features not
found in the equipment reported previously. Specific differences
are:

1. The path length can be varied from 0.05 mm to 20.0 mm.

2. It has large windows which eliminate the need for beam
condensers.

3. The cell is separate and can be isolated from the
reactor to minimize damage from spraying solutions should the
windows fail.

4. It is sufficiently compact to fit into a variety of
commercial Infrared and UV-Visible spectrophotometers without
modification of the instrument.

5. Assembly is relatively easy with little damage to brittle
window material.

6. The construction metal is resistent to corrosive solutions at high temperature.

7. The volume of solution needed is relatively small (< 5 ml).

The Catalysts

For the sake of clarity we must define several terms which will be used subsequently.

1. The "charged catalyst" or "catalyst precursor" is the form of the metal complex which is added (charged) to the reaction solution.

2. The "active catalyst" is any metal complex which is involved in the catalytic cycle.

3. The "predominate form of the catalyst" is the principle metal species present at reaction conditions.

The determination of the IR spectrum of the active catalyst would give the most direct evidence regarding a particular reaction mechanism. However, it is likely that the active catalyst may be present in very low concentration and, therefore, its spectrum may not be directly measurable. Under these conditions only the predominate form of the catalyst would be observed. Although not as satisfying as a direct observation of the active catalyst, the identification of the predominate form of the catalyst is no less essential to any kinetic study of the reaction system. This is true because kinetic experiments are designed to elucidate the molecularity of the transition state of a reaction but do so only in relation to the molecularity of the predominate form of the catalyst and not in relation to the charged catalyst unless the two are identical. This statement should become clear in terms of the following example.

The hydroformylation of olefins (Eq. 1) is catalyzed by rhodium chlorotris(triphenylphosphine), $RhCl(Ph_3P)_3$.

$$R\text{-}CH{=}CH_2 + CO + H_2 \xrightarrow{\quad RhCl(Ph_3P)_3 \quad} R\text{-}CH_2\text{-}CH_2\text{-}CHO + R\text{-}CH(CHO)\text{-}CH_3 \quad (1)$$

Assume, for the sake of discussion, the rate law for this reaction is

$$-\frac{d[R\text{-}CH{=}CH_2]}{dt} = k[R\text{-}CH{=}CH_2][H_2][RhCl(Ph_3P)_3] \qquad (2)$$

The rate of reaction is independent of [CO] and the mechanism must involve a non-rate-determining insertion of CO into an alkyl-rhodium bond. However, spectral studies show that the predominate form of the catalyst is RhClCO(Ph$_3$P)$_2$ formed according to reaction 3.

$$RhCl(Ph_3P)_3 + CO \longrightarrow RhClCO(Ph_3P)_2 + Ph_3P \qquad (3)$$

Although the rate law does not indicate the presence of CO in the transition state, spectral studies in conjunction with kinetic experiments clearly demonstrate that CO is involved in the molecularity of the reaction.

DESIGN AND CONSTRUCTION OF THE CELL

The Spectrophotometer Cell

An assembled and exploded cross section view of the cell is reproduced in Figure 1. The body of the cell is Hastelloy C, a particularly non-corrosive form of stainless steel. The cell is relatively compact measuring 9.50 cm x 8.00 cm x 6.35 cm and has been used in a variety of commercially available UV-Visible and IR spectrophotometers (e.g., Perkin Elmer 221; Beckmann IR 4, 7, and 12; Cary 14; and a Perkin Elmer Hitachi 124) without the need for special instrument modification.

The cell is heated with four - 50 watt tubular heaters inserted in the holes drilled at points A. Temperature changes are detected by a thermister inserted in hole B and the temperature of the cell is maintained at $\pm$ 5°C of the set temperature using a thermistor temperature controller [14]. When employing the cell for studies below 200°C, we have not used insulation between the cell and the spectrophotometer.

The windows (C) are held apart by a Hastelloy spacer (D) until pressure is applied. The o-rings (E) encompass the windows and are compressed between the spacer (D) and the soft copper seats (F) by the screw plugs (G) to form the initial seal. The reaction solutions are introduced into the sample cavity (H) through a hole in the cell body (I) and removed through a similar hole at the top of the cell (J). The openings at I and J tapered for 1/4" Aminco high-pressure fittings.

We have designed a variety of spacers which can be used to produce path lengths of from 2.0 cm down to 0.05 mm. When a spacer smaller than 2.0 cm is employed, washers of the appropriate thickness are inserted between the windows (C) and the screw plugs (G) to extend the depth of these back-up plugs (G).

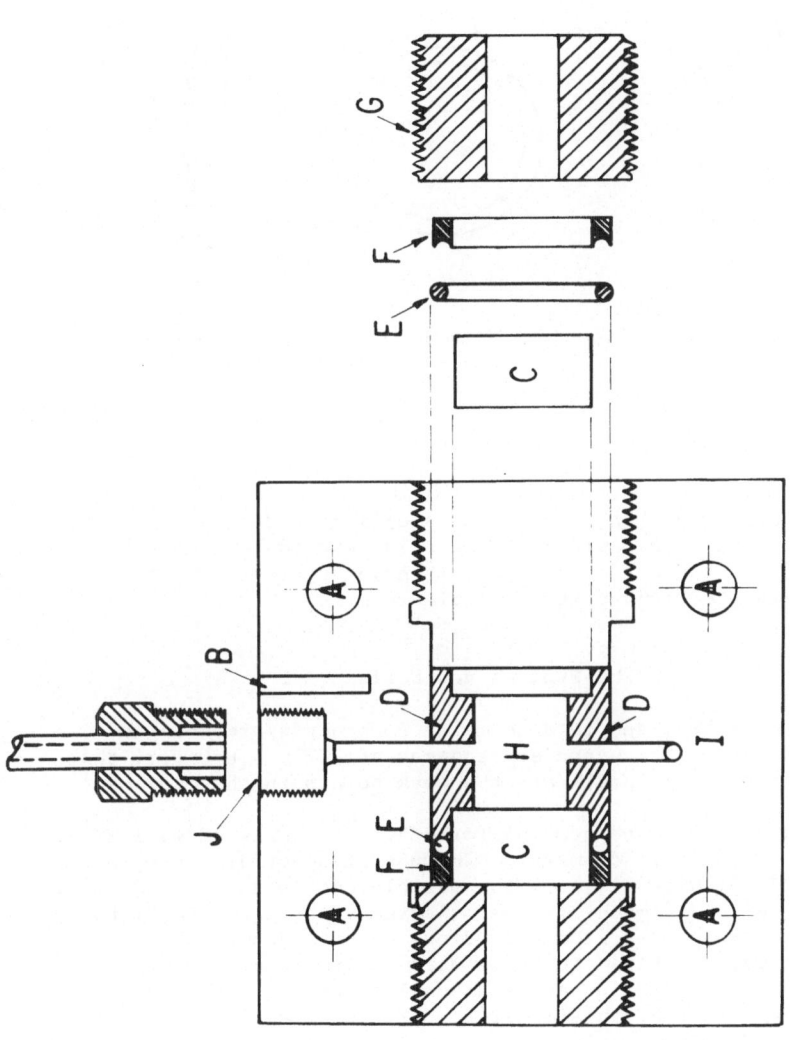

Fig. 1   Cross section view of high-pressure, high-temperature spectrophotometer cell.

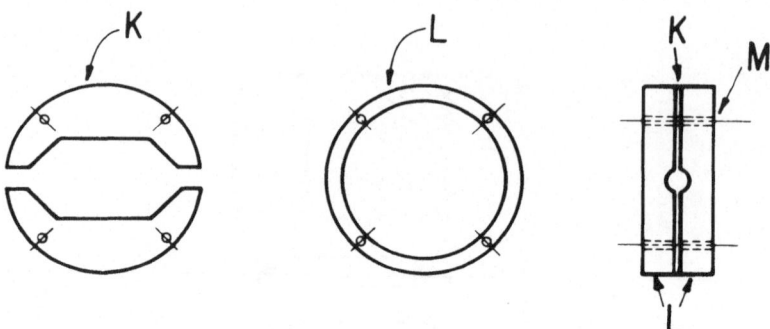

Fig. 2   Spacer sub-assembly for path lengths of 1.0 mm or less

Spacers less than 1 mm are of a different design as shown in Figure 2. The spacer plates (K) are sandwiched between two rings (L) and the entire assembly is held together with four pins (M) as shown in the right portion of Figure 2.

The Windows and Seals

The cell body and related metal components are able to withstand several thousand atmospheres at 250° C. However, the window material and seals are the weak points in the equipment.

Most studies have been performed using CaF_2 windows 1.27 cm thick and 2.54 cm in diameter. The useful spectral range for CaF_2 (1.27 cm thick) is from the visible out to about 1400 cm^{-1}. Sapphire, which is much stronger than CaF_2 and less likely to break, is transparent from the UV out only to about 2300 cm^{-1}. Consequently, sapphire cannot be used for the study of most transition metal-hydride and -carbonyl compounds. Pressed MgF_2 has mechanical strength similar to CaF_2 but transmits only about 1/2 as much energy. KRS-5 (a TlBr-TlI mixture) has mechanical strength comparable to CaF_2 and is transparent out to about 300 cm^{-1}.

The maximum pressure the window will withstand is related to the thickness and unsupported diameter of window by equation 4 [15].

$$\frac{T}{D} = 1.06 \ (p/Fa)^{0.5} \tag{4}$$

T = Window Thickness, 1.27 cm
D = Unsupported Diameter, 1.27 cm
p = Pressure (psig)
Fa = Apparent Elastic Limits (psig)

The equation is valid at 25°C and contains a safety factor of 4. The Fa for calcium fluoride is 5.3×10^3 psig consequently at 25°C the cell as designed will withstand at least 5000 psig. If the safety factor of 4 is not considered, the cell should withstand ~ 20,000 psig.

The upper pressure limit could be increased by decreasing the unsupported area of the windows. However, this would involve a concomitant decrease in the quantity of transmitted radiation and, at some diameter, as the cell opening decreases, a beam condenser and optical bench would be required to insure that sufficient energy would reach the detector to produce satisfactory spectra.

No systematic study has been performed to determine the rupture point of the windows or seals. Therefore, no general statement can be made regarding the maxima of temperature and pressure the cell will withstand. However, employing calcium fluoride windows, the cell has been used at 2000 psi and 25°C and at 1500 psi and 200°C with no evidence of impending window failure.

Teflon, viton, and ethylene-propylene o-rings have been used in the cell. Teflon, although desirable because it is chemically inert, is very stiff and tends to chip all windows other than sapphire. Ethylene-propylene o-rings are attacked by olefins and halocarbons and viton o-rings are reactive toward aldehydes. However, both were used to study the hydroformylation of 1-hexene to isomeric heptanals in 1,2-dichloro ethane without any effect on the quality of the spectra. This suggests that there is very little solvent transfer from the sample compartment to the o-rings. On the other hand, if the cell is disassembled after ~ 20 hours of use, the o-rings are swollen and must be replaced before the cell can be reassembled.

APPLICATION

This spectrophotometer cell may be used to examine a wide
variety of reactions at elevated pressures and/or temperatures;
however, initial efforts at our laboratory have been concerned
with the rhodium-catalyzed hydroformylation of olefins [16]
(Eq. 1). In a typical reaction study, the solvent, rhodium
catalyst, and CO/H_2 gas blend are added to a 300 ml stainless
steel autoclave [17]. The autoclave, its contents, the
spectrophotometer cell, and transfer tube are equilibrated at
reaction pressure and temperature. The reaction is started by
pressure-injecting olefins into the clave. After appropriate time
intervals samples are transferred to the cell and the spectrum is
recorded using a Perkin-Elmer 221 Infrared Spectrophotometer.

The physical relationship among the autoclave (at the left);
the high-pressure, high-temperature spectrophotometer cell; and
the Perkin-Elmer 221 spectrophotometer are depicted in Figure 3.

Since the solutions are transferred from the autoclave to the
cell, the objection can be raised that the studies are not truly
in situ. However, the cell is maintained at a temperature which
is within a few degrees of the reactor temperature and spectral
results clearly demonstrate that a large quantity of CO remains
dissolved in the reaction solution during the spectral measurement.
For these reasons we feel that the studies, although not truly
in situ, are virtually in situ.

Several different rhodium complexes have been examined under
pressures of CO and H_2 separately, of CO/H_2 blends, and under
reaction conditions, i.e., 100°C, 250 psi of CO, and 250 psi of H_2.
The detailed chemistry of the rhodium complexes under these
conditions will not be discussed but rather the results will be
noted as examples of the use of the cell.

Rhodium Chloro Dicarbonyl Dimer

The rhodium chloro dicarbonyl dimer, $[RhCl(CO)_2]_2$, is a very
active catalyst for the hydroformylation reaction [17]. The
spectra reproduced in Figure 4 show the effect of CO pressure on
the spectrum of this complex. Under 200 psi of H_2 and 25°C, the
spectrum is identical with that observed at room temperature and
pressure, i.e., carbonyl bands at 2110, 2093, and 2037 cm^{-1}.
However, under CO pressure the spectrum changes (Figure 4) to 2
bands at 2097 and 2060 cm^{-1}. The band at 2137 cm^{-1} is due to CO
dissolved in the reaction medium. Although not yet demonstrated,
this spectral change appears to be due to the formation of
$Rh(CO)_3Cl$ which results from CO attack on the chloride bridge
(Eq. 5).

Fig. 3. Autoclave, spectrophotometer, and spectrophotometer cell

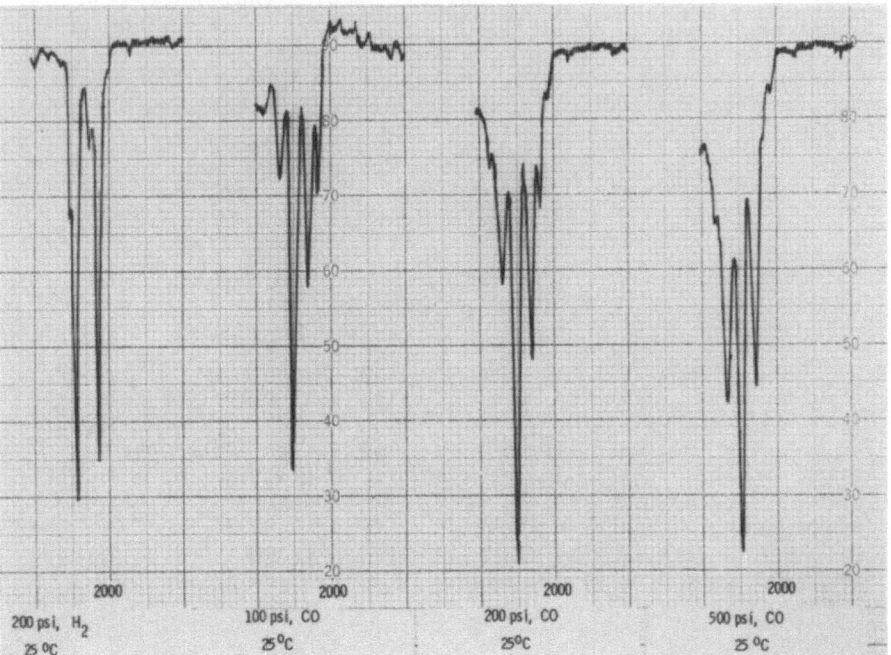

Fig. 4. Effect of CO pressure on the spectrum of $[RhCl(CO)_2]_2$.

$$\begin{matrix} CO & & Cl & & CO \\ & \diagdown Rh \diagup & & \diagdown Rh \diagup \\ CO \diagup & & \diagdown Cl \diagup & & \diagdown CO \end{matrix} + 2CO \; \rightleftharpoons \; 2RhCl(CO)_3 \qquad\qquad (5)$$

When CO pressure is released, the initial spectrum returns and no compound other than the initial dimer can be recovered from the reaction solution. At reaction conditions, the spectrum is identical to the spectrum at ambient conditions, i.e., 3 carbonyl bands at 2110, 2093, and 2037 cm^{-1}. The predominate form of the catalyst and the charged catalyst are the same.

Rhodium Chloro Carbonyl bis(triphenylphosphine)

Another effective catalyst for olefin hydroformylation is $RhClCO(Ph_3P)_2$ (ν_{CO} = 1975 cm^{-1}). The spectra reproduced in Figure 5 show that high pressures of CO alter the complex and a new species with two carbonyl bands (ν_{CO} = 1975 and 1950 cm^{-1}) is observed. Presumably, this is a dicarbonyl complex formed by the reaction illustrated by equation 6. Unfortunately, the dicarbonyl

$$RhClCO(Ph_3P)_2 + CO \; \rightleftharpoons \; RhCl(CO)_2(Ph_3P)_2 \qquad\qquad (6)$$

reverts to the monocarbonyl starting material when CO pressure is removed; therefore, only the initial complex can be isolated. A similar reaction of $IrClCO(Ph_3P)_2$ (ν_{CO} = 1967 cm^{-1}) with CO at atmospheric pressure to form the stable, isolable iridium dicarbonyl complex, $IrCl(CO)_2(Ph_3P)_2$ (ν_{CO} = 1988 and 1934 cm^{-1}), suggests that reaction 6 is responsible for the spectral change (Figure 5).

At 100°C, $RhClCO(Ph_3P)_2$ is unaffected by CO and/or H_2 pressure and the charged catalyst is the predominate form of the catalyst.

Rhodium Chloro Carbonyl bis(triphenylphosphine) Promoted by Hydroperoxide

The rate of hydroformylation of cyclohexene (80°C, 500 psi of CO/H_2 blend) catalyzed by $RhClCO(Ph_3P)_2$ is greatly accelerated by the addition of cyclohexyl hydroperoxide (a factor of ~ 20). The increase in rate followed a short induction period and was accompanied by a change in the spectrum of the reacting solution (Figure 6). The new species, i.e., the predominate form of the catalyst, exhibited carbonyl bands at 2094 and 2012 cm^{-1}. Further studies demonstrated that this new complex is $RhCl(CO)_2Ph_3P$ which resulted from the oxidation of one Ph_3P to Ph_3PO by cyclohexyl hydroperoxide (Eq. 7).

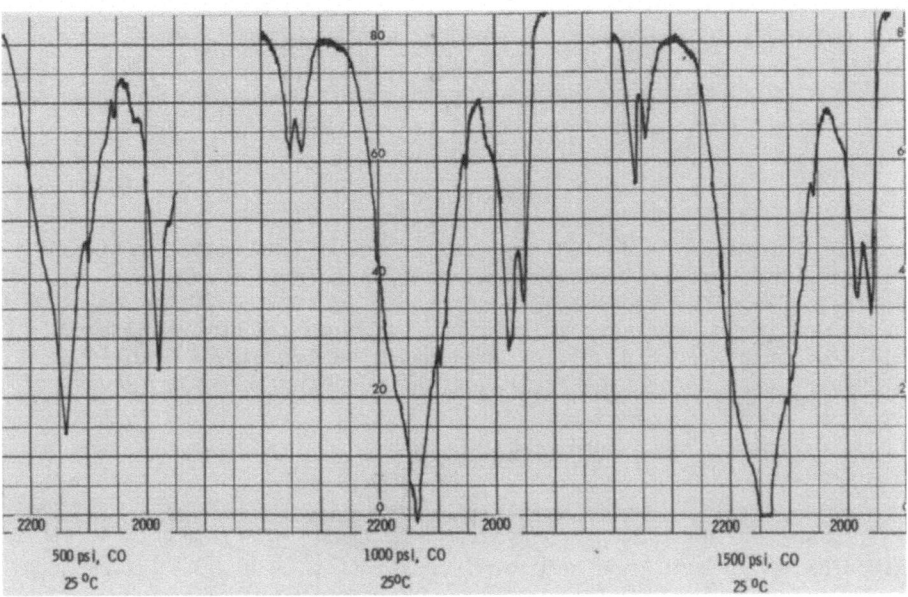

Fig. 5. Effect of CO pressure on spectrum of RhClCO(Ph$_3$P)$_2$

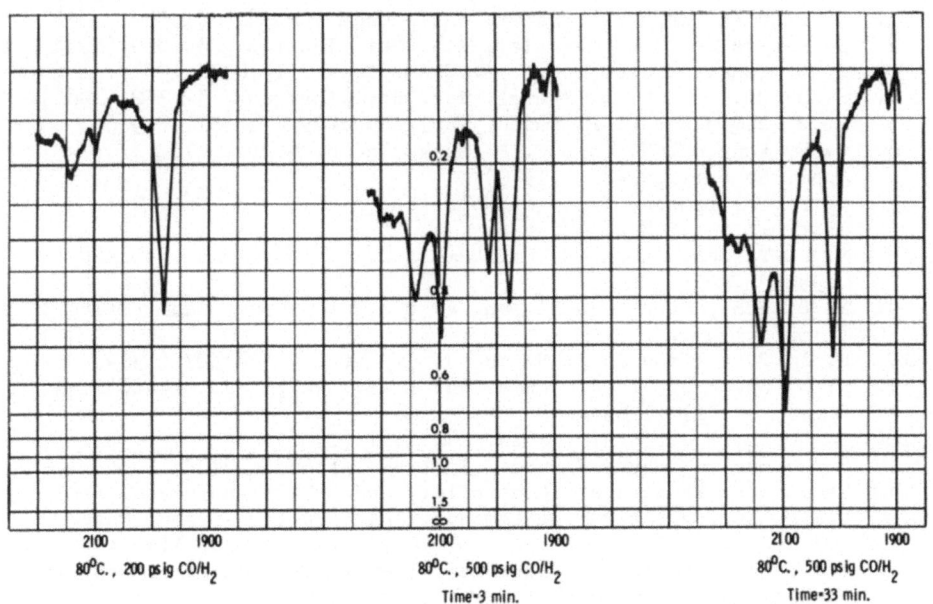

Fig. 6. Spectrum of RhClCO(Ph$_3$P)$_2$ and cyclohexylhydroperoxide as a function of time.

$$RhClCO(Ph_3P)_2 + CO + \langle\bigcirc\rangle OOH \rightarrow RhCl(CO)_2Ph_3P + Ph_3PO \quad (7)$$

+

organic products

SUMMARY

The high-pressure, high-temperature spectrophotometer cell can be used to examine the spectra of reaction solutions and to characterize metal complexes at reaction conditions. Using this technique, the predominate form of the catalyst can be established. The predominate form of the catalyst must be determined before the mechanism of the reaction can be established from kinetic results.

ACKNOWLEDGMENT

We gratefully acknowledge the helpful suggestions and technical assistance of the Instrument Development Laboratory, Central Research Department, and helpful discussions with Dr. J. H. Craddock.

REFERENCES

1. E. Fishman and H. Drickamer, Anal. Chem., 28:804 (1956).
2. E. Peters and J. J. Byerley, Rev. Sci. Instrum., 34:819 (1963).
3. S. S. Penner and D. Weber, J. Chem. Phys., 19:807 (1951).
4. S. J. Gill and W. D. Rummel, Rev. Sci. Instrum., 32:752 (1961).
5. D. W. James and G. P. Smith, Appl. Spectry., 20:317 (1966).
6. C. F. Fong, J. V. Fox, C. E. Mauk, and H. W. Prengle, Jr., Appl. Spectry., 24:21 (1970).
7. D. E. Williamson, I. A. Nichols, and B. Schurin, Rev. Sci. Instrum., 31:528 (1960).
8. R. W. Parsons and H. G. Drickamer, J. Opt. Soc. Am., 46:464 (1956).
9. M. Bianchi, E. Benedetti, and F. Piacenti, Chem. Ind. (Milan), 51:613 (1969).
10. R. L. Pruett and J. A. Smith, J. Org. Chem., 34:327 (1969).
11. W. W. Spooncer, A. C. Jones, and L. H. Slaugh, J. Organometal. Chem., 18:327 (1969).
12. K. Noack, Spectrochim. Acta, 24A:1917 (1968).
13. R. Whyman, J. Phys. E: Sci. Instrum., 3:573 (1970).
14. Thermistor Temperature Controller, Model 7115, purchased from Yellow Spring Instrument Company, Inc., Yellow Spring, Ohio, and modified by installing a 100 Ω resistor in parallel with the probe in order to expand the temperature-range. Four 50 watt heaters were sufficient to maintain 200° C.

15. Harshaw Optical Crystals, Harshaw Chemical Company, 1967, p. 15-17.

16. D. E. Morris and H. B. Tinker, presented at the Second North American Catalysis Society Meeting, Houston, Texas, February 24-26, 1971; C & E News, p. 53, March 8, 1971.

17. This equipment has been described in detail elsewhere. J. H. Craddock, A. Hershman, F. E. Paulik, and J. F. Roth, Ind. Eng. Chem. Prod. Res. Develop., 8:291 (1969).

RECENT DEVELOPMENTS IN HIGH RESOLUTION RAMAN SPECTROSCOPY OF GASES

Alfons Weber

Physics Department

Fordham University, Bronx, N.Y., 10458

Several recently developed experimental methods for obtaining Raman spectra of gases with laser excitation are reviewed. These developments are placed in the context of the accomplishments achieved with the pre-laser technique which employed mercury arc lamps as the primary light source. An apparatus is described with which the rotational and rotation-vibrational Raman spectra of a number of heavy polyatomic molecules have been observed with a practical limit of resolution of approximately $0.07 \mathrm{cm}^{-1}$. The use of the Fabry-Perot interferometer in high resolution Raman studies of gases is briefly discussed.

INTRODUCTION

The field of high resolution Raman spectroscopy of gases has come to denote the study of the rotational fine structure of the rotation-vibrational Raman bands of polyatomic molecules. The rotation-centrifugal distortion and Coriolis coupling constants as well as the frequencies of the fundamental and combination vibrations derived from the Raman spectrum take their place alongside the information provided by infrared and microwave absorption spectroscopy to provide the data that are necessary for an understanding of the structure of molecules.

Prior to the invention of the laser modern Raman spectroscopy of gases was carried out with the experimental techniques developed by H.L. Welsh and his collaborators [1] and notable advances were made in this field which had lain dormant since the early 1930's.

137

B.P. Stoicheff and his collaborators[2] studied the pure rotational spectra of a number of non-polar symmetric top and slightly asymmetric top molecules from which they derived a set of highly accurate bond lengths. H.L. Welsh and his collaborators studied the rotation-vibration spectra of ethane and its isotopic variants and thereby provided the first spectroscopic proof for the "staggered" structure of the ethane molecule.

By now nearly 50 diatomic, linear, and non-linear polyatomic molecules, including spherical, symmetric and asymmetric top rotators have been studied in the Raman effect under high resolution. However, of the 18 symmetric top molecules studied thus far (not counting their isotopic variants) only three (ethane, cyclopropane, allene) have been studied in the rotation-vibration region (in addition to the study of their pure rotation spectra), whereas for the other 15 molecules only the pure rotation spectra have been studied. For asymmetric top molecules the score is even worse. Of a total of 13 such molecules studied thus far there is only one, ethylene, for which there has been a study of rotation-vibration spectrum. The reasons for this state of affairs are well understood: Rotation-vibration spectra are very much weaker than pure rotation spectra. The whole body of the Raman spectroscopic findings (up to 1971) and their correlation with the results obtained from infrared spectra and electron diffraction studies is reviewed elsewhere[3].

Since Raman spectroscopy of gases is done mostly in the visible region of the spectrum, "high resolution" means spectroscopic resolving powers of at least 100,000. At $\lambda = 5000\text{Å}$ the resolving limit is then 0.2cm^{-1}. Most of the Raman spectroscopic studies that have been reported thus far in the literature have been performed with the now outdated mercury arc method of exciting the spectra. The best that has been achieved with this old technique is a limit of resolution of 0.245cm^{-1} observed in the pure rotation spectrum of cycloheptatriene. The development of the laser technique during the past few years has very considerably changed the outlook on the usefulness of Raman spectroscopy of gases. The limit of resolution has been lowered to somewhat less than 0.07cm^{-1} with the 90° scattering configuration[4] and a limit of somewhat less than 0.05cm^{-1} has been observed using the "forward" scattering configuration[5]. However, in spite of these improvements a lot remains to be done before one can say that the potential of the laser as the exciting source for high resolution Raman spectroscopy has been sensibly exploited.

While the emphasis thus far has been on the resolution of the rotational fine structure of Raman bands, the techniques of high resolution spectroscopy are also of importance to the study and analysis of unresolved, or partially resolved bands.

EXPERIMENTAL TECHNIQUES

Before the laser could usefully be employed in Raman spectroscopy of gases new techniques for illuminating a volume of gas and for efficient collection of the scattered radiation had to be designed. The simplest method is of course to focus a laser beam in the gas and to send the scattered radiation that emanates from the focal region into a spectrograph or a spectrometer. This works very nicely if a fast spectrometer with a small focal ratio is employed.

The whole Raman apparatus is quite small in size and many such "table top" installations are now widely used. Figure 1 shows the optical arrangement developed by Claassen et al.(6) which was very effectively used in the study of the highly corrosive spherical top hexafluorides, as well as F_2 (6,7). Barrett and Adams (8) have demonstrated the efficacy of the focussed beam technique. Their experimental arrangement is shown in Figure 2. With mirror M_2 removed, the Raman sample area is located inside the laser cavity. The scattered radiation is collected by a fast (f/0.95) lens L_2 and is transferred to the grating spectrometer (Perkin-Elmer E1) by way of the Dove prism image rotator P_2. With lens L_1 having a focal length of 3cm the size of the scattering volume is ca. $10^{-8} cm^3$. At atmospheric pressure this corresponds to 10^{11} scatterers. With this apparatus the rotation-vibration spectra of N_2, O_2, and CO_2 at atmospheric pressure were recorded in only a few (i.e. 1-3 hours).

These two methods are based on the $90°$ scattering configuration and are thereby limited to small scattering volumes and consequent low intensities of scattered radiation. This limitation is circumvented through the use of the principle of the light pipe. Figure 3 shows two arrangements suggested by Chapput et al. (9). In (a) the laser beam passes through a hole in a diagonally placed mirror and travels coaxially down the length of the Raman tube. The nearly "back scattered" radiation is guided by internal reflection to the diagonal mirror which directs it out of the Raman tube and thus to the spectrometer. In the second arrangement shown in (b) it is the nearly "forward scattered" radiation which is fed into the spectrometer. The direct beam is intercepted by a light trap and is thus prevented from flooding the spectrometer. The "forward scattering" configuration possesses the very considerable advantage of minimizing the Doppler broadening of the Raman lines.

These "table top" size arrangements using fast (i.e. small f/D ratio) recording spectrometers are incorporated in commercial apparatus and sample spectra of O_2, N_2 and CO_2 are now routinely shown in the advertisements. However,insofar as

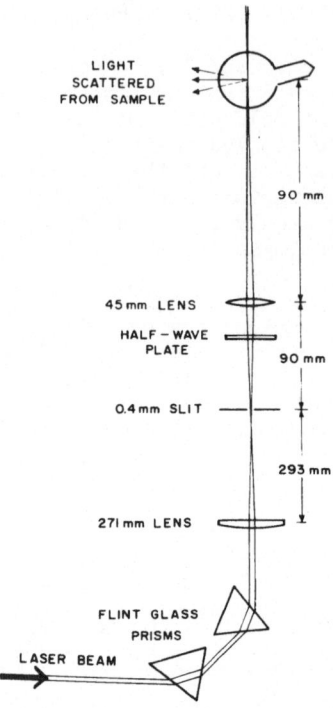

Figure 1.   Optical system used by Claassen et al. (6). To remove
             the unwanted Ne lines emitted by a He-Ne laser the
             laser radiation is filtered with a double prism
             spectrograph. The 6328Å radiation that emerges from
             the 0.4mm wide slit is focussed in the scattering
             cell. The half wave plate serves to suitably adjust
             the plane of polarization of the exciting light.

the research program of high resolution Raman spectroscopy is
concerned, according to which a resolving limit of less than
0.2cm^{-1} together with accurately determined wavenumber shifts
is desired, the photographic methods of recording the spectra is
preferred. Several attempts have been described to utilize the
multiple mirror Raman tube (1,2) with the laser light source.

    The principle of the four-mirror Raman tube is described with
the aid of Figure 4. The mirror pair C,D constitutes the "front"
end of the Raman tube and the scattered radiation is generated
in the volume contained between the front (C,D) and rear (A,B)
pairs. The mode of operation of this mirror system is understood by having the light from the spectrograph enter the Raman

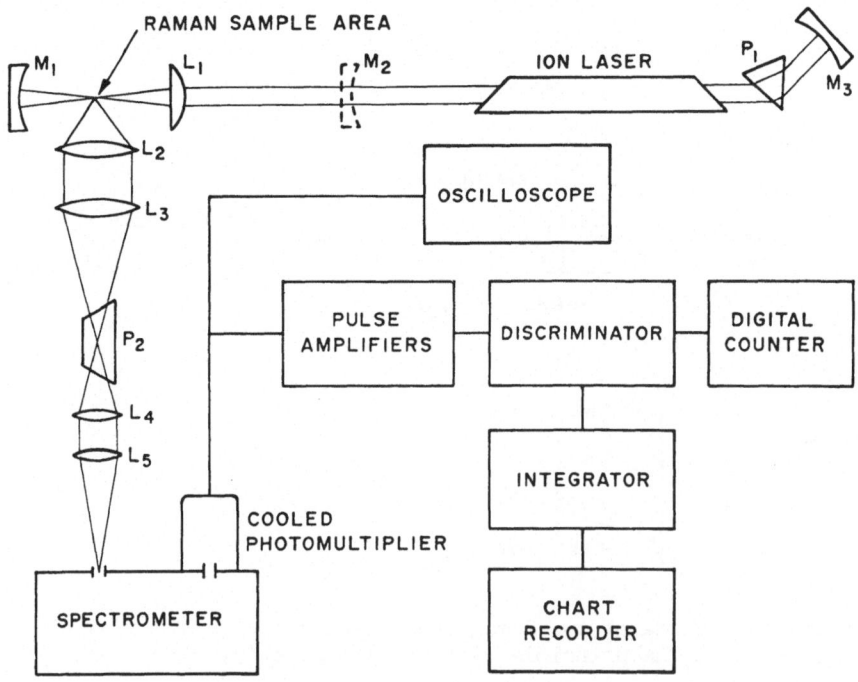

Figure 2.   Schematic diagram of the Raman apparatus developed
by Barrett and Adams (8).

tube in reverse through the slot between C and D and fill the
two rear mirrors A,B. With the centers of curvature of the
mirrors as shown, the rear mirrors will form a series of slot
images on the front mirrors and thereby fill the volume with
light. In the forward operation of the system scattered light
that is generated within the volume will then be fed by multiple
reflections onto the slot between C and D and thus leave the
Raman tube. External transfer optics then pass this radiation
into the spectrograph.

Two methods of coupling a laser beam into a multiple mirror
system are shown in Figure 5. In (a) one of the Raman tube
front mirrors also serves as a laser resonator mirror with the
laser beam passing into the tube through the gap between the
rear mirrors A,B (10). In (b) the laser cavity is formed by
two independent mirrors and stray light that is generated in
method (a) through scattering in the window that closes off the

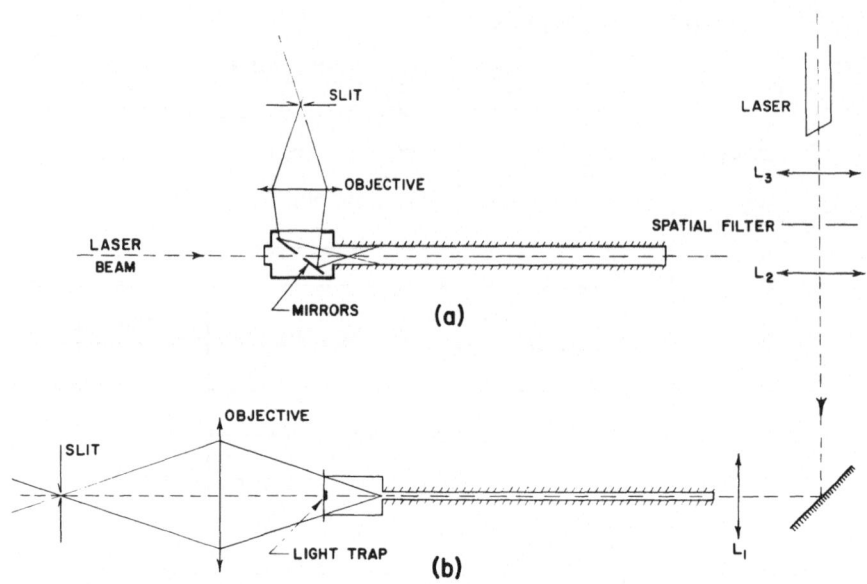

Figure 3.   Schema of two Raman tube illumination systems based
            on the principle of the light pipe (9).

rear of the Raman tube as well as in the coating of mirror D is
avoided. A similar method is used by Moret-Bailly and Berger (11)
who use the three-mirror Raman tube described by White et al.(12).

    A more direct method of coupling a laser beam into a four-
mirror tube has been devised by Rich and Welsh (13). In this
method the laser beam is focussed into the middle of the slot
between the front mirrors C,D (see Figure 4) from which the beam
diverges and fills the two rear mirrors A,B. Figure 6 shows the
coupling arrangement. A short 90° prism deflects the laser beam
onto the slot. A series of images of the focal spot is then
formed on the front mirrors by the "reverse" operation of the
mirror system. The scattered radiation that is produced by the
gas in the tube leaves the tube through the slot as shown, with
the prism occulting only a very short portion of the slot length.
As described by Rich and Welsh this system also generates stray
light which can be "filtered out" by a linear sheet polarizer

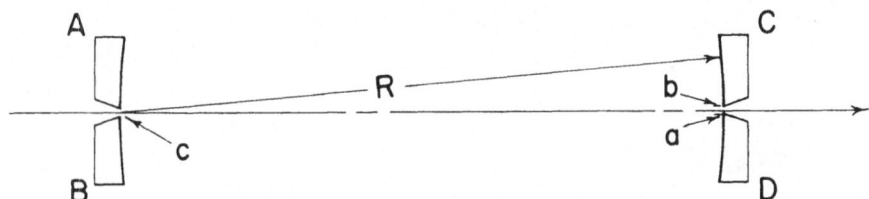

Figure 4.   Four-mirror Raman tube for gases. A,B,C and D are
semi-circular mirrors, all with the same radius of
curvature. The separation of the pairs A,B and C,D
is equal to the radius of curvature, R. Point C is
the common center of curvature of the pair C,D,
which thus act as a single spherical mirror. Mirrors
A and B are "crossed", with centers of curvature at
a and b respectively.

which is "crossed" with respect to the plane of polarization of
the (linearly polarized) exciting radiation and which is placed
in front of the spectrograph slit. More bothersome, however, is
the fluorescence in the quartz prism and the degradation of the
front mirror coatings due to the high energy density at the focal
spot images.

    The multiple-reflection multiple-pass Raman tube originally
devised by Weber et al.(10) has been incorporated into a 90°
scattering apparatus (4) depicted in Figure 7. The Raman tube is
in the shape of a cross and is located inside the laser cavity

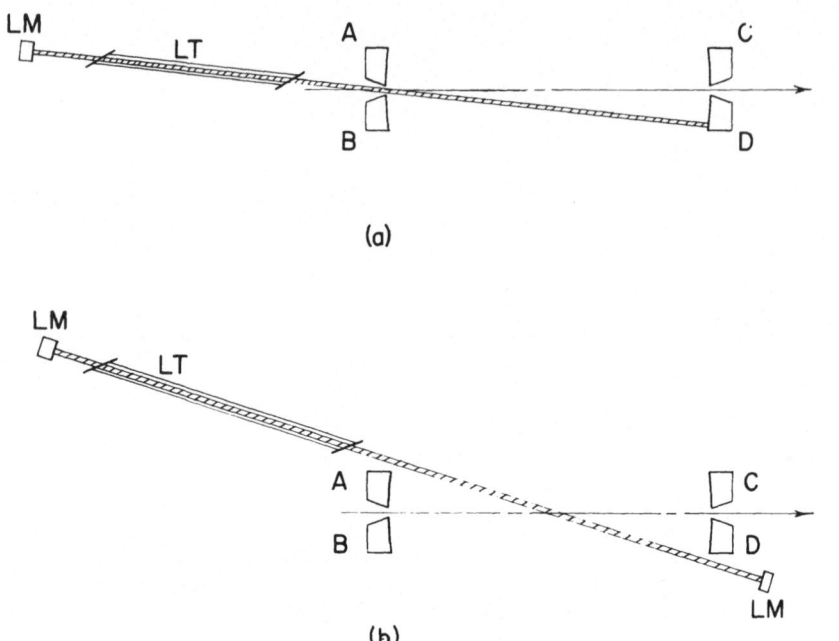

(a)

(b)

Figure 5.   Method of coupling a laser beam into a multiple
             mirror Raman tube (10).

which is defined by the fully reflecting mirror LM and the
Littrow Prism Mirror combination LP+M. The laser beam is caused
to multiply traverse the Raman tube by means of the plane folding
mirrors FM. The scattered radiation is collected by four spherical
collecting mirrors CM (see Figure 4) and is transferred to the
spectrograph slit S by the lenses L_1, L_2 and if necessary, a
Dove prism image rotator DP. The laser beam polarization is
adjusted to be in the scattering plane by means of the half wave
plate, a filter F serves to prevent harmful ultraviolet radiation
from entering the Raman tube and a solid Fabry-Perot etalon
serves to produce a single mode laser beam whose spectral line
width is $\sim$ 0.001 cm^{-1}. With this apparatus and Kodak IIIa-J
plates which have been hypersensitized by baking a series of pure
rotation as well as rotation-vibration spectra have been successfully photographed. A few of these are presented in Figures 8-13.

    The pure rotation spectra (Stokes side only) of n-propane,

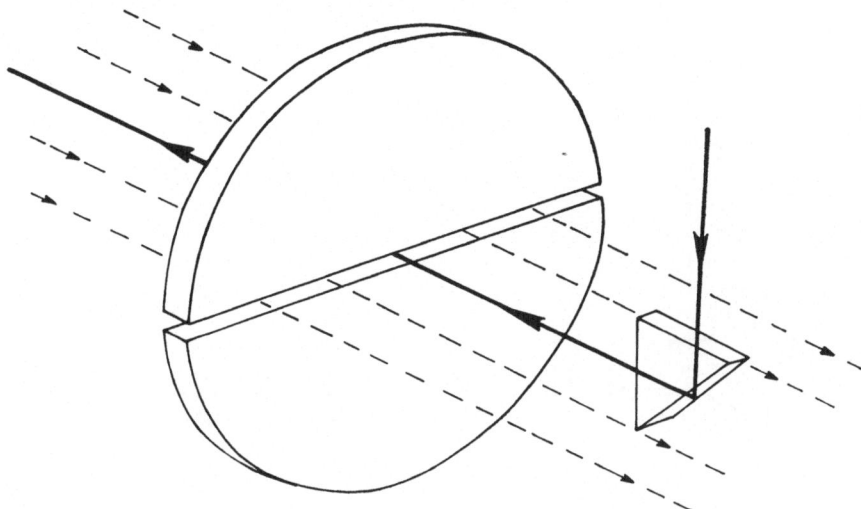

Figure 6.   Coupling of laser beam to a four-mirror Raman tube
according to Rich and Welsh (12).

n-pentane and n-butane are shown in Figure 8. The gas pressures
and exposure times were, respectively, for propane: 400 torr,
$7^3/4$ hr; for butane: 400 torr, 10 hr; for pentane: 164 torr, $3^3/4$hr.
These molecules are relatively weak scatterers and earlier attempts to obtain the spectra of propane and butane with the old
mercury arc technique of excitation were not successful (14).
These spectra resemble those of a symmetric top rotator. The
intensity distribution in the spectrum of n-pentane shows beats
which suggests the superposition of two series of lines which
are slightly out of step with one another. Since the presence of
isotopic species in the pentane sample may be safely ruled out
these two series of lines must be due to the fact that pentane
is a slightly asymmetric top or that two structural isomers
(of n-pentane) are present which have only slightly different
rotation constants, or both.

Figure 9 shows the pure rotation spectra of trans-difluoro-

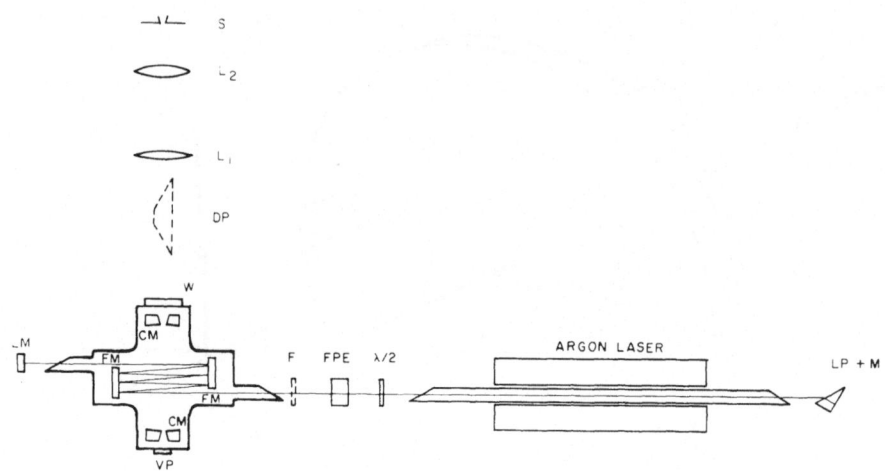

Figure 7.   Raman scattering apparatus according to Weber and
             Schlupf (4).

ethylene (100 torr, 7 hr) and trans-dichloroethylene (74 torr,
1½ hr), both of which are asymmetric top rotators. An intensity
beat is seen on the original spectrogram of trans-difluoroethylene and this must be ascribed to the asymmetric top character
of the molecule. For trans-dichloroethylene as many as four intensity beats are seen on the original spectrograms. A microphotometer trace of the spectrum in Figure 9 is shown in Figure 10.
This shows three intensity beats, with the first two particularly
well developed. It is clear, although the isotope effect plays an
important part the asymmetric top character of the molecule is
intrinsically responsible for the intensity modulation.

    The unresolved rotation spectrum of trans-dichloroethylene
was observed with the old mercury arc excitation technique (14).
Similarly, the unresolved spectrum of borontrichloride was observed prior to the availability of lasers (14). This spectrum is
now obtained with good resolution and is shown in Figure 11. Due
to the isotope effect the observed spectrum is a superposition

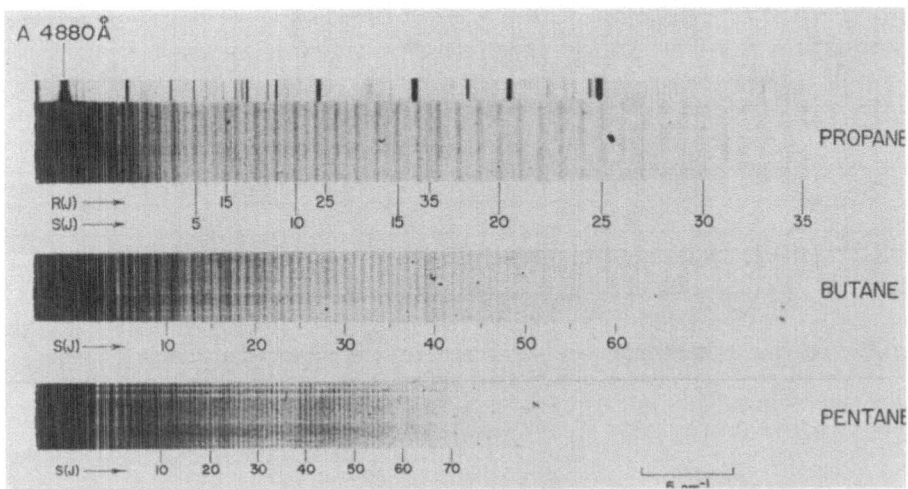

Figure 8. Rotational Raman spectra of some normal alkanes.

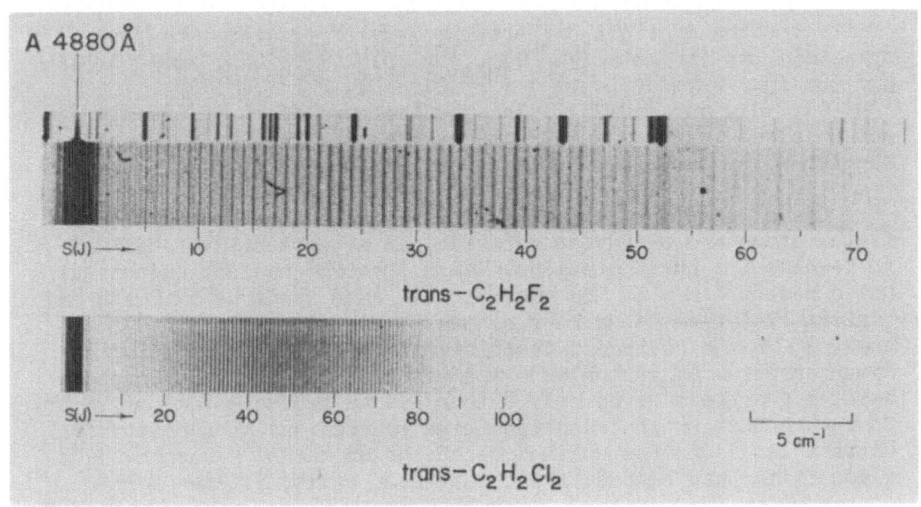

Figure 9.   Rotational Raman spectra of trans-difluoroethylene
and trans-dichloroethylene.

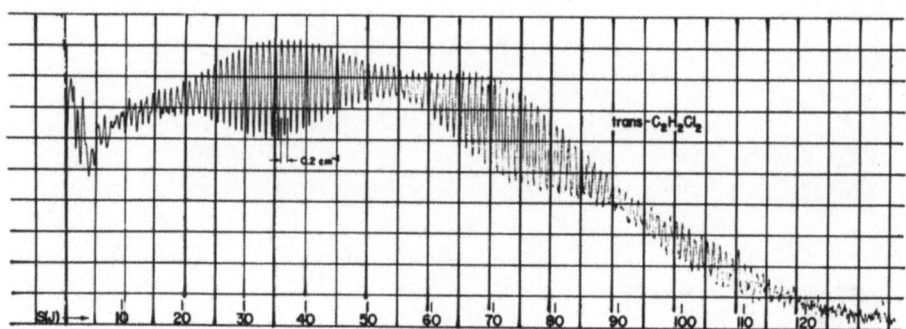

Figure 10.   Microphotometer trace of the trans-dichloroethylene
spectrum shown in Figure 9.

of the spectra of eight different molecular species, the four
symmetric top variants $^{10}B^{35}Cl_3$, $^{10}B^{37}Cl_3$, $^{11}B^{35}Cl_3$, and $^{11}B^{37}Cl_3$
and the four asymmetric tops $^{10}B^{35}Cl_2^{37}Cl$, $^{10}B^{35}Cl^{37}Cl_2$,
$^{11}B^{35}Cl_2^{37}Cl$, and $^{11}B^{35}Cl^{37}Cl_2$. The isotope effect is clearly
not resolved in the spectrum and manifests itself only through
the intensity distribution in the spectrum.

One of the tantalizing problems of molecular spectroscopy
is the structure of carbon suboxide. An attempt by Stoicheff (15)
to resolve its pure rotational Raman spectrum was not successful
but a recent study of its vibrational Raman spectrum by Smith and
Barrett (16) cleared up many of the heretofore unexplained features. With the equipment described in Figure 7 the rotational
Raman spectrum of carbon suboxide was successfully photographed
using a gas pressure of only 8 torr and an exposure time of $4\frac{1}{2}$ hr.
To insure that no photodecomposition occured due to the intense
ultraviolet radiation emitted by the argon plasma a Corning 3-73
glass filter was placed inside the laser cavity (filter F in

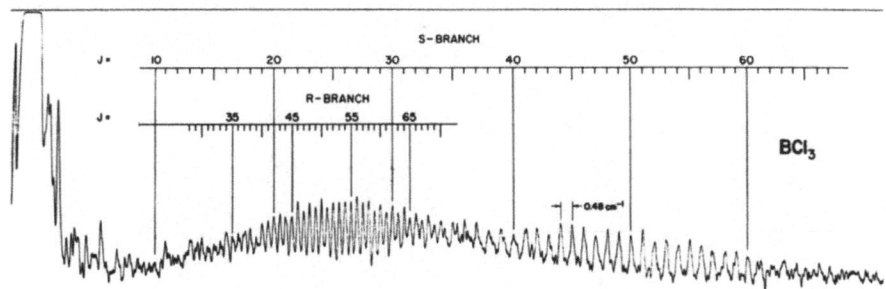

Figure 11.   Rotational Raman spectrum of BCl_3.

Figure 7). Figure 12 shows the microphotometer trace of the spect-
um. The observed spectrum is the result of the superposition of
the vibrational ground state spectrum and a large number of pure
rotational "hot bands" i.e. pure rotational transitions in excited
vibrational states. At low J-values these transitions nearly
coincide to give two series of lines of alternating intensity.
For increasing J-values the different rotational constants as well
as the different values of the vibrational angular momentum quantum number for the various vibrational states cause these series
first to go out of step and then go partially back into step. As
seen from the microphotometer trace the spectrum above J = 20 is
rather more complex than was initially believed and it is not
surprising that the original attempt to resolve it was not
successful (15).

    In general the presence of "hot bands" which are not resolved
from the ground state pure rotation spectrum afflicts any "observed
pure rotational Raman spectrum". The degree to which the hot bands

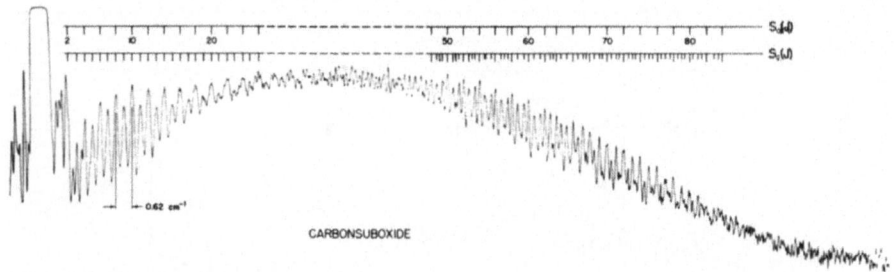

CARBONSUBOXIDE

Figure 12.   Rotational Raman spectrum of carbon suboxide. The
assignment is tentative.

contribute to the intensity and thereby cause erroneous molecular
constants (B_0 and D_J values, say) can be answered only for each
specific case. In the few instances where a comparison of the Raman
B_0 and D_J values with the infrared and microwave determined values
has been carried out the agreement is quite satisfactory even though
hot bands are present. TABLE I shows such a comparison for the
linear molecules CO_2, CS_2 and OCS. To be sure, the presence of
hot bands in these cases is of lesser severity than in the case of
carbon suboxide.

In addition to the spectra presented above the spectra of
borazole ($B_3N_3H_6$), cyanuric fluoride ($C_3N_3F_3$), sym-trifluorobenz-
ene ($C_6H_3F_3$) , and hexafuorobenzene (C_6F_6) have been obtained and
are now being analyzed (24). In the spectrum of hexafluorobenzene
the J = odd lines of the R-branch are resolved from the neighboring
S-branch lines. With an R-branch line spacing of 0.068 cm^{-1} this
represents the best degree of resolution obtained to date for the
$90°$ scattering configuration.

We briefly turn to the rotation-vibration spectra of polyatomic molecules. These are very much weaker than pure rotation
spectra and in only a very few cases (ethane, acetylene, ethylene,
allene, methane, carbondioxide, carbondisulfide) has it been
possible to resolve the rotational fine structure of the vibrational bands with the old mercury arc technique. Even with the laser
technique the work in the rotation-vibration region is arduous.

TABLE I

Molecular Constants of CO_2, CS_2, and OCS[a]

Molecule	Constant	Infrared	Ref.	Raman	Ref.
CO_2	B_{00^00}	39021 ± 4	19	39027 ± 0.7	17
	D_{00^00}	13.5 ± 0.5	19	12.9 ± 0.3	17
	$B^c_{01^10}$	39063.5	20	39065 ± 4	17
	$D^c_{01^10}$	14.1	20	8.2 ± 4	17
CS_2	B_{00^00}	10912.3 ± 1.0	21	10912 ± 0.7	18
	D_{00^00}	1.23 ± 0.12	21	0.83 ± 0.18	18
	$B^c_{01^10}$	10932.1	22	10935 ± 2	18
	$D^c_{01^10}$	1.1	22	1.5 ± 0.6	18
OCS	B_{00^00}	20285.6[b]	23	20287 ± 2	14
	D_{00^00}	4.37[b]	23	4.28 ± 0.55	14

[a] B- and D-values are in units of 10^{-5} cm^{-1} and 10^{-8} cm^{-1} respectively.
[b] Microwave determined values, converted to cm^{-1} with c = 2.997925×10^{10} cm/sec.

As an example , the triply degenerate γ_4 band of CF_4 at 631 cm^{-1} is shown in Figure 13. An earlier gas phase Raman spectrum shows this band as unresolved (25). This spectrum was obtained in 168 hr at a gas pressure of 1 atm. The $\gamma_2(e)$ = 435 cm^{-1} band was also observed (in 68 hr) but with less resolution. Since CF_4 is not one of the strong Raman scatterers the long exposure required for the observation of these bands is not too forbidding.

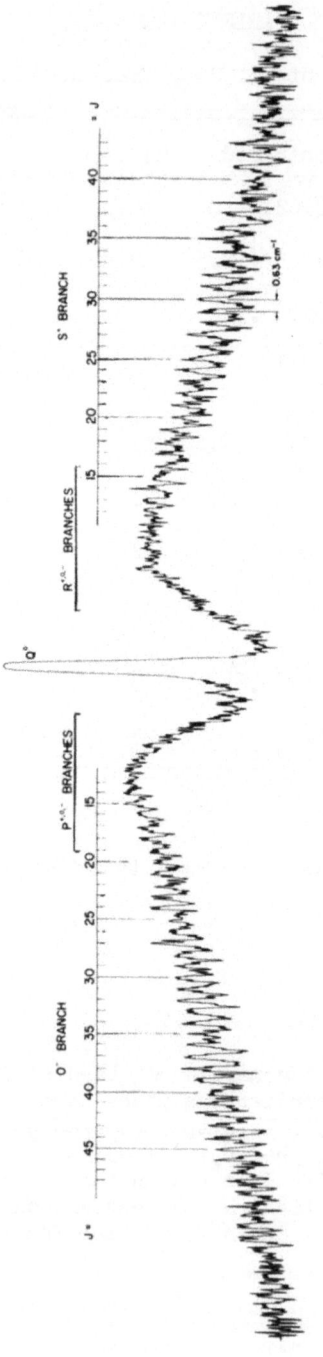

Figure 13.   The $\nu_4(f_2) = 631$ cm^{-1} Raman band of CF$_4$.

USE OF THE FABRY-PEROT INTERFEROMETER

The traditional spectroscopic technique used in high resolution
Raman spectroscopy has relied on the use of large plane and concave
grating spectrographs. The use of interferometer techniques was
precluded until the invention of the laser due to the weakness of
the Raman spectra and the width and hyperfine structure of the
mercury (natural abundance) $\lambda = 4358\text{Å}$ radiation that was used nearly
exclusively until that time. The availability of laser sources
has stimulated investigations into the use of the Fabry-Perot
interferometer in the study of the Raman effect. Figure 14 shows
the classical scanning arrangement utilized by Barrett and Myers (26).

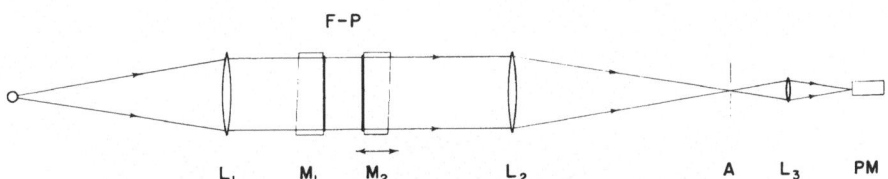

Figure 14.   Schematic diagram of Fabry-Perot interference
spectrometer used by Barrett and Myers (26).

The Fabry-Perot mirror M_1 is held stationary while mirror M_2 is
moved along accurate ways so as to either increase or decrease the
gap between them in a smooth, linear manner. The Haidinger fringes
are formed by lens L_2 on the plane of a pin hole aperture A, with
the center of the ring system focussed on the pin hole. As the
mirror M_2 is translated the signal picked up by the photomultiplier

PM is modulated in intensity. A spectrum that consists of regularly
spaced lines (i.e. the pure rotation spectrum of a linear molecule
if the effects of centrifugal distortion are ignored) is then
fully transmitted by the interferometer when its free spectral
range is commensurate to the frequency difference between adjacent
spectral lines. The interferogram then consists of regularly
spaced maxima and the deduction of the rotation constant is then
a simple task. The presence of centrifugal distortion effects only
slightly modifies the analysis but for a spectrum that consists
of non-regularly spaced lines (i.e. asymmetric top rotation
spectrum) the Fourier transform of the interferogram would have to
be computed. In this case, however, it was shown by Röseler (27)
that it would be more advantageous to employ a Michelson interferometer. This is so since the amplitude of the Fourier components
is greater for the interferogram generated by the Fabry-Perot
interferometer and a larger number of terms must be computed than
is the case with the interferogram obtained with the Michelson
interferometer.

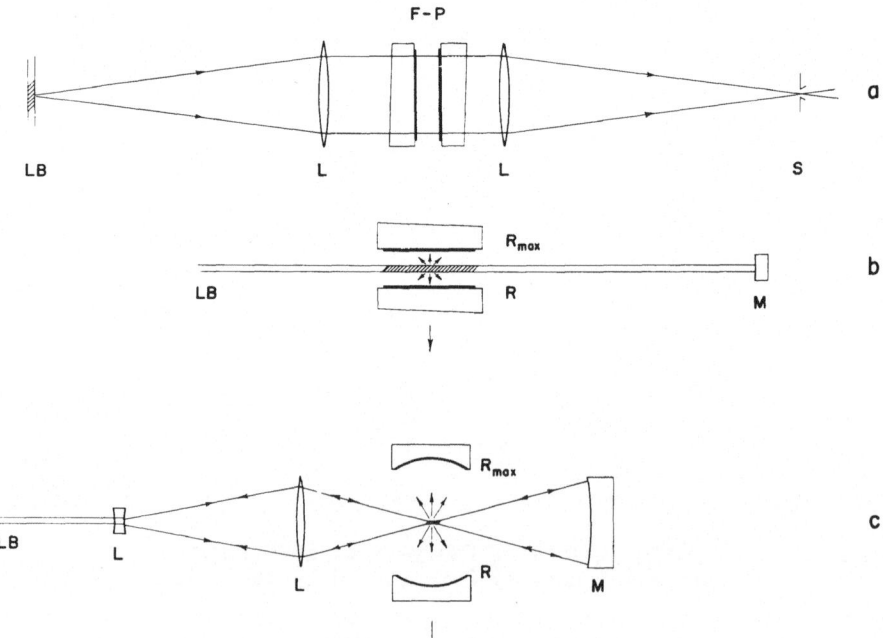

Figure 15.   Fabry-Perot interferometer in Raman spectroscopy. (a)
              Traditional arrangement with external source and Haidinger
              rings focussed on slit plane of spectrograph; (b)
              Internally illuminated plane-parallel interferometer;
              (c) Internally illuminated spherical interferometer.

Somewhat more classical arrangements of the Fabry-Perot
interferometer are shown in Figure 15. The traditional arrangement
is shown in (a). Here the plane parallel interferometer is placed
in the collimated beam between the Raman source (either the focal
volume in Figures 1 and 2 or the exit slot between the front
mirrors, Figures 4-7) and the spectrograph slit S. Internally
illuminated Fabry-Perot interferometers (28) are depicted in (b)
and (c). Here the scattering gas fills the space between the
interferometer mirrors, one of which is coated to maximum
reflectance whereas the other (the "exit" mirror) is coated to
give the desired transmission and resolving power for a given
plate separation. The laser beam traverses the gas and is scattered
by it. Using the He-Ne laser as a source of exciting radiation
the internally illuminated spherical Fabry-Perot interferometer
shown in (c) has been used by Schlupf and Weber (29) in a study
of the fine structure of the rotational lines of O_2. The schemes
shown in Figure 15 a and b have been tested with photographic
recording of the interferograms. Figure 16 shows the interference

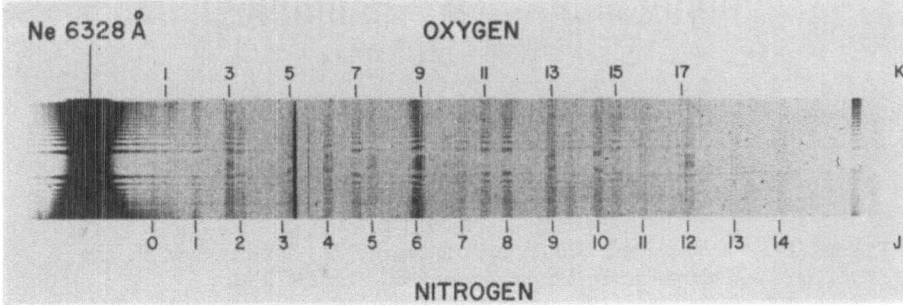

Figure 16. Rotational Raman spectrum of air at atmospheric pressure
obtained with an internally illuminated plane-parallel
Fabry-Perot interferometer (mirror spacing = 12.5mm)
and He-Ne laser excitation.

spectrum of air obtained in 1966 with the arrangement of Figure 15b
and He-Ne laser excitation (14). The interferometer mirrors were
positioned inside the Raman cell and took the place of the
spherical collecting mirrors CM (see Figure 7). With the air at
atmospheric pressure and a He-Ne laser rated at 20 mw the spectrum
was generated in 36 hr. The spectrum is seen to be rather "dirty"
due to the strong Rayleigh line produced by scattering off dust
particles (the Raman cell was not closed). The two sharp lines
between J = 3 and J = 4 are due to mercury whose spectrum was
directly photographed without passing through the interferometer.

Greater success is had with the argon laser. Figure 17 shows

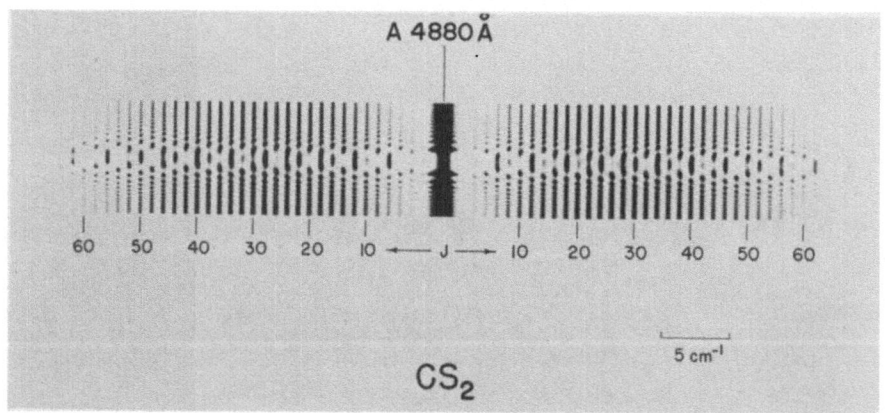

Figure 17.   Rotational Raman spectrum of CS_2 obtained with the
Fabry-Perot interferometer, Figure 15a.

the rotational Raman spectrum of CS_2 obtained with the arrangement
shown in Figure 7. The interferometer (plate separation = 15 mm)
was placed at the aperture stop between the lenses L_1 and L_2.
The gas pressure was 275 torr and the exposure time on a baked

Kodak IIIa-J plate was 2 hrs. The rotation spectrum in the $01^{1}0$ vibrational state (J = odd transitions, S-branch) is easily recognized on the original interferogram.

Using a He-Ne laser, photoelectric recording and high gas pressures Clements and Stoicheff (5) have similarly demonstrated the feasibility of using the Fabry-Perot in high resolution Raman spectroscopy.

Whether or not interferometric techniques are especially useful in Raman spectroscopy of gases depends on the specific problems. The required instrumental resolving power depends on the width of the Raman lines, this being determined primarily by Doppler and pressure broadening effects. The Doppler width of a Raman line (full width at half height) is given by (3)

$$\Delta\nu^{D_0}_{FWHH} = \frac{2}{c}\sqrt{2\frac{kT}{m}\ln2}\sqrt{4(\nu_0^2+\nu_0\nu_R)\sin^2(\theta/2)+\nu_R^2}$$

where ν_0 is the exciting frequency and $\nu_R=\frac{\Delta E}{h}$ is the Raman frequency shift, both in cycles/sec, θ is the scattering angle and the other symbols have their usual meaning. For Rayleigh scattering $\nu_R = 0$ and the Doppler width of the Rayleigh line is given by

$$\Delta\nu^{D_0}_{FWHH} = \frac{4\nu_0}{c}\sqrt{2\frac{kT}{m}\ln2}\cdot\sin(\theta/2)$$

This expression also represents the Doppler width of the pure rotational Raman lines to a good approximation.

The line broadenings due to the Doppler effect are thus seen to depend on the scattering angle, with the "forward" scattering ($\theta \approx 0$) giving the minimum width. As shown by Clements and Stoicheff (5) the improvement in the resolving power due to the reduction of the Doppler line broadening through the forward scattering technique is rather remarkable.

SPECIAL EFFECTS

In addition to the Doppler effect molecular interactions contribute to the line broadenings. This broadening effect can be reduced through the use of low gas pressures (for given temperature) and this is now done routinely in high resolution studies. On the other hand pressure broadening of Raman lines is of intrinsic interest in itself since it reveals the effects of those intermolecular interactions which are usually not observed with the methods of infrared and microwave absorption spectroscopy.

Pressure broadening of Raman lines has been intensively studied
both experimentally and theoretically by Jammu et al.(30),
May et al. (31), Fiutak and Van Kranendonk (32), Gray and
Van Kranendonk (33), and by Gordon (34). A pressure induced
narrowing of Raman lines has also been observed (35) and has been
investigated most recently by May et al.(36).

Since its discovery in 1928 the study of the Raman effect
has been restricted mainly to stationary, stable systems. It is
therefore of particular interest to note that the availability
of lasers and fast recording spectrometers have allowed an
extension of the Raman spectroscopic technique to the study of
gases which carry an electric discharge. Barrett and Weber (17)
have obtained the pure rotational Raman spectrum of CO_2 gas at
40torr pressure, carrying an electric discharge. The intensity
distribution in the spectrum is seen to be severly modified. Most
noticeable is the appearance of lines with J = odd whose intensity
relative to the lines with J = even is much greater than that for
the unexcited gas. Moreover, the intensity distribution suggests
that vibrational levels other than the 01^10 level also contribute
to the pure rotation spectrum. Nelson et al. (37) report the
vibrational Raman spectrum of N_2 gas (70-200 torr) carrying an
electric discharge. Vibrational hot bands (Q-branches)
$v' - v'' = 1-0, 2-1, 3-2$ were observed on both the Stokes and
anti-Stokes sides of the exciting frequency. These initial
experiments demonstrate the feasibility of using the Raman
effect in the study of gas phase kinetics, energy transfer in
gases and other problems of molecular dynamics.

The vapor phase spectra of Cl_2 and CCl_4 have been reported by
Hochenbleicher and Schrötter (38) who use a novel method of
illuminating the gas sample. The isotope structure of the v_1
vibration of (liquid) CCl_4 has been the traditional touchstone
for demonstrating the resolution performance of Raman spectrometers
and the non-zero value of the measured depolarization factor of
this vibration has always been a slightly disturbing feature
of the spectrum. Hochenbleicher and Schrötter's work on CCl_4 was
performed over the temperature range 25-120°C and the results
suggest that the hot bands found in the vapor spectrum are useful
for the explanation of the v_1 band structure in the liquid.

Lastly, we wish to mention the work of Masri and Fletcher (39)
on the contours of Raman bands. As shown by Edgell and Moynihan (40)
for the case of infrared bands, the analysis of the contours of
unresolved bands is capable of providing accurate values of the
Coriolis ζ - constants for the triply degenerate vibrations of
spherical top molecules. This method has been expanded by Masri
and Fletcher to the Raman spectra of spherical top molecules.
An earlier attempt by Mathai et al. (41) to derive the ζ -constants

for the doubly degenerate E' and E" vibrations of cyclopropane
from an analysis of the contours of the unresolved Raman bands
was only partially successful. It is very likely that this was
due to insufficient resolution that was available to these
investigators and that with the improved techniques based on laser
excitation more reliable and conclusive results can be had from
unresolved Raman band contours.

ACKNOWLEDGMENTS

This work was supported in part by a grant from the National
Science Foundation. Thanks are also due to Mr. Joseph Schlupf for
assisting in many phases of the author's work and for obtaining
the spectrogram shown in Figure 13.

REFERENCES

1. H.L. Welsh, E.J. Stansbury, J. Romanko, and T. Feldman,
    J. Opt. Soc. Am. 45, 338 (1955).
2. B.P. Stoicheff, "High Resolution Raman Spectroscopy of Gases",
    in Advances in Spectroscopy, H.W. Thompson, Ed. (Academic
    Press, New York, 1959), vol.1, pp.91-174.
3. A. Weber, "High Resolution Raman Studies of Gases", in The
    Raman Effect, A. Anderson, Ed. (Marcel Dekker, Inc., New
    York, 1972) vol.2, chapter 10.
4. A. Weber and J. Schlupf, J. Opt. Soc. Am., (in print), 1972.
5. W.R.L. Clements and B.P. Stoicheff, J. Mol. Spectry. 33, 183
    (1970).
6. H.H. Claassen, H. Selig, and J. Shamir, Appl. Spectroscopy,
    23, 8 (1969).
7. H.H. Claassen, G.L. Goodman, J.H. Holloway, and H. Selig,
    J. Chem. Phys. 53, 341 (1970).
8. J.J. Barrett and N.I. Adams,III, J.Opt.Soc.Am. 58,311 (1968).
9. A. Chapput, M.Delhaye, and J.Wrobel, Compt.Rend.272B,461(1971).
10. A. Weber, S.P.S. Porto, L.E. Cheesman, and J.J. Barrett,
    J. Opt. Soc. Am. 57, 19 (1967).
11. J.Moret-Bailly and H. Berger, Compt.Rend. 269B, 416 (1969).
12. J.U. White, N.L. Alpert, and E.G. DeBell, J. Opt. Soc.
    Am. 45, 154 (1955).
13. N.H.Rich and H.L.Welsh, J.Opt.Soc.Am. 61, 977 (1971).
14. A. Weber, unpublished work.
15. B. P. Stoicheff, ref. 2, pp.124-126.
16. W.H. Smith and J.J. Barrett, J. Chem. Phys. 51, 1475 (1969).
17. J.J. Barrett and A. Weber, J. Opt. Soc. Am. 60, 70 (1970).
18. W.J. Walker and A. Weber, J. Mol. Spectry. 39, 57 (1971).
19. C.P. Courtoy, Ann.Soc. Sci. Bruxelles 73, 5 (1959).
20. H.R.Gordon and T.K.McCubbin, J. Mol. Spectry. 19, 137 (1966).
21. G.Blanquet and C.P.Courtoy, Ann.Soc.Sci.Bruxelles,84,293 (1970).

22. A.H. Guenther, T.A. Wiggins and D.H. Rank, J.Chem.Phys. 28, 682 (1958).
23. W.C.King and W.Gordy, Phys.Rev. 90, 319 (1953).
24. J.Schlupf and A.Weber, to be published.
25. B.Monostori and A.Weber, J.Chem. Phys. 33,1867 (1960).
26. J.J. Barrett and S.A. Myers, J.Opt.Soc. Am. 61,1246 (1971).
27. A.Röseler, Optik, 24,606 (1966-1967).
28. A.Kastler, Appl. Optics 1,17 (1962).
29. J.Schlupf and A.Weber, unpublished work.
30. K.S.Jammu, G.E.St.John and H.L.Welsh, Can.J.Phys. 44, 797 (1966).
31. A.D.May, G.Varghese, J.C.Stryland and H.L.Welsh, Can.J.Phys. 42, 1058 (1964).
32. J.Fiutak and J.Van Kranendonk, Can.J.Phys. 41,21 (1963).
33. C.G.Gray and J.Van Kranendonk, Can.J.Phys. 44,2411 (1966).
34. R.G.Gordon, J.Chem.Phys. 42,3658 (1965); 44,3083 (1966); 45, 1649 (1966).
35. For a summary and references to this work see ref. 3.
36. A.D.May, J.C. Stryland and G. Varghese, Can.J.Phys. 48,2333, (1970).
37. L.Y.Nelson, A.W. Saunders, Jr., A.B.Harvey, and G.O. Neely, J.Chem.Phys. 55,5127 (1971).
38. G.Hochenbleicher and H.W. Schrötter, Appl.Spectroscopy, 25, 360 (1971).
39. F.N. Masri and W.H. Fletcher, J.Chem.Phys. 52,5759 (1970).
40. W.F.Edgell and R.E. Moynihan, J. Chem.Phys. 27,155 (1957).
41. P.M. Mathai, G.G. Shepherd and H.L.Welsh, Can.J.Phys. 34, 1448 (1956).

ATMOSPHERIC AND SPACE SPECTROSCOPY

MEASUREMENT OF METEOROLOGICAL STATE VARIABLES

BY REMOTE PROBING

John A. Cooney

Physics Department, Drexel University

Philadelphia, Pennsylvania 19104

The measurement of meteorological state variables includes a measurement of the space-time distribution of temperature, pressure and humidity. In what follows, a detailed description of a number of experiments is given which provides measurements of temperature, density and humidity.

The notion of remote probing, in atmospheric physics, has advanced markedly over the last decade. The description below relates to the observation of Raman backscatter from the atmosphere when it is illuminated by very high powered lasers. The Raman scatter is distinguished from the Rayleigh scatter in that the scattered light is shifted in frequency from the incident light by a specific amount of frequency corresponding to a energy level difference in the scatterer itself. In addition, as with the Rayleigh, the Raman intensity is proportional to the local molecular density. The cross-section for the Rayleigh process is larger than the non-resonant Raman process; ranging anywhere from a factor 10 to a factor of a thousand greater. In the interesting case of nitrogen, if the incident energy were launched at a wavelength of 6943Å as in the ruby laser, we would expect to observe the nitrogen vibrational-rotational Raman transitions centered at 8283Å. It should be noted that there are rotational band structures super-imposed on this particular vibrational line and in figure 1 the Raman return for the nitrogen vibrational levels shows

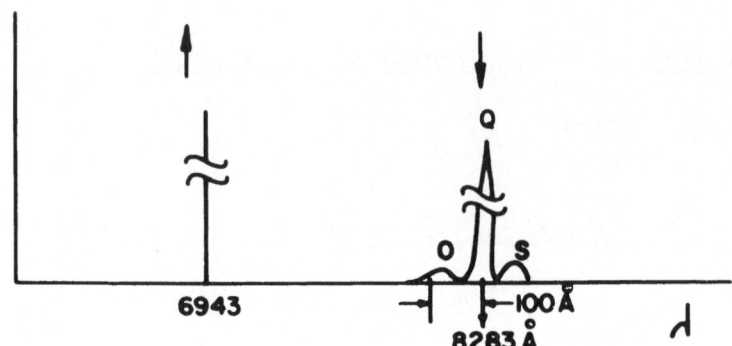

Figure 1.   Raman return of N_2 vibrational level.

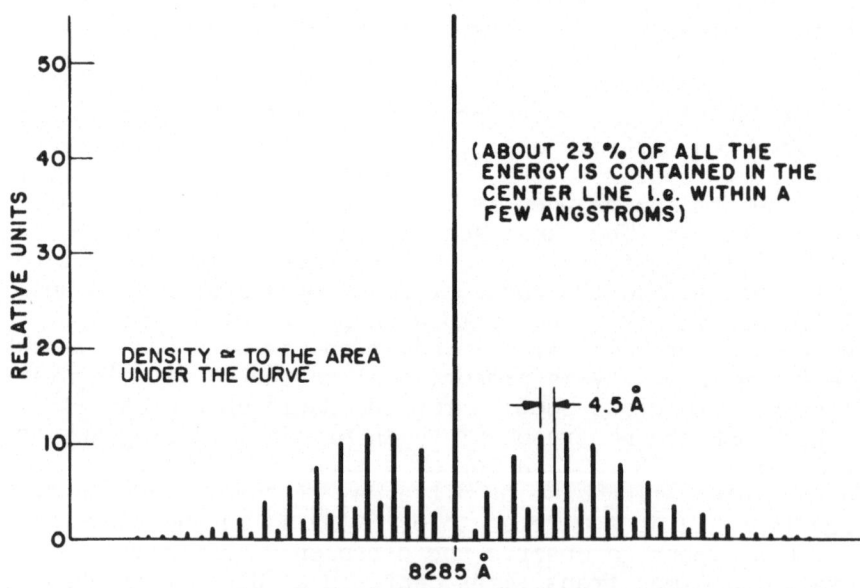

Figure 2.   Rotational-vibrational Raman spectra of N_2.

such a structure including Stokes and anti-Stokes bands.
Figure 2 shows the more detailed structure of these bands.
As is well known, it is composed of a series of discreet
lines and the branches are, in the direction of increasing
wavelength, respectively O, Q and S.

These preliminary remarks on N_2 lead into a description
of the first of four separate experiments which gives rise
to the measurement of profiles of number density. There
are, to be sure, four primary meteorological variables.
However, the wind field is a dynamical variable and is
excluded from present considerations. Humidity, density
and temperature are measured directly and, as a result,
pressure as a derivative measurement can be obtained.
However, the presence of particulate matter in the atmosphere can influence the accuracy of density measurements,
and so the scattering arising from these materials has to
be measured and this constitutes the fourth measurement.

Starting with a description of the hardware, the
details of the first of the four measurements is given. If
one looks at figure 3, one sees a drawing of the angular
aperatures of the launch beam and the optical receiver beam.
This provides an effective scattering volume which is essentially the overlap volume of the transmitter and receiver.
This is a monostatic arrangement whereby it is meant that
the receiver and transmitter are co-mounted and share a
common optic axis. Figure 4 is a picture of some of the
earlier field hardware. On the right, on the bench is a
ruby laser system with the cover off and to the left is
some rack mounted subsidiary electronics. The mirror used
is shown in figure 5. Notice the post to the left of the
mirror. Mounted on its top at the end of the arm is a
filter wheel-photomultiplier housing combination. The flat
disk part contains the filter wheel (see figure 5A) which
is a means of selecting narrow band filters to interpose
in the optical channel. The cylinder atop the disk contains
the photomultiplier. The arm position controls the multiplier filter wheel combination in order that the multiplier
cathode is placed in the focal plane of the mirror. The
optical filters can be interposed between the mirror and
its focal point. Then the light is transduced to an electrical signal. The cross-like attachment on the right is
the mount by the laser which is co-mounted with the
receiver.

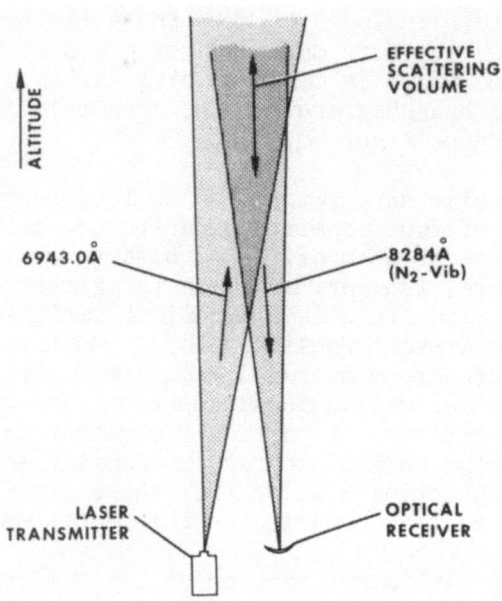

Figure 3. Geometry of laser probing systems.

Figure 4. Laser transmitter and associated electronics.

Figure 5. Laser transmitter-receiver mount.

Figure 5A. Laser system filter wheel.

Because the system is constructed to operate in a radar
like fashion, the basic information coming back is in the
form of an altitude profile.

An actual case of a return signal can be seen in
figure 6. Notice that there is a strong rising component
to the signal followed by a falloff. The rising portion
is due to the fact that although the strength of the scattering per unit volume is decreasing with distance and
therefore ought to become less in intensity, the overlapping
of the volumes is increasing with distance at such a rate
that the diminution of signal does not begin to take hold
until it reaches that signal apex, at which time the volumes,
as shown in figure 3 earlier, are overlapped pretty much.
After this, one has effectively a $(1/r^2)$ tail off. Thus,
the basic form of the signal results from these two factors.
This basic signal configuration really has little to do, in
a sense, with the basic information which is to be extracted
from the curves. (r is the altitude coordinate).

The first experiments consisted of observing the
nitrogen density as a function of altitude. This is done
simply by observing the signal return at 8283Å due to the
6943Å launch pulse. With the system constants and altitude
dependencies eliminated from the raw data, one can construct
a curve of nitrogen density as a function of altitude. Of
course, the nitrogen density is very well known, and one
needn't worry so much about being able to measure the nitrogen
density "per se". As is seen below, however, it does provide
an extremely important subsidiary function for measurements
of water vapor. The description of the N_2 profile measurements is given in more detail in (Cooney, App.Phy. Lettr.,1/68).

In the second set of experiments, measurements were
made separating the gaseous and aerosol components of laser
backscatter. In this connection it should be noted that the
atmosphere, in addition to consisting of molecular constituents, quite often has a significant amount of particulate
matter suspended in it. This solid matter scatters
radiation with intensity which quite often exceeds that of
the so-called Rayleigh scatter. As with the Rayleigh return,
this so-called Mie scatter is monitored at the incident
frequency. We see that the backscatter return at the on
frequency or incident frequency must, therefore, contain
not only molecular scatter, but aerosol or particulate
scatter as well. If, however, one has a subsidiary

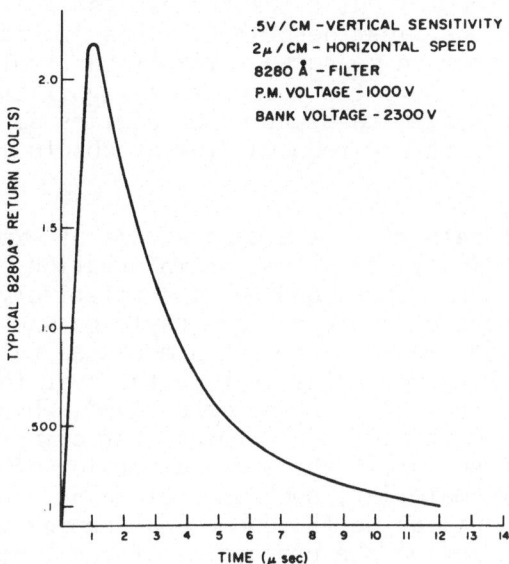

Figure 6. Typical form of a laser radar return signal.

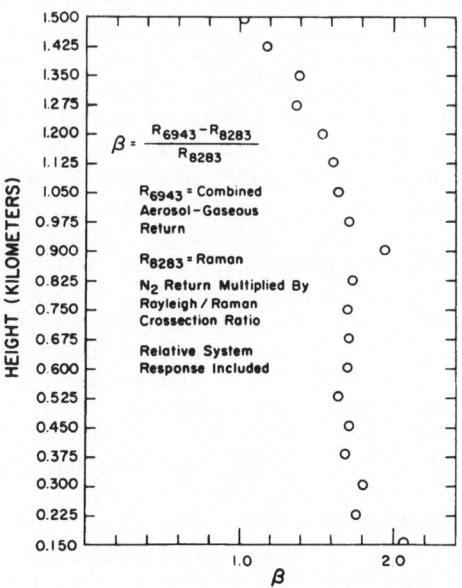

Figure 7. Normalized aerosol back-scatter cross section.

measurement of the molecular scatter available with the Raman
nitrogen measurement just described. This provides, with the
aerosol backscatter, a measured ratio of particulate to
gaseous scatter. The ratio measurement significantly reduces
the errors from system constant uncertainties as the calibration requirements are reduced from an absolute to a
relative one.

The actual data of this second series of experiments
gave rise to basically two kinds of information. When the
visibility was high corresponding to a relatively low
aerosol content, the departure from Rayleigh was relatively
small and the data was used to get a measured value of the
ratio of aerosol to molecular backscatter over the normal
range of the instrument. On the other hand, when the ratio
of particulate scattering to Rayleigh scattering was a
factor of 10 or so, or if there was a portion of the path
occupied by some material, such as a cloud or a haze layer,
of significant optical depth, one could then gather only
special information on the properties of the layer such as
the optical thickness of the layer. In such a case the loss
of signal strength when the beam passed through an area of
significant optical thickness was such that one could not
simply interpret the return data. Such things as multiple
scattering confused the interpretation of this type of return data. Nonetheless, data of this kind, as will be
indicated below, can have a reassuring effect in one specific
regard. To show this, I would like to go back to the data
corresponding to a relatively clear or a relatively sparse
aerosol content and ask you to look at figure 7. One sees
there a measured, and I emphasize measured, ratio of aerosol
to molecular backscatter and then by knowing what the molecular backscatter function ought to be from theory (this
is known quite well), one can compute the real or the actual
aerosol backscatter. It should be realized, of course, that
the aerosol scatter is highly anisotropic in character and
that a knowledge of the backscatter is insufficient to determine the total cross-section. These cross-sections depend
very much on such things as the kinds of material that the
particulates are composed of as well as the size distribution
of the particles themselves. So that the knowledge of the
backscatter from aerosols will simply help us eliminate the
error in the backscattered return from the Rayleigh signal
or from the Raman signal. This measurement is also a verification, a way of helping to determine what the actual
density profile is. However, it is re-emphasized, that one

already knows fairly well the nitrogen density. And so, in this connection the process basically becomes one of upgrading of the accuracy of the laser radar measurement of atmospheric molecular density.

The second type of return referred to in connection with these aerosol measurements corresponds to data shown in figure 8. In the experiment not only is the Rayleigh return monitored, but also the Raman return. This particular experiment shows the data taken of a simple vertical profile at both the Raman and Rayleigh wavelengths. The target of opportunity which manifest itself on the 6943 return was a semi-transparent haze layer which drifted over the optical path of the measuring instrument. In this case the haze layer gives rise to the significant bump seen on the Rayleigh return. What is interesting about this particular measurement is the total absence of a return on the Raman return. There were, to be sure, a large number of traces taken and some curves show a small bump on the 8283Å at the same place as is given on the 6943Å return, but these could not be statistically validated due to the presence of noise. The bump has a maximum of about a factor of four above the base of the return. Thus we see that with a high degree of confidence it is possible to assume that transmission losses, even when operating thru significant haze, do not affect the Raman return very much as long as transmission losses are not large enough to diminish the beam intensity.

These first two measurements which essentially provide for a measurement of nitrogen atmospheric density, while not of overriding or practical importance in themselves, are quite interesting scientifically because it becomes possible to study the fluctuations in density. However, as a practical matter they do provide a means whereby some remote probing profiles of some highly interesting, highly variable atmospheric constitutents can be measured. Thus we come to the third series of measurements involving atmospheric water vapor. Basically, the atmospheric water vapor measurement is made by monitoring the return at the Raman vibrational shifted wavelength of the water vapor molecule. For example, using the doubled output frequency of the ruby laser for which the launch pulse is at 3472Å, observations of water vapor return occur at 3976Å. Basically, the water vapor measurements are made with the hardware essentially as described earlier. The principal exception, of course, is that the

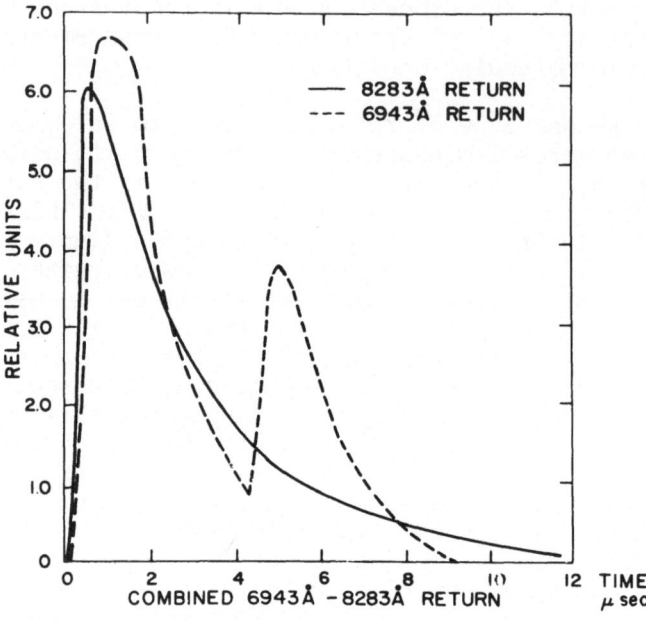

Figure 8.

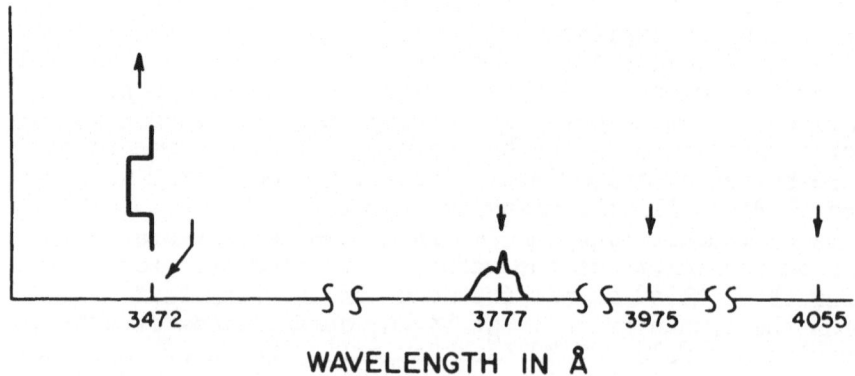

Figure 9.   Lines monitored by laser receiver for
the water vapor measurement.

filter wheel contains four separate narrow band filters or
optical channels at the specifically needed wavelengths.
Thus, there is an optical channel at the output or launch
frequency, a second at 3777Å corresponding to the nitrogen
vibrational rotational transition, a third channel (the
water vapor channel 5-10Å wide channel) centered at 3976Å
and finally a channel at 4055Å to monitor spurious signals.
In the latter case if the instrument is not working correctly
or if there were signals coming back from spurious sources
imposing themselves upon the nitrogen return at 3777 or the
water vapor return at 3976, some manifestation of this would
also occur in this so-called dead-band channel. The wavelength disposition is shown in figure 9.

Some preliminary measurements were made in Boulder,
Colorado, in the Spring of 1969 and figures 10 and 11 show
vertical profiles of absolute humidity when taken in April 17
and the other taken May 28. Basically, both the water vapor
and nitrogen returns were taken essentially simultaneously.
The time difference being that required to shift the optical
channel to the appropriate wavelength. When making the
nitrogen returns, the filter wheel was shifted to the 3777Å
optical channel and when water vapor returns were required,
the filter wheel was shifted to 3976Å. A number of checks
were periodically made at 4055Å to make sure that our signals
were not arising from spurious sources. The question of
why one also makes nitrogen measurements in the case of
humidity measurements has not really been fully discussed.
If we attempted to make a water vapor profile based on the
water vapor measurements alone, we would require what amounts
to an absolute calibration of an optical system at the
received wavelength 3976. This is an especially difficult
thing to do. By making, in addition to the water vapor
measurements, measurements of the nitrogen density and
assuming that the nitrogen density is otherwise known, we
are then able to reduce the calibration requirements to those
of making a relative calibration between a 3976 channel and
3777 channel. This significantly reduces the error associated
with the water vapor measurement. Much of the error that
enters into these measurements, enters in, in essentially the
same way in both channels. Thus, what we do is to take the
ratio of the water vapor returns to that of the nitrogen
returns and so make, in essence, a measure of the mixing
ratio. However, to repeat, we assume that we know the amount
of nitrogen from temperature and pressure measurements and

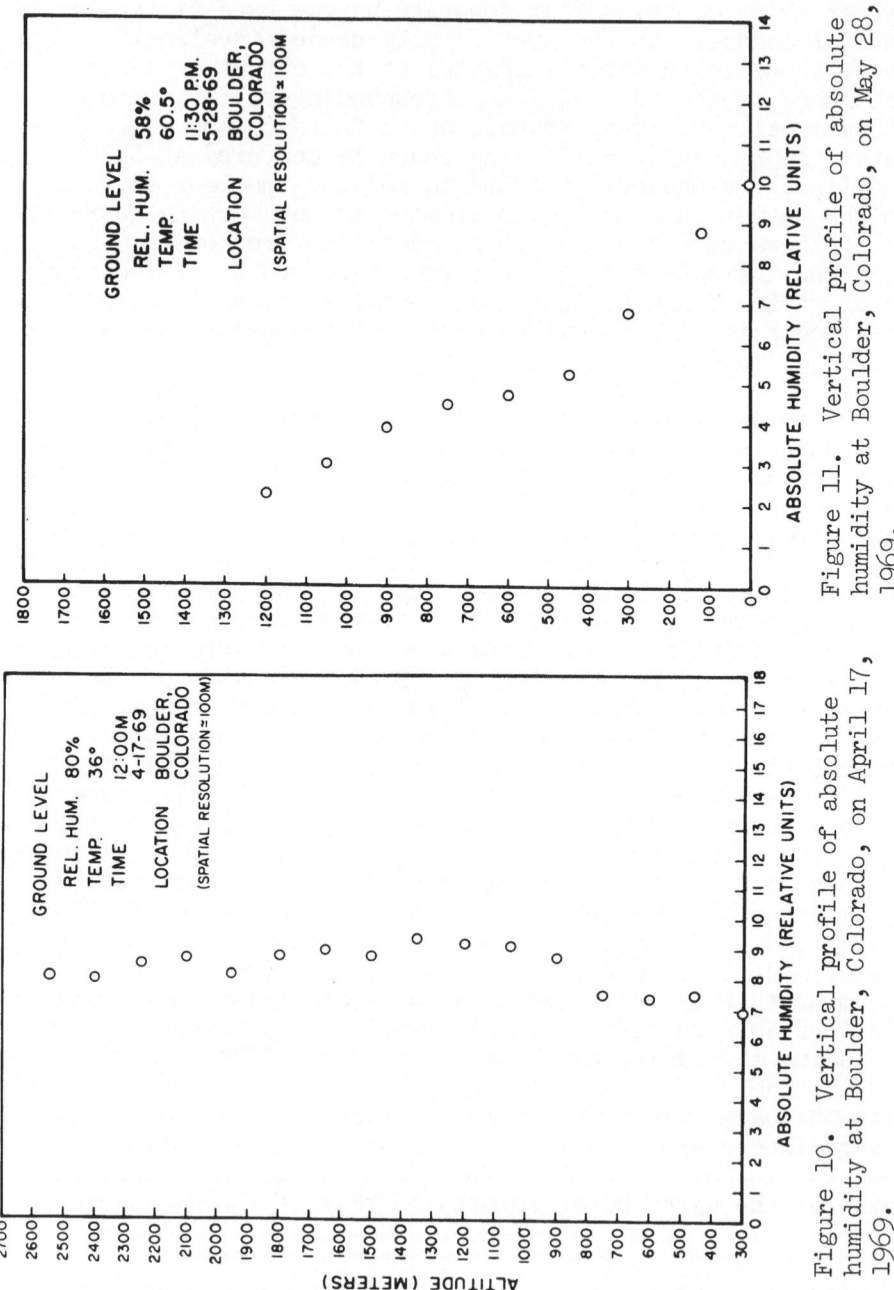

Figure 11. Vertical profile of absolute humidity at Boulder, Colorado, on May 28, 1969.

Figure 10. Vertical profile of absolute humidity at Boulder, Colorado, on April 17, 1969.

assumptions of lapse rates, and are able, therefore, to come up with curves of absolute water vapor density or absolute humidity content.

The correlation of the measurements made at Boulder with the orthodox techniques for making these measurements, namely the radiosonde, provided by the Denver Weather Bureau, were such as to encourage a more complete program and thus it was that a more complete program was carried on at Wallops Island, Virginia. Thus, although there was a very promising correlation between the Denver Weather Bureau and Boulder measurements, the space time correlation was not sufficiently good. For example, the laser measurements were made at midnight and the Denver Weather Bureau radiosonde were taken at 5:00 P.M. and compared. There was a seven hour time separation and a 22 mile spatial separation. One of the chief reasons for going to Wallops Island, Virginia to the NASA station was that the meteorological station facilities were very complete, and that we had available radiosonde measurements within literally minutes of the time we wanted them as well as detailed subsidiary meteorological information. After a considerable amount of setting up, data was recorded corresponding to about 150 profiles on 2 successive evenings in 1969, on October 22 and 23. Because we desired to obtain as tight a spatial correlation as possible, we attached a radiosonde package to a line and suspended it about 250 feet below a helicopter so that we could fly precisely through the optical path of the laser beam just before and just after the laser shots so that the space time correlation would be its tightest. Thus it was that we not only had balloon flights, the orthodox means of flying the radiosonde package, but we had helicopter flights of the radiosonde packages as well. It should be noted that prior tests on the usefulness of helicopter mounting of radiosonde packages were conducted.

In Table I one sees a timetable of radiosonde and laser profiles taken on the two nights mentioned earlier. The reduced profiles are shown in figure 12 for the laser and in figure 13 for the radiosonde and the helicopter-sonde. How well did these measurements inter-compare? This can be seen in figure 14. Here we take the average of a series of profiles both nitrogen and water vapor shots (15 to 25) and compare these to the average of the two orthodox radiosonde contiguous in time flights, to be laser flights. For example, laser B is compared with the average of flight 1 and flight 2.

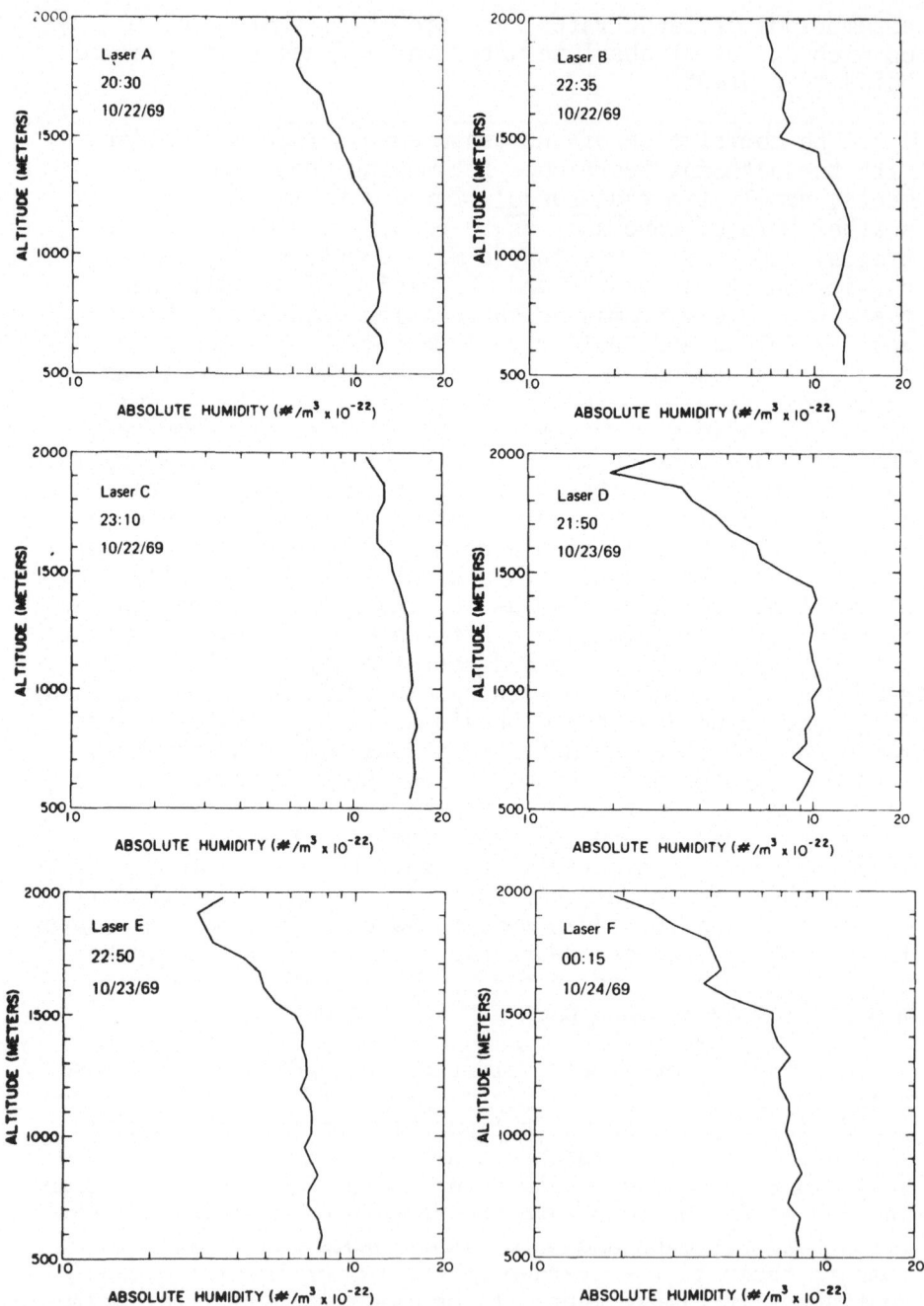

Figure 12.   Six averaged laser profiles with time and date.

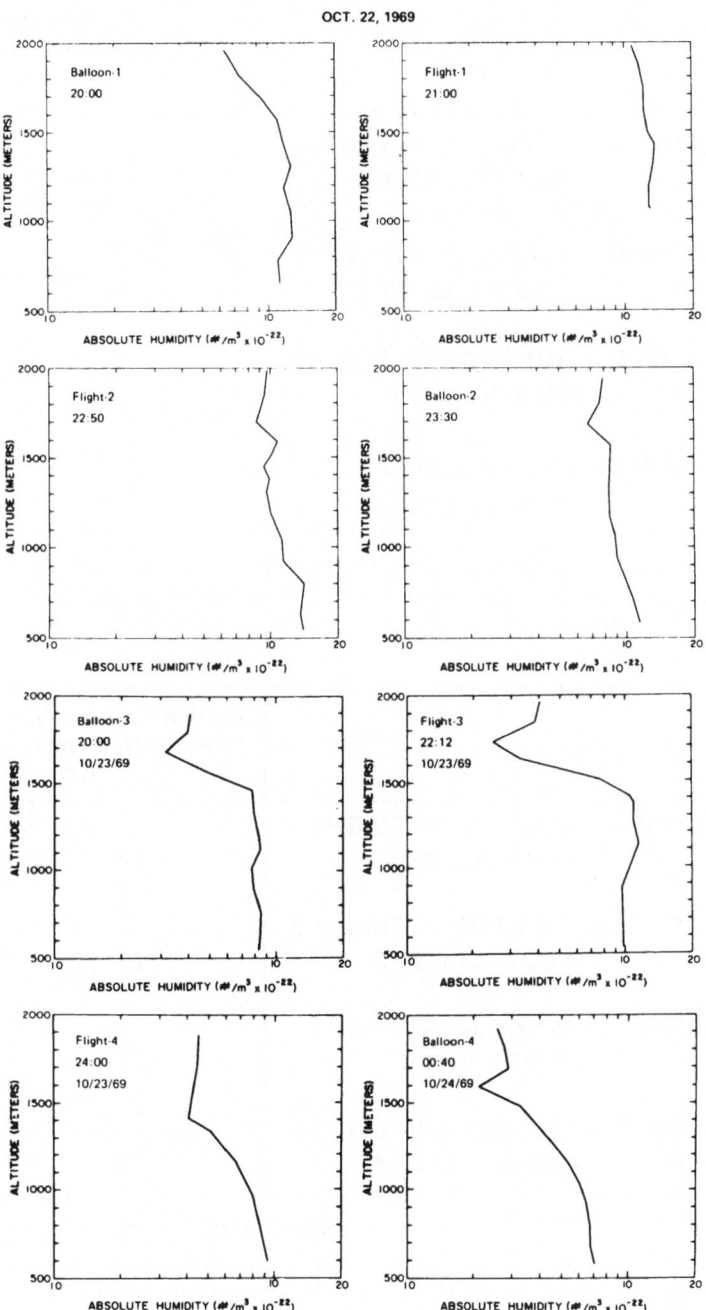

Figure 13.   Eight sonde profiles with time and date.

	Relative Difference in Integrated H_2O Vapor Profile
LASER B – AVERAGE OF FLIGHT 1 AND FLIGHT 2	.11
LASER C – AVERAGE OF FLIGHT 2 AND BALLOON 2	.11
LASER E – AVERAGE OF FLIGHT 3 AND FLIGHT 4	.12
LASER F – AVERAGE OF FLIGHT 4 AND BALLOON 4	.15

Figure 14. Fractional relative differences in water vapor content in a column of air extending from 500 m to 2000 m. Profiles compared given on left.

	Absolute Difference in Integrated H_2O Vapor Profile
LASER B – AVERAGE OF FLIGHT 1 AND FLIGHT 2	.09
LASER C – AVERAGE OF FLIGHT 2 AND BALLOON 2	.73
LASER E – AVERAGE OF FLIGHT 3 AND FLIGHT 4	.22
LASER F – AVERAGE OF FLIGHT 4 AND BALLOON 4	.09
AVERAGE	.28

Figure 15. Fractional absolute differences in water vapor content in a column of air extending from 500 m to 2000 m. Profiles compared given on left.

As can be seen in the earlier figure, laser B is surrounded
in time by flight 1 and flight 2. Laser B occurred at 22:35
and flight 1 occurred at 21:00 and flight 2 at 22:50. In
addition, we took the average of flight 2 and balloon 2 and
compared it with laser C. On the following night we took
the average of flight 3 and flight 4 and compared it with
laser E; and flight 4 and balloon F and compared it with
laser F. Notice that we did not use laser A and laser D
in these comparisons. It will become clear why below.

What variable did we use to make these inter-comparisons?
The profiles were integrated to get the total water vapor
content in the column between 500 meters and 2 kilometers
and this datum as obtained both by laser soundings and by
orthodox radiosondes was compared using the averages mentioned
above. In figure 14 the relative differences of laser, radio
and sound profiles are variously 11%, 12% and 15%. What do
we mean by relative difference? Here is meant that we take
and arbitrarily place the 500 meter altitude value of the
profile obtained by laser B at the same humidity value as
the 500 meter reading obtained from the average of flight 1
and flight 2. Thus, the measurement shows that the difference ·between the integrated value of the humidity between
500 meters and 2000 meters of the two types of measurement,
varies only by 11% etc. Similarly, this occurred with later
and earlier readings. This is what is meant by a relative
difference. In addition to this, what is called an absolute
difference of laser and radiosonde profiles was obtained.
In this case the idea was to take the first balloon flight
and take the first laser series of measurements, and arbitrarily place the laser A profile with respect to the balloon
1 profile in such a way that the difference of the integrated
water vapor profile was a minimum. This effectively cali-
brates the laser system. That is to say that we use a balloon
profile and the first laser profile to perform a calibration
on the system. Now we allow the other balloon and helicopter
radiosonde data as well as the other laser flights to take
the positions wherever they might as shown in figure 15.
We did not adjust the relationship between laser A and laser
B. Laser B was placed where its measurement said it should.
Similarly, this was done with the average of the humidity
data from the balloon and helicopter flights. So that when
we compare laser B with the average of flight 1 and 2 as
seen in figure 15 this, in a sense, is an absolute measurement. As can be seen for those absolute measurements, we

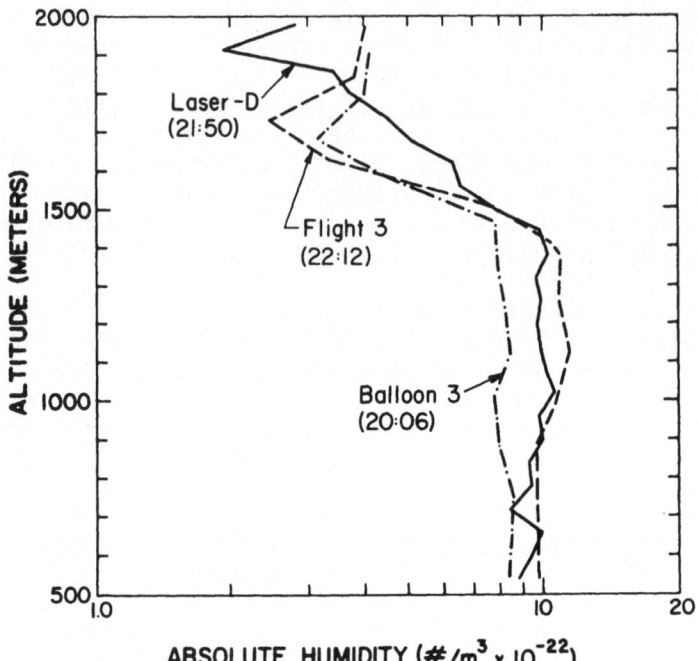

Figure 16. Humidity profiles.

**Difference in Integrated Water Vapor Content
from Sonde to Sonde**

			ABSOLUTE	RELATIVE
Balloon 1	—	Flight 1	.161	.111
Flight 1	—	Flight 2	.013	.173
Flight 2	—	Balloon 2	.182	.029
		AV	.119	AV .104
Balloon 3	—	Flight 3	.144	.005
Flight 3	—	Flight 4	.282	.241
Flight 4	—	Balloon 4	.315	.063
		AV	.247	AV .103

Figure 17. Sonde to sonde comparison of relative and absolute differences in water vapor
content in a 500 m to 2000 m column.

had 9%, 73%, 22%. The 73% is a fairly wide variation, and this might really have occurred because there was an enormous fluctuation in the water vapor content before the time the laser C was taken. However, such a fluctuation is highly unlikely and, in addition, we found that there was a saturation of an amplifier signal during this particular set of measurements. From this latter fact, it seems quite clear that the system will take quite good calibrations and hold them for at least one evening. What is required to make them, or whether they in fact can hold them for quite some much longer period of time is, of course, a matter to be determined by further experiment. Up to this point, measured comparisons between laser and balloon have been made for the integrated data.

One can ask how much fine detail can be recorded. Figure 16 shows separate rate profiles: one of balloon 3, one of laser D and one of flight 3. It is to be seen that all 3 pickup the sharp break in the moist layer at 1500 meters, all within a 100 meters of each other. It should be mentioned that these systems integrate in a slightly different way and therefore a very fine structure for these particular instruments cannot be precisely compared. There is no question however, of course, that the general trends as measured by radiosonde are followed very nicely by the laser. In fact it may very well be that the laser is doing a better job of measuring humidity but, of course, there is no way of proving this at this point.

We noticed the variations of the laser readings from the radiosonde readings in both the absolute and relative comparisons. If we now intercompare the variations between the various radiosonde measurements themselves, we notice as in figure 17 that on an absolute basis balloon 1 and flight 1 are 16% apart. That is to say that 16% difference in the water vapor content between the measurement made by balloon 1 and that made by flight 1 and there is an 11% difference in the relative measurement. The interesting point to be learned about such comparisons is that the variations in the measured values of water vapor content from one radiosonde to the next was almost as large as the variation from the radiosonde to the laser. This suggests that the laser is already on an accuracy parity with the radiosonde in terms of making such measurements and to repeat the point made above, namely that the laser is very possibly doing a more accurate measurement already. These measurements by the nature of the hardware used were restricted to night time measurements. However, there are at least two promising techniques for overcoming the handicap of making laser measurements at night. One of them will be alluded to in the last of four measurements which will be described now.

Up to this point we have described the measurement of nitrogen density and of humidity and of the measurements separating aerosols and gaseous scatter. I should mention before going on that humidity is considered to be a minor constituent and that one has something of the order of 1% relative composition. Thus, the ratio of water vapor molecules to nitrogen molecules is lower by a factor of approximately 100.

Subsidiary measurement, by which I mean independent calibrations of the transmitter and receiver allow us to turn the measurements of water vapor around and come up with a cross section for the ν_1 Raman transition in water vapor. What was not known, of course, at the outset of these humidity measurements is the size of this Raman transition. However, if we assume that we know the water vapor content and if we assume we know the nitrogen content, we can measure the ratio of the cross section of ν_1 transition to the nitrogen vibrational rotational transition. This was done and has been reported in Spectroscopy Letters in Dec. 1970.

The final of the four measurements concerns the measurement of atmospheric temperature profiles. This is also done by the observation of the detection of Raman backscatter. In this case, however, the Raman nitrogen rotational bands are used. These are Raman returns close by the exciting line. They consist essentially of two bands placed symmetrically on either side of the exciting line and, of course, are due to the rotational energy level structure of the molecule. In the case of interest (rotational structure of nitrogen) the first Raman line is removed a little more than 12Å from the exciting line. If we are using the ruby laser, one would look for the bands to begin within 10Å or so of the 6943Å exciting line, and extending from 50 to 100Å to either side. In principle, of course, the lines extend into infinity. In practice one goes to about $J = 21$ (rotational quantum number) and has already covered the overwhelming fraction of the energy involved in these bands.

The fact that the Raman spectrum contains temperature information is well known. However, although measurements began on nitrogen density and water vapor density within a year or so of the advent of the O-switched laser, the measurement of temperature has really yet to be made. The prime reason for this delay is that the Raman signal is relatively insensitive to temperature. Thus, one calculates a change of .2% in signal intensity of the nitrogen rotational spectrum at $300^{\circ}K$ at the wavelength of maximum sensitivity. On the other hand, the measurement technique described below can increase the signal sensitivity by close to two orders of magnitude. Basically, the suggested method is a differencing scheme in which two separate input signals are inserted into a differential amplifier. The separate signals are generated from one prime signal and therefore many of the characteristics are the same. The two separate signals come from slightly different portions of the N_2 rotational spectrum. If we examine figure 18, we see what is meant by the temperature change of the signals. It gives the intensity as a function of rotational quantum number J. The envelope of the band peaks at lower J values as the temperature decreases. Thus, for example, the maximum J which is J = 8 for nitrogen at $300^{\circ}K$, decreases to lower values of quantum number as the temperature decreases.

If one examines figure 19, a scheme is shown for separating or breaking into two components the incoming signal. Figure 19 shows an incoming signal on the left being split by a pellicle and the separated signals, each going into an optical filter. One is designated F1 and the other designated F2. F1 and F2 are slightly different in wavelength transmissions. As we see in figure 20 their relative position in respect that the Raman rotational intensity spectrum is given. The signals pass through the filters and go into individual photomultipliers. They are then transduced to electrical signals and are fed into the input of a differential amplifier. The output is fed into the "scope". It is the "scope" output that constitutes the temperature profile. If we look at figure 20, we see the relative dispositions of the filters F1 and F2 as a function of wavelength and then we can appreciate, viewing figure 18 and 20 together, what is happening. We've gotten the relative magnitudes of the two signals coming through their respective channels which are changed in magnitude as a function of temperature. The signals are fed into the differential amplifier and the output of the differential

amplifier which is fed to the scope, is the fundamental
temperature profile. In figure 21 one has the filter
properties for F1 and F2. F1 is 14Å full width at half maximum and the center wavelength is 6979Å whereas in F2 the
full width at half maximum is 20Å and center wavelength is
7008Å. Transmission maximum is 30% in both cases. These
characteristics were arrived at on the basis of an optimizing
computing program. Several constraints were involved. Namely,
primarily that the Rayleigh return be very low and so one
had to have a substantial blocking action for both filters
at the incident or launch frequency.

     In order to really determine what level of temperature
resolution is available from the signal, one has to make
signal to noise calculations in order to see what the change
in signal level per degree change in temperature there is
in relationship to the mean noise level. If we assume that
the main source of noise is shot noise which in most practical cases is to be expected, we can go ahead and calculate
the temperature sensitivity. Figure 22 contains a table of
equations. The last one on the right is a measure of the
temperature sensitivity. These equations are used to compute
the various columns shown in figure 23. The three equations
in the lefthand column compute the light signal intensity
both before and after the optical filters. The two far
equations in the right hand column compute the differential
amplifier input and output. α is the photomultiplier conver-
sion factor and B is the amplifier gain. (Note that I_{F1} =
I_1 and I_{F2} = I_2 in figure 23.) The last equation on the
lower right compute the temperature sensitivity of the output of the amplifier.

     Note also at the bottom of figure 23 the Rayleigh re-
turn is listed for both channels. Basically the reason we
have to restrict the size of the Rayleigh signal is because
the Rayleigh signal will tail off in intensity at about the
anticipated rate that the temperature signal tails off. As
a consequence of which if the Rayleigh signal comes through
in any appreciable magnitude, the tail off of signal from
the temperature above will be confused because it will be
due in part to the Rayleigh signal. So the way to eliminate
this source of error is simply to make the Rayleigh signal
very small. In fact we set a criterion that it be approximately 1% of the signal difference which arises per degree
temperature change and would therefore add a relatively

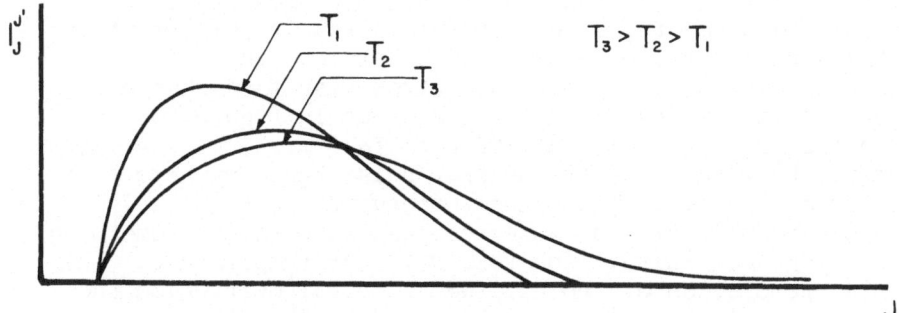

Figure 18. Graph of Equation (1) depicting shape changes of Raman rotational spectral envelope arising from temperature changes.

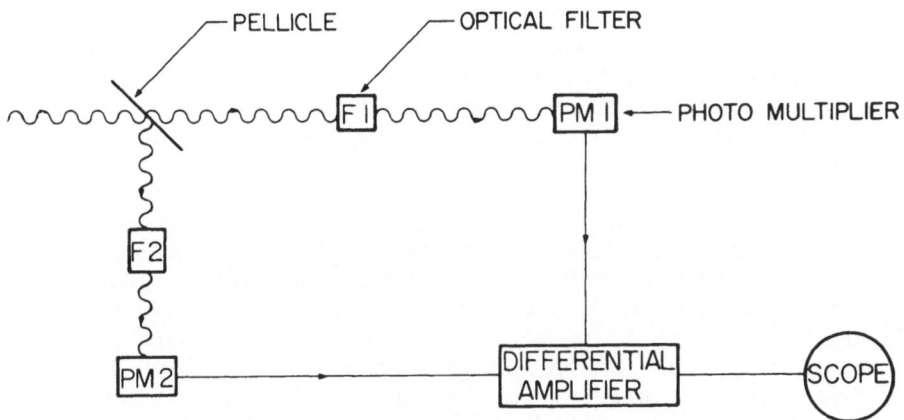

Figure 19. Schematic diagram of electro-optics of receiver system of remote temperature measuring apparatus.

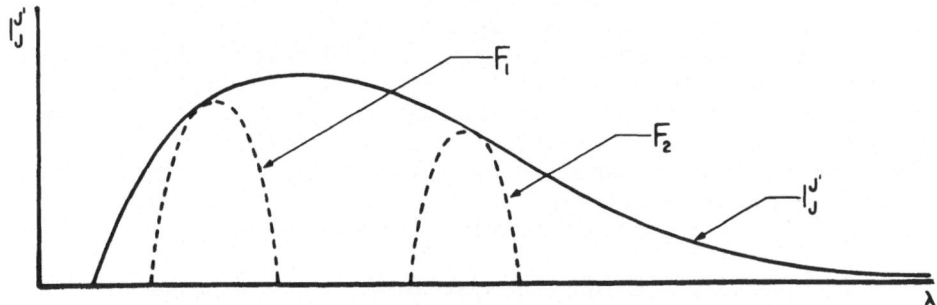

Figure 20. Disposition of optical filters relative to Raman rotational spectrum.

small error to the estimate of temperature resolvability.
The shot noise on the other hand, is a measureable fraction
of the signal difference per degree Kelvin. In fact, we
find the temperature sensitivity gives rise to a change,
using typical system parameters, as seen in figure 23, anywhere from 28 millivolts at $262^{0}K$ to 18.7 millivolts at $222^{0}K$.
Whereas the mean value of the shot noise runs anywhere from
9 to 20 millivolts. Thus, the temperature resolution is
at best about a third of a degree, and at worst at something
like a degree. This is the resolution of temperature. The
spatial resolution or the altitude interval over which the
signal averaging is done is determined by the electronic
receiver bandwidth and that is set, for convenience, at
100 meters corresponding to a bandwidth of approximately
1.5 MHZ.

    It should be emphasized that in contradistinction to
the first three measurements no field measurement has been
made of temperature, although plans have been made to do so.
It appears to be possible to get to an altitude of 3 or 4 km
with the existing hardware for the temperature measurement
with about $1^{0}K$ resolution.

    To repeat a point made above, spurious signals coming
in from the outside will be largely cancelled. That is to
say the differencing scheme will create two light signals
and the difference of these signals will be cancelled in
the difference amplifier to within a very high order. For
example, a typical common mode rejection ratio (which is
the parameter of interest for the operational amplifier)
is of the order 70 to 80 db. This suggests that one can
measure the difference between 1000 millivolts and 1001
millivolts to a tenth of a microvolt. However, realisitic-
ally these amplifiers have input noise levels of the order
of 10 microvolts. Thus, to exploit the full 70 db. capability one would need an input signal of 100 volts. Interestingly
enough, this could easily be the case with the rotational
structure as the cross-section associated with the rotational
structure is fairly substantial. The cross-section for the
nitrogen rotational Raman is about 10% of the Rayleigh signal.
Thus, one could expect from the very low altitude regime
from 2 to 4 km a very substantial signal.

    In closing, I think the most interesting aspects of the
four measurements resides, for extrapolation into the future

	F_1	F_2
F.W.H.M.	14.0Å	20.0Å
λ_{cent}	6979Å	7008Å
TRANS. (Max)	0.3	0.3

FILTER PROPERTIES

Figure 21.   Computer-optimized realizable filter properties.

$$I = \int_0^\infty I_J' \, dJ = 1.0 \qquad\qquad I_\Delta = a_1 I_1 - a_2 I_2$$

$$I_1 = \int_0^\infty I_J' F_1 \, dJ \qquad\qquad I_{\Delta 0} = Ba(I_1 - I_2)$$

$$\text{where } a_1 = a_2 = a$$

$$I_2 = \int_0^\infty I_J' F_2 \, dJ \qquad\qquad I_{\Delta T} = \frac{I_{\Delta 0}(T + \Delta T) - I_{\Delta 0}(T)}{\frac{1}{2}\{I_{\Delta 0}(T + \Delta T) + I_{\Delta 0}(T)\}}$$

F_1 and F_2 are filter responses of filters one and two.

REVIEW OF PERTINENT EQUATIONS

Figure 22.   Review of pertinent equations.

Temp.	$I_{F_1}^{\dagger}$	$I_{F_2}^{\dagger\dagger}$	$I_{\Delta}(\times 10^{-2})$	$I_{\Delta T}(\times 10^{-3})$
262	.091347	.088556	.27912	.28775
263	.091185	.088681	.25035	.28572
264	.091023	.088805	.22178	.28376
299	.085537	.092147	.66105	.22123
300	.085386	.092218	.68317	.21965
301	.085235	.092287	.70514	.21802
320	.082446	.093382	.10936$(\times 10^{-1})$	.18999
321	.082303	.093428	.11126$(\times 10^{-1})$	.18862
322	.082160	.093474	.11315$(\times 10^{-1})$	.18723

[†]Rayleigh contribution is 6.2×10^{-6} [††]Rayleigh contribution is 5.27×10^{-6}

Figure 23. Table of signal computations.

OCT. 22, 1969	LOCAL TIME	OCT. 23, 1969	LOCAL TIME
Balloon 1	20:00	Balloon 3	20:00
Laser A	20:30	Laser D	21:50
Flight 1	21:00	Flight 3	22:12
Laser B	22:35	Laser E	22:50
Flight 2	22:50	Flight 4	24:00
Laser C	23:10	Laser F	00:15
Balloon 2	23:30	Balloon 4	00:40

Figure 24. Time table of radiosonde and laser profiles.

in the accuracy with which the water vapor was measured on
a first try. It is to be recalled that the accuracy is
relative to existing measurements, but when one takes into
account the normal difficulties one has in systems of the
kind described, this becomes all the more interesting. It
has to be said, I think, that normalizing effect of the
nitrogen measurement is perhaps the prime component in
giving rise to the relatively good accuracies in the water
vapor measurement. There seems to be little doubt that this
is a technique which can be useful.

What would one look for down stream? To begin with, in
the temperature measurement which is pretty much signal/
noise limited, we can anticipate the use of some innovations
that will appear from laser development. For example, narrow
band optical amplifiers are on the way, and will be available
in the 3 to 5 year time span. With such a development the
shot noise problem can be eliminated. As a consequence we
could anticipate going to greater altitudes where we would
then be limited by other noise considerations. Indeed, the
development of practical narrow band optical amplifiers will
solve a broad spectrum of technological problems. I think
also that one can safely look forward to a very fruitful use
of remote probing. Of course, the primary thing that one
does with remote probing in respect of what is presently
being done by the more orthodox "in situ" techniques is to
measure the instantaneous three dimensional space-time configuration of the various scalar fields. In practice, of course,
one is going to have to back off because of the finite scan
rates of laser systems and so one would not quite get an
instantaneous space-time profile. Some small time interval
would be required to scan the 2π solid angle over which the
laser can make observations.

Further along in time other, more exotic schemes seem
possible. Thus, one can envision a system with multiple
launch beams where, for example, one beam could be used to
"tailor" a given atmospheric species while a 2nd beam inter-
rogates it. Possibilities of this type should provide a
"field day" for the spectroscopists interested in atmospheric
problems.

SOLAR SPECTROSCOPY FROM SPACE VEHICLES

R. Tousey

E. O. Hulburt Center for Space Research

Naval Research Laboratory, Washington, D.C. 20390

This is the Silver Anniversary year of the commencement of space research, and to-day's Symposium comes exactly nine days after the first observation of the sun's ul-traviolet spectrum, made by the Naval Research Laboratory [1] from a V-2 rocket launched on October 10, 1946. Until then, the solar spectrum below 2900 Å had been hidden by the earth's ozone layer. This opened a new field for research, which increased rapidly and continuously until it now forms a major portion of the work in science and engineering in the United States and in other countries too. At NRL we considered having a celebration on the very day, but space research has now struck such a pace that not one of us had the time to celebrate, at least in proper fashion.

Where did ultraviolet astronomy stand before 1946? Observations were made from the earth's surface, including high mountains, and a few experiments were carried out from aircraft and high flying balloons. The altitudes attained were not sufficiently great to be of much assistance. Almost the entire atmosphere still lay above, a hard and fast limitation that could only be overcome by placing instruments higher above the earth than was possible in those days.

The constraints imposed by the atmosphere were of three kinds. First, there is turbulence; the atmosphere is never completely stagnant. The passage it gives light waves is anything but smooth, so that it is rare indeed to resolve solar details as fine as one arc-second. In recent years, however, some locations on the earth have been dis-covered where one arc-second seeing is not too uncommon, and better than this is sometimes achieved. The second constraint is imposed by the light of the sky itself. When the sky is not cloudy, and when it is not polluted with haze, which is becoming rarer day by day, we have left only the dark blue sky produced by Rayleigh scattering of sunlight by the molecules of pure air. But even the purest and darkest blue sky pre-vents seeing the sun's white light corona, except during the moments of a total solar eclipse. The third and the most serious problem arises from the extreme opacity of the atmospheric gases themselves. Far above the earth's surface all radiation at wavelengths shorter than about 2900Å is absorbed by the gases always present in the atmosphere.

In 1946 with a single stroke the rocket swept away all these constraints. There was nothing left to prevent observing the sun over the entire spectrum and in full detail, except for problems of a technological character. These problems were indeed severe, but during the first 25 years tremendous progress has been made.

Progress comes dearly. Now we are realizing that the cost of space research rises ever more rapidly as astronomers strive to see more and more. How much more will have been discovered by the Golden Anniversary than today I cannot predict. But on the Silver Anniversary the field has grown so broad that in a short talk I can touch only the highlights.

This anniversary would not be complete without telling once again the story of the beginning. In 1949 U.S. scientists and engineers rushed to Europe and followed the retreating German armies to gather up everything new and interesting they could find. In Nordhausen, just ahead of the Russians, they came upon an underground cache of V-2 rockets and equipment. This is how we happened to get nearly 100 of the V-2's that had been intended for London.

The V-2 rockets were taken to the White Sands Proving Grounds in New Mexico, and many of the top German engineers accompanied them to nearby Fort Bliss. In January 1946, NRL and several other laboratories were invited to place scientific instruments in a number of these rockets. Needless to say we were extremely enthusiastic and went to work as though never again would there be such a wonderful chance for space research, once all the V-2's had been expended. With the assistance of Baird Associates, Inc., we built an extreme ultraviolet (XUV) spectrograph, designed to fit into the nose of the former V-2 warhead. In that first year working conditions at White Sands were rugged. As shown in Figure 1, we installed the instruments by working at the tops of fire ladders, the very ones that had been used with the V-2's in Germany.

The first V-2 launched in the U.S.A. flew on April 16, 1946 with success. We installed our first spectrograph in the nose of the first V-2 to carry instrumentation for scientific research. Launched on June 28, 1946, this V-2 remained streamlined, came down in one piece at a tremendous speed, and vanished underneath a huge crater. Digging equipment was brought out by the Army, as shown with the crater in Figure 2. It was a valiant attempt, but all that was recovered was a bushel of scraps, Figure 3.

But on October 10, 1946, our next attempt, the instrument was mounted in one of the large tail fins rather than in the nose. The rocket was blown apart after reaching peak, and with its streamlined character destroyed, went into a flat spin. This caused the large tail fins to break off, and the one with the spectrograph fell gently back to earth with the parachuting action of a seed-key from a maple tree. Both the instrument and the spectra were recovered, and we saw for the first time the sun's ultraviolet spectrum no longer hidden by the earth's atmosphere. As can be seen from Figure 4, while the rocket rose through the ozone layer the spectrum became extended to shorter wavelengths, and at 55 km altitude, which was nearly the peak altitude reached by this particular V-2, it was well above all but a vestige of the ozone layer.

Fig. 1—The nose of a V-2 rocket, where the first spectrograph was installed. In 1946 the installation was made from fire ladders.

Fig. 2—The crater resulting from the impact of the V-2 rocket flown on June 28, 1946, carrying the first rocket spectrograph.

Fig. 3—All that was recovered from the June 28, 1946 crater after several weeks of digging.

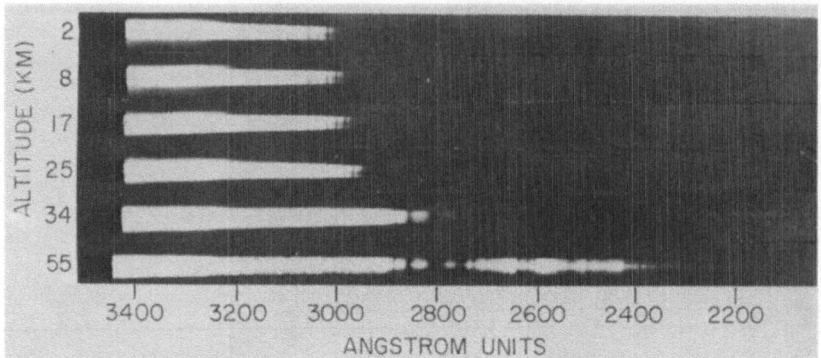

Fig. 4—The first photographs of the uv spectrum of the sun beyond the limit reached from the ground. Several exposures are reproduced from a sequence made by the Naval Research Laboratory with a spectrograph flown in a V-2 rocket on 10 October 1946.

The spectrum was recorded to 2400Å or a little beyond, which was about all that we could expect at this altitude with such a simple instrument.

This opened the way to two fields of research: One, on the nature of the sun itself; The other on the composition of the earth's atmosphere by absorption spectroscopy, using the sun as the light source.

The first atmospheric problem that we tackled was the determination of the vertical distribution of ozone by making careful measurements of the spectra exposed at different altitudes, using more or less standard methods of photographic photometry. Improved spectra (Figure 5) were obtained three years later with several instruments carried in an Aerobee rocket reaching a peak altitude of 112 km. This rocket was launched close to sunset in order to make use of the maximum slant optical path through the ozone layer. As a result we [2] measured the ozone distribution to an altitude of 70 km. The results are shown in Figure 6 which gives the ozone concentration in millimeters per km of height for three flights. These are the first measurements of the vertical distribution of ozone in the earth's upper atmosphere made by an accurate direct method. It was shown that the ozone concentration decreased to 2.5 x 10^{-5} mm (STP)/km at 70 km.

Recently there has been a revival of interest in the earth's ozone layer in connection with the SST. There has raged a controversy, whether the combustion products of the engines of many SST's might somehow modify the ozone layer sufficiently to allow more of the sun's ultraviolet to reach the earth. Our old data have suddenly become important, for at high altitudes there are no others. With great thoroughness, Harold Johnston [3] has examined the question of the interaction with ozone of the nitrogen oxides produced by the SST exhaust, listing 30 different chemical reactions. He concludes that nitrogen oxides from the SST pose a much greater threat to the ozone than does the water from the SST exhaust, and that the earth's ozone shield

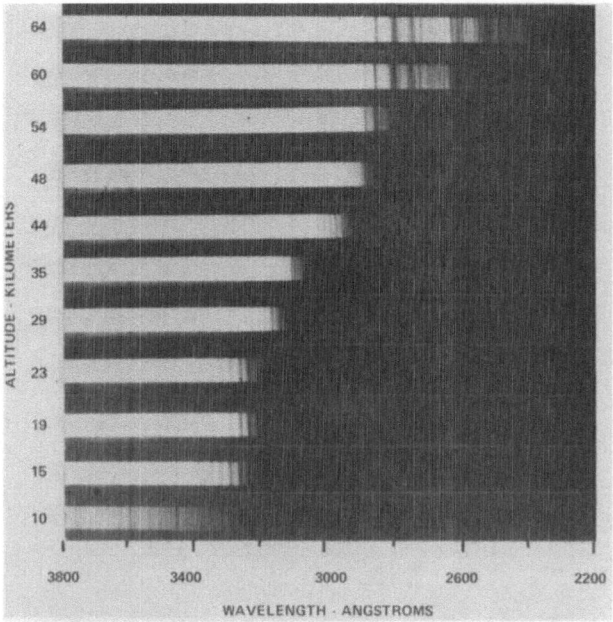

Fig. 5—A series of exposures made on June 14, 1949, with a spectrograph flown in an Aerobee rocket. The spectrum became extended as the rocket ascended through the ozone layer. The effect of ozone absorption was greatly enhanced because the rocket was flown at sunset.

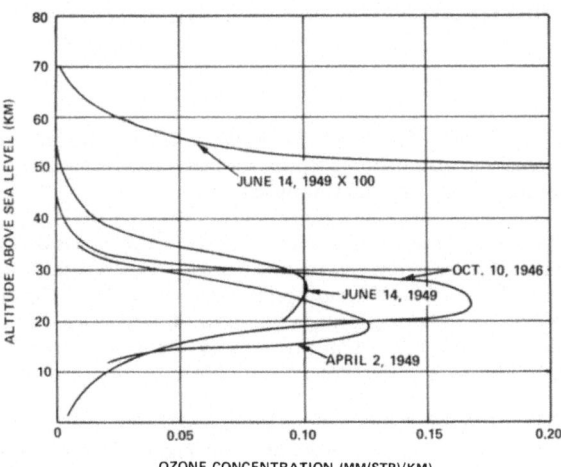

Fig. 6—Ozone determinations from rockets. The determination which extended above 50 km is shown on an expanded scale.

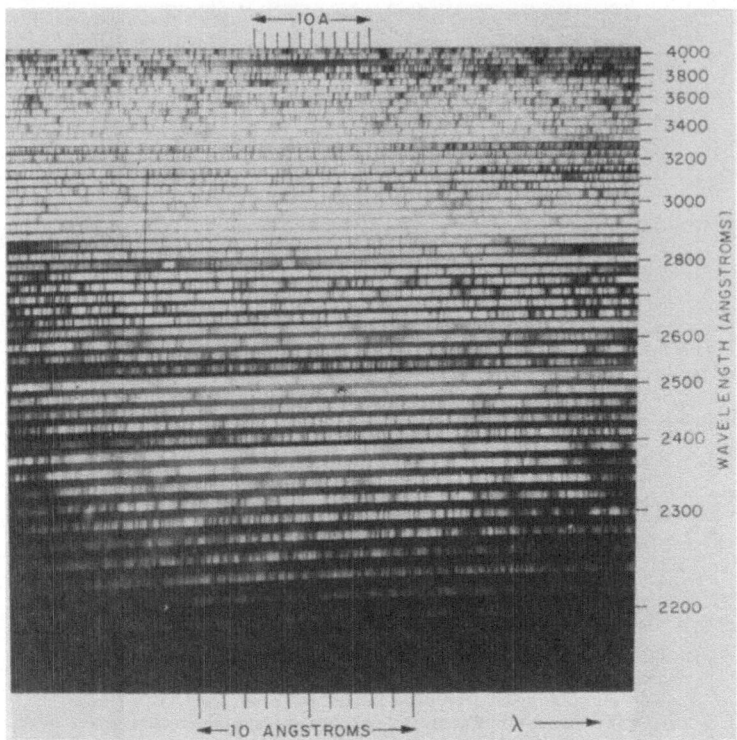

Fig. 7—The uv spectrum of the sun, photographed with an echelle grating spectrograph by the Naval Research Laboratory on 29 August, 1961, from an Aerobee-Hi Rocket.

could be reduced by a factor of two, which would certainly have an effect on our complexions if we stayed long out of doors. I am sure that experiments will be carried out to determine whether or not his conclusions are correct.

For the first decade progress in solar rocket spectroscopy was slow, but by 1961 we [4] had obtained the spectrum shown in Figure 7. This was made with one of Professor G. R. Harrison's early echelle gratings, replicated by David Richardson of Bausch and Lomb. It covers the wavelength range down to 2200 Å with 30 mA resolution, and is still the best spectrum in existence over most of this spectral range. It contains some 4000 Fraunhofer (absorption) lines below 3000 Å.

The early spectra of Figure 4 showed a great gap at 2800 Å, which was the most conspicious feature of the new solar ultraviolet. In Figure 7 it can be seen that this gap is actually a broad, deep depression, with two emission lines, seven Angstroms apart, appearing inside. This region can be seen better in Figure 8, an enlargement of that

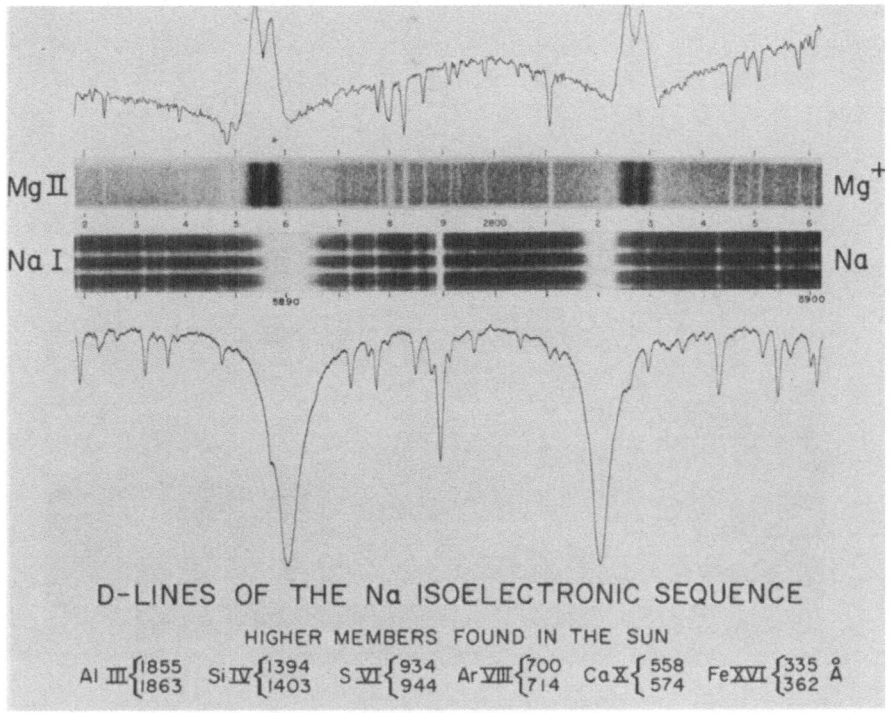

D-LINES OF THE Na ISOELECTRONIC SEQUENCE

HIGHER MEMBERS FOUND IN THE SUN

$Al\ III \begin{cases} 1855 \\ 1863 \end{cases}$ $Si\ IV \begin{cases} 1394 \\ 1403 \end{cases}$ $S\ VI \begin{cases} 934 \\ 944 \end{cases}$ $Ar\ VIII \begin{cases} 700 \\ 714 \end{cases}$ $Ca\ X \begin{cases} 558 \\ 574 \end{cases}$ $Fe\ XVI \begin{cases} 335 \\ 362 \end{cases} Å$

Fig. 8—The H&K, or D-lines of ionized Mg, photographed with the NRL echelle spectrograph, compared with the D-lines of Na, photographed by the Mt. Wilson Observatory.

section of Figure 7. The great absorption feature is caused by the Mg II resonance lines at 2795 and 2803 Å. They correspond to the D-lines of sodium, the best known absorption lines in the solar spectrum, and shown in the same figure. Together with H-α, the red line of hydrogen, the sodium D-lines are the most prominent absorption features in the visible solar spectrum.

Figure 8 shows the beginning of the sodium-like one-electron iso-electronic sequence. The second member is Mg II, spectroscopists designation for the ion Mg^+ having one electron removed; Thus it is a one-electron atom just like sodium, but with a heavier nucleus. The D-lines of Magnesium are slightly farther apart than those of sodium.

The effect of the D-lines of ionized Magnesium on the solar spectrum is very great. In appearance they are much like the Calcium H and K lines, produced by once-ionized calcium, a one-electron potassium-like atom. The emission features at the centers of the broad absorption lines come from the outer layers of the sun's atmosphere, the chromosphere, and in turn they show a second self-reversal at their centers. The explanation is closely connected with the sun's atmosphere and is far from complete.

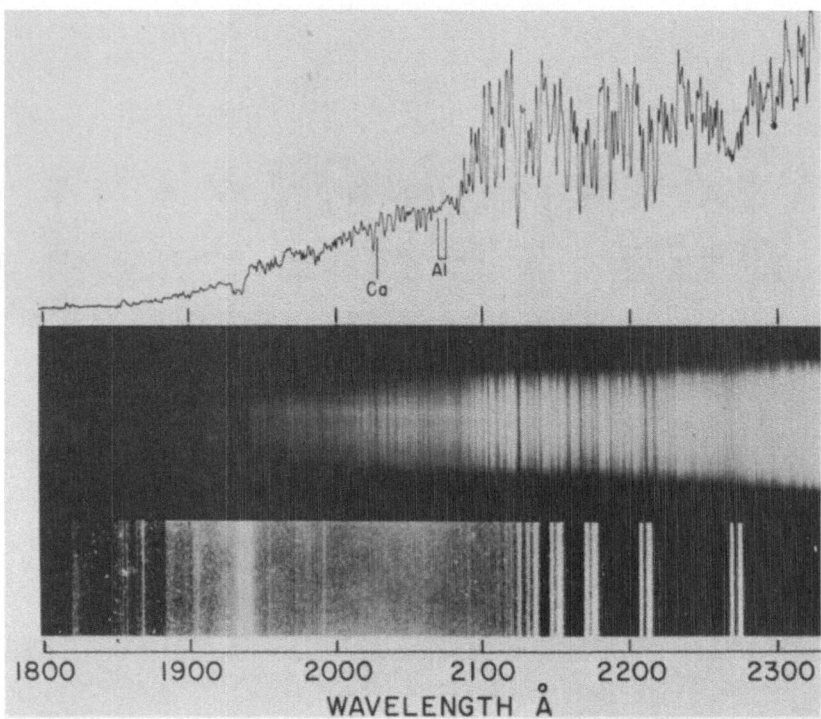

Fig. 9—The sun's spectrum in the region where its character changes; the Fraunhofer lines suddenly become faint. At the bottom an emission spectrum of an aluminum arc is reproduced from [6].

We have followed the sodium iso-electronic sequence far down in the solar spectrum, as listed at the bottom of the figure. The D-lines from ion to ion lie stretched out in an orderly fashion progressing down the spectrum to shorter wavelengths, and Fe XVI, which will be seen in a later figure, comes in the XUV near 350 Å. This listing is no longer quite complete. Ni XVIII and perhaps one or two others have now been recorded.

Down to 2085 Å the sun's spectrum continues without any striking changes. As seen in Figure 9, it is a continuum from the hot photosphere, packed densely with Fraunhofer lines, produced in the cooler, outer part of the photosphere. But at 2085 Å a sudden change takes place. Continuous absorption sets in, slices off the continuum so that it can no longer be seen, and leaves exposed only the bottoms of the Fraunhofer lines. This absorption is produced by aluminum for the most part. The curve and the spectrum just below it are from the sun [5]; the spectrum at the bottom

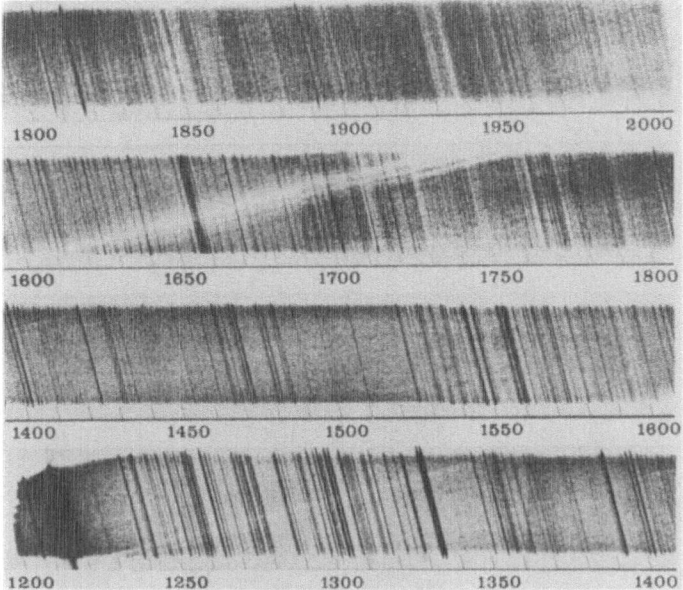

Fig. 10—The sun's spectrum photographed on 22 August 1962 with the NRL
crossed-dispersion spectrograph. Dirt in the slit caused the streaks.

is the emission spectrum of aluminum [6]. The effect of the continuous absorption
of aluminum beyond its series limits is striking. Everything fits together reasonably
well, and the auto-ionization lines of aluminum at 1935 Å appear strongly in absorption in the solar spectrum, and in emission in the laboratory spectrum.

To shorter wavelengths the spectrum changes completely. Figure 10 shows the
range 2000 to 1200 Å, photographed by NRL [7] in 1962 with about 0.2 Å resolution. Fraunhofer lines can be seen faintly in the top section and over the right hand
half of the second section. They disappear below about 1700 Å. Emission lines soon
replace the Fraunhofer lines toward shorter wavelengths. The first, at 1893 Å, is the
intersystem resonance line of Si III. The next two at about 1810 Å are the resonance
lines of Si II. There is much Fe I in absorption and Fe II in emission, but they do not
overlap to any great extent. Si I is last seen in absorption at about 1690 Å, but re-
appears in emission at shorter wavelengths. The smooth spectrum from 1530 to
1400 Å is the emission continuum of Si, produced by the capture of an electron by a
silicon ion [5]. To shorter wavelengths there are many other features of interest of
which the most important is the intense, broad emission line at 1216 Å, the Lyman
alpha line of hydrogen, which is the strongest line in the entire solar spectrum.

In recent years we have obtained spectra covering this wavelength range with much
higher spectral resolution. A short section of a spectrum [8] obtained on August 13,

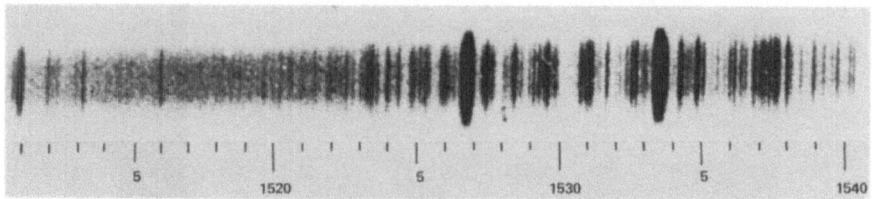

Fig. 11—A section of the NRL high spectral and spatial resolution solar spectrum
exposed with the 2 arc sec slit positioned 7 arc sec inside the limb.

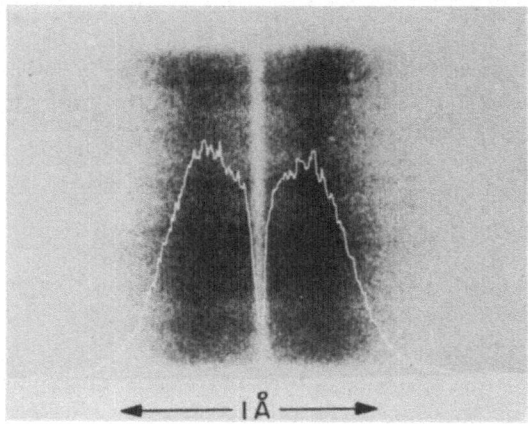

Fig. 12—The profile of hydrogen Lyman-α showing the outer
atmospheric hydrogen absorption core, from an NRL rocket
flown on July 21, 1959.

1970 is shown in Figure 11. This has approximately 6 mA resolution, the highest as
yet obtained in this range. The Si I ^{3}P ionization limits lie at 1521.0, 1522.7, and
1526.1 Å; just about all the Si I lines known from laboratory spectra that lead up
to these limits are present, and many new Si I lines can be identified in the solar
spectrum by extrapolating the series [9]. Beyond the sun's limb, which we see by eye,
the atmosphere becomes cooler for a short distance, then starts to increase rapidly in
temperature as the chromosphere is entered. In the gap at 1630 Å, one is probably
looking at the coolest part of the sun's outer atmosphere. But below 1530 Å, the
silicon continuum arises from the hotter chromosphere beyond.

The hydrogen Lyman-alpha emission line is both extremely intense, and also broad.
The profile of the line is shown in Figure 12 from a spectrum obtained with a high-
dispersion spectrograph [10] flown by NRL in 1959 and 1960. The resolution in this
image is about 30 mA. The line is one Angstrom wide at half maximum and is deeply

self-reversed. Part of the self-reversal arises in the sun itself, just because there is so much hydrogen, as explained by Morton and Widing [11], using the theory of non-coherent scattering developed by Jefferies and Thomas [12].

The narrow Lyman-alpha absorption core in Figure 12 is produced by atomic hydrogen between the earth and the sun, mostly in the earth's outer atmosphere. By making use of the profile of this narrow core it was possible to determine by absorption spectroscopy, the amount of hydrogen between the earth and the sun [13]. This is the most direct way to measure the total atomic hydrogen. Between 100 and 200 km altitude, the region where we obtained exposures from the rocket, there was a total of about 10^{-12} ground state atoms per cm^2 column, and above 200 km altitude there was also about 10^{-12} atoms per cm^2 column.

Now I should like to pose a question. Why should we be so interested in the sun per se, and not simply as a tool for investigating our atmosphere? The answer is simple and obvious. It is the sun that has brought life to the planet Earth through the creation of its atmosphere. It is the sun that regulates our climate. It is the sun that causes fundamental changes in the earth, most of them so slow that they go un-noticed, but extremely important in the long run. I refer to the ice ages. Other changes that the sun produces are spectacular, for example the aurora borealis. Still others are important to man's activities, in particular the ionospheric storms that disrupt radio communication. Changes in the sun's output may also to some extent cause short term weather changes, but this of course, is still highly controversial.

All these changes in the earth occur because the sun itself is changing. Every 11 years the sun goes through its cycle, increasing in activity, pouring out more of the short wavelength radiation that affects our upper atmosphere, then subsiding, and repeating it all, over and over again. But no two solar cycles are exactly alike. Furthermore the reason why the sun has cycles is still completely unknown. It is one of the great scientific challenges. Once we have learned more about the sun, expecially by observing more completely its changing character at the short wavelengths accessible from rockets and orbiting vehicles, we should be able to explain the cycling, and perhaps even predict the nature of an upcoming solar cycle and the solar phenomena to follow.

Until about 30 years ago the sun's atmosphere was considered fairly well understood—just a hot layer through which heat was transmitted outward by convection and then radiated away to the earth. Where the white light we see originates, its temperature is about 6000°K. From there outward its temperature decreases as the white light energy is radiated away. But formerly it was believed that the temperature kept right on decreasing; the absorption lines were simply produced by abundant elements in the cooler outer part of the solar atmosphere.

One of the many unsolved solar problems at that time was a strange unidentified line which had been observed for many years in the sun's corona that is seen during the brief moments of a total solar eclipse. In the 1930's Lyot [14] invented and constructed the coronagraph, an instrument with which this strange line in the sun's inner coronal could be seen at any time and without a total eclipse. Figure 13 shows a spectrum of the corona photographed by Lyot. The very bright line at 5303 Å is the unidentified line, previously known only from eclipse spectra. In desperation astronomers proposed that this line was emitted by an unknown element which they dubbed coronium.

Fig. 13—A portion of the spectrum of the sun's inner corona, photographed by Lyot
with his coronagraph. This shows the intense 5303 Å coronal emission line, identified
by Edlèn as produced by Fe XIV.

The explanation for the 5303 Å line was given in about 1940 by the Swedish spec-
troscopist Edlén [15]; following a suggestion of Grotrian [16], he realized and proved
that this line was produced by FeXIV, that is 13-times ionized iron. This he did as a
result of extensive and precise extreme ultraviolet spectroscopic researches in his laboratory. The point of this is that Edlén's proof that 13-times ionized iron exists in the
sun's atmosphere, meant that astronomers had to change completely their picture of
the sun. The hot center is where nuclear reactions produced the heat, which seeps
outward, passing by convection through the photosphere, then radiating away and
according to the old theory causing the temperature of the sun's atmosphere to fall from
about 6000° K to a value near 4500°K. But from here out, Edlén's discovery changed
the picture completely, because high in the atmosphere Fe^{+13} was shown to be present, and a temperature of one million degrees is required to produce this ion. It is
now believed that the temperature rises from the minimum near 4500°K, at first with
extreme rapidity through the chromosphere, and then more gradually into the corona
where temperatures well over one million degrees are reached.

This new picture posed a question which is still not completely answered. How can
energy be transmitted outward through the solar atmosphere against an uphill temperature gradient and in such vast quantities? The explanation lies in transmission by shock
waves, but the details are still not understood.

Among the greatest of solar mysteries is the solar flare. This is a tremendous brighten-
ing occurring in a very small region located close to a group of cool sunspots in the
midst of intense and complicated magnetic field structures. During the active part of
the solar cycle flares take place quite frequently but during solar minimum they are
rare. Flares emit x-rays which upset the ionosphere and cause disturbances that disrupt radio communications. After large flares electrons and protons reach the earth
and cause severe disturbances in its magnetic field. Because flares are so closely associated with the solar cycle, when we understand flares we may be well on the road to
understanding the solar cycle itself.

Fig. 14—The solar spectrum photographed by NRL on May 10, 1963 with a grazing-incidence spectrograph. Most of the intense lines from 171 to 220 Å could not be identified at that time.

Much that is new about solar flares has been learned from space research, especially from the nature of the solar spectrum at still shorter wavelengths in the extreme ultraviolet and x-rays. To wavelengths shorter than about 500 Å grazing-incidence spectrographs are usually employed because of their great speed, whereas normal-incidence spectrographs are slow and can be used in special applications only. A spectrum [17] obtained by NRL in 1963 with a grazing-incidence spectrograph is reproduced in Figure 14; the range 500 to 300 Å is shown in the lower section and in the upper section the spectrum is continued to about 150 Å. The Lyman-alpha line of ionized helium, HeII 304 Å, is extremely prominent; it is one of the most intense lines in this region. This figure shows the range where the resonance lines of many highly ionized atoms appear; we see the sodium D-lines of sodium-like FeXIV [18], 335 Å and 361 Å in the isoelectronic sequence shown in Figure 8; the resonance line of FeXV, 284 Å, a magnesium-like ion with two electrons in the outer shell. These lines were relatively easy to identify and follow directly the conclusion reached from Edlén's identification of FeXIV, 5303 Å that led to the realization that million degree temperatures must prevail in the corona. The difference is that FeXIV 5303 Å is a forbidden transition, but these XUV lines are resonance lines.

The upper section of Figure 14 contains many lines of great intensity which for several years were largely unidentified. Identification of most of these lines was extremely difficult. The greatest progress was made by the Culham Laboratories in England, using the high temperature plasma machine called ZETA. Wilson [19] and collaborators decided to observe the spectrum which the plasma in the ZETA machine

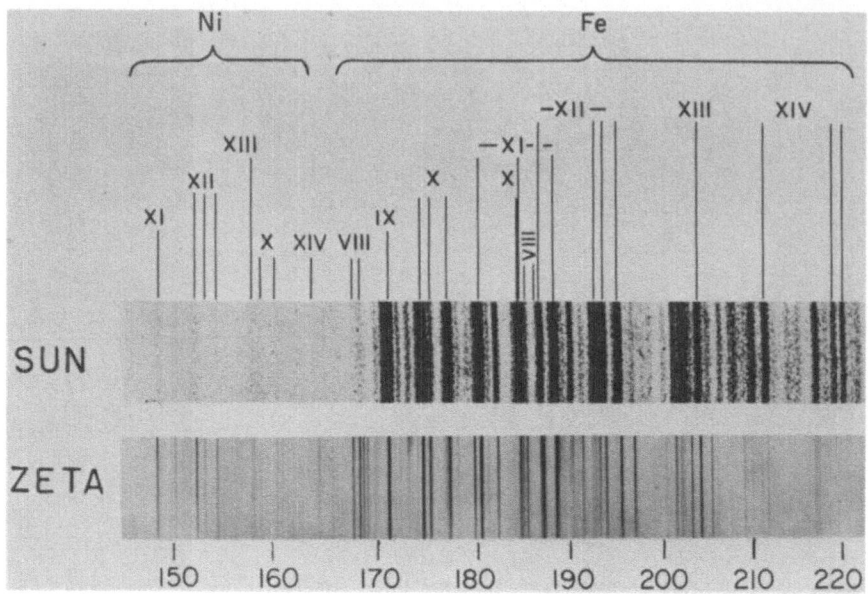

Fig. 15—The NRL solar spectrum of Figure 14 compared with the spectrum of the high temperature plasma produced by the Zeta discharge at Harwell, England. The unknown lines were identified largely by the Culham Laboratory through laboratory spectroscopy using a Theta pinch.

would radiate in the extreme ultraviolet. Figure 15 shows the results. They obtained
the spectrum of ZETA, and on comparing it with the spectrum of the sun, found that
the two were close to identical. At first this seemed a great surprise, but soon it was
realized that it was to be expected, because the temperatures of the two plasmas were
not very different, and the walls of the ZETA machine contained iron and nickel which
are abundant elements in the sun. Following this, a large amount of laboratory work
was required, but most of the lines have now been identified. In Figure 15 some of the
most prominent are noted; Fe VIII, IX, X, XI, XII, XIII and XIV form the tremen-
dously intense group of lines from 171 to 220 Å. The corresponding lines from nickel,
which is less abundant than iron in the sun, are present at slightly shorter wavelengths.
As a practical result of this work, the ZETA plasma was much more completely understood, and some energy which had been missing was accounted for by the radiation in
these XUV lines. The conclusion is that study of the sun's natural plasma physics laboratory by means of space research technology is a valuable adjunct to the great plasma
laboratory devices constructed on the earth in connection with the development of
nuclear power.

My next topic is the detailed appearance of the sun when viewed with radiation of
different wavelengths. To the unaided human eye the sun is white, extremely bright,
and has a sharp edge. With the aid of a piece of smoked glass one can see the dark sunspots. What does the sun look like in certain monochromatic spectral lines, and most
important, what does it look like in lines in the extreme ultraviolet?

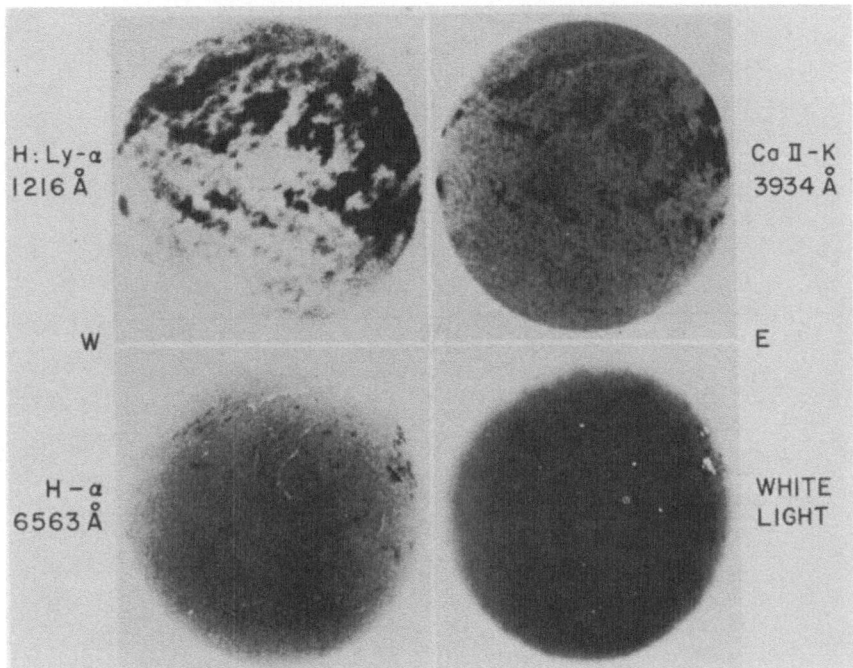

Fig. 16—The sun photographed by NRL on March 13, 1959 in hydrogen
Lyman-α, compared with its images in CaK, H-α, and white light.

Four solar images, produced with radiation of different wavelengths, are shown in
Figure 16. These are negative images such as one obtains by direct photography on
negative material. The white light sun image photographs as a black disk, except for
the sunspots. With a spectroheliograph, or a narrow-bandpass filter set for the H-alpha
line of hydrogen, it is by no means uniform; it shows great detail of all sorts, including bright patches called active regions or plages, and cool dark streaks called filaments.
The H-alpha radiation comes from the low chromosphere where the temperature has
just begun to rise. The upper right image was made in one of the resonance lines of
ionized calcium, the K line. This is still different; it shows more contrast and less detail than H-alpha, and comes from still higher in the chromosphere. The upper left
image was obtained by NRL [20] in 1959 using emission in the Lyman-alpha line of
hydrogen, 1216 Å. This pattern is the coarsest of all, and shows very intense regions
high in the chromosphere where the sun is still hotter yet. All these images, however,
really come from not far above the temperature minimum, just a few hundred kilometers above it, but over this part of the sun's atmosphere the temperature is rising
very rapidly, from 4500°K to a value some ten times greater. It is in this region that
the shock waves originate and carry outward the energy against the rising temperature
gradient.

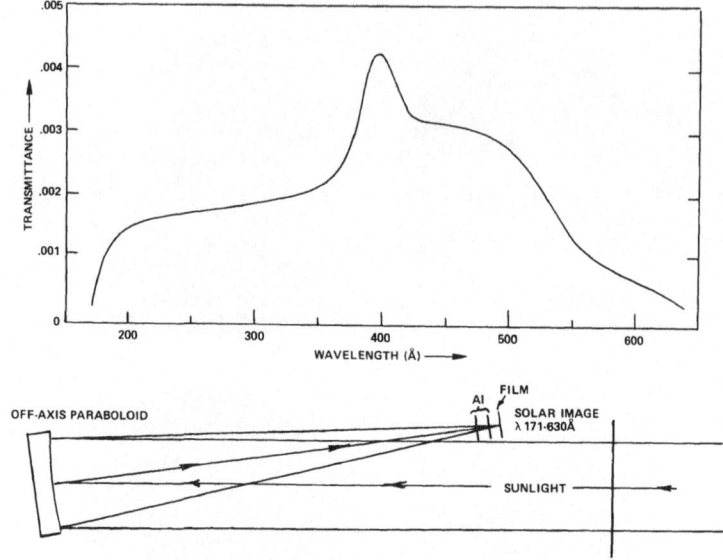

Fig. 17—Optical diagram of the NRL XUV heliograph which produces photographic solar images in the broad band 171-630 Å.

To obtain images of the sun from levels in its atmosphere where the temperature is
still higher it is necessary to employ radiation at even shorter wavelengths. For this
purpose we [21] have used a simple instrument which we call an XUV heliograph. In
its simplest form shown schematically in Figure 17, an off-axis paraboloidal mirror is
used to form by reflection an ordinary white light image, including all wavelengths
from the visible through the ultraviolet and into the extreme ultraviolet. But a special
filter is introduced consisting of three films of aluminum, each 1000 Å thick which
together with the platinum-coated mirror produces the transmittance curve shown at
the top of the figure. Aluminum itself begins to transmit at 835 Å, reaches peak transmittance at 171 Å, the Al $L_{2,3}$ edge, and then drops to a low value [22]. This filter-
mirror combination excludes all visible and near ultraviolet but allows the instrument
to transmit the sun's radiation in a band from 171 to 630 Å. The second form of this
instrument uses a concave grating rather than a mirror, and converts the instrument
into a slitless spectrograph, or spectroheliograph. One aluminum filter is still used to
eliminate the long wavelength stray light scattered from the grating. A spectrum of
monochromatic solar disk images is produced, exactly as are the lines in a normal slit-
type spectrograph.

In Figure 18, a broad-band image, or heliogram, made with the off-axis paraboloid
is reproduced [23]. This includes for the most part emission from the sun's corona.
The spatial resolution is film-limited to about 10 arc sec, because the image diameter
was only 2.3 mm. The pattern is extremely different from the images in Figure 16.
Although there is some coronal emission from the entire sun, it is enhanced enormously

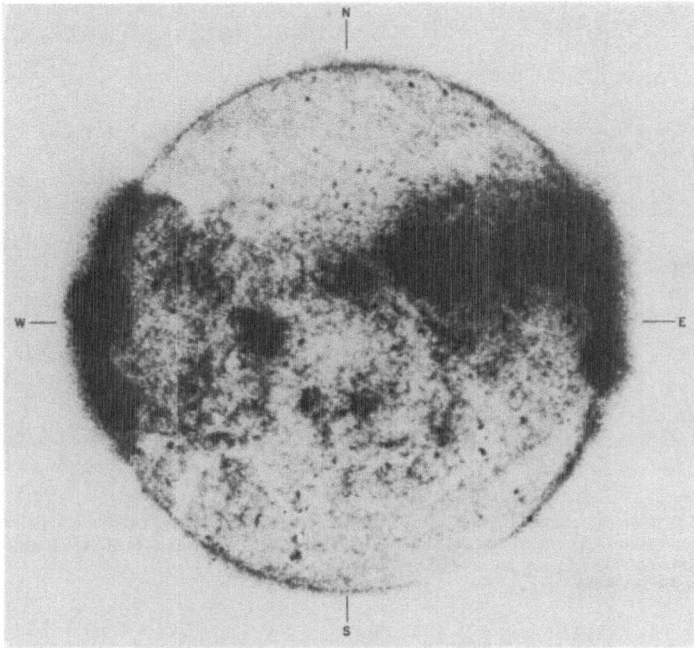

Fig. 18—An XUV heliogram photographed on August 13, 1970
using the heliograph of Figure 17.

over the active regions. The corona is optically thin, therefore, looking edgewise at the
corona around the sun's limb, we see a bright ring. Also, above the active regions close
to the limb, the corona is greatly enhanced because it is seen edgewise, and shows
emission extending far above the visible limb.

The image in Figure 18 contains contributions from many XUV emission lines. An
important question is to determine whether the image is the same or is very different
in character in the various emission lines included in this pass band, and originating
from different levels in the solar atmosphere.

In Figure 19 are shown spectroheliograms [24] made with the diffraction grating
version of the instrument shown in Figure 17. The reproduction is a positive and a
long and a short exposure are shown. The intense disk, filled with detail, is produced
by the Lyman-alpha line of ionized helium, 304 Å. To the left is a strange ring, produced by FeXV, 284 Å, and at the very right is a hoop produced by MgIX, 369 Å.
Still different are the images produced by the sodium-like lines FeXVI, 335 and 361 Å.
Each of these lines is different and has its own signature.

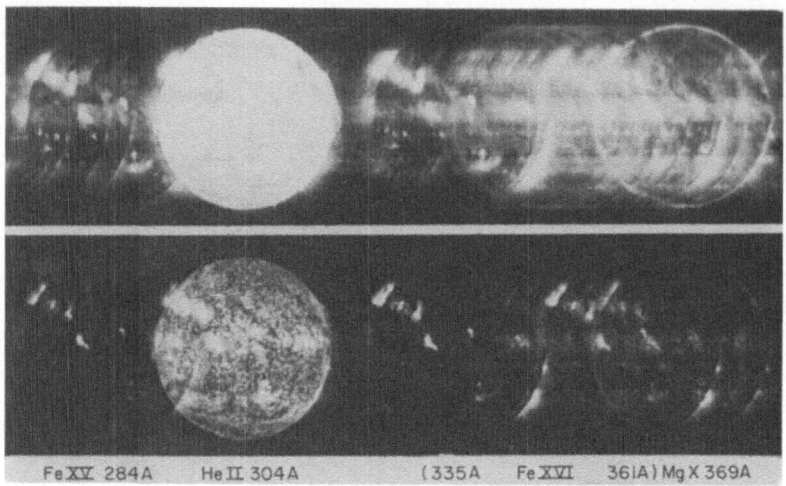

Fig. 19—Two spectroheliograms photographed on September 22, 1968 with different exposure times. The NRL rocket was launched from the White Sands Missile Range on the same day that a total solar eclipse took place in Siberia.

The collection of solar images recorded in the various emission lines, produced in the chromosphere and corona at different temperatures all the way from a few tens of thousands of degrees into the millions, form a set of maps of the sun's atmosphere for different altitudes above the sun's surface. For example, Figure 20 shows enlarged the same part of the sun in HeII, 304 Å and O V, 630 Å. In HeII there is a tremendous arch prominence, such as one sees occasionally in hydrogen H-alpha. But in O V there is a strange absence of emission in place of the prominence. Somehow, the prominence has disturbed the atmosphere in such a way that O V radiation is not emitted. Probably it has just lowered the temperature of the atmosphere in this region so much that little O V is present.

The best image that we have obtained of the sun in the Lyman-alpha line of HeII is shown in Figure 21. The resolution is about 5 arc sec, and in some short exposures reaches 3 arc sec, just about the resolution that is obtained from the earth's surface under normal observing conditions. This image was obtained while a solar flare of importance 2N was in progress. To accomplish this was difficult and we are indebted to personnel from NOAA, Boulder, Colorado for watching the sun continuously and informing us when a flare took place. The flare spectrum over this range shows as a series of small dots, shown enlarged in a later figure.

There is present in Figure 21 a tremendous, complicated active region which has the shape of a butterfly, one of the most striking objects we have ever seen. This

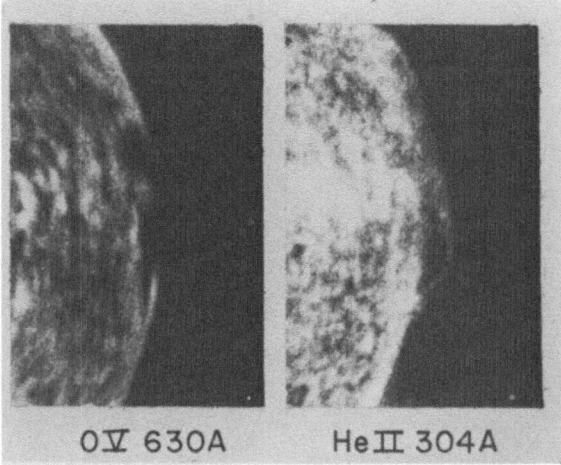

Fig. 20—The sun photographed on September 22, 1968 with
NRL's spectroheliograph in a low chromosphere emission line,
He II 304 Å, and in O V 630 Å which is produced much higher
in the sun's atmosphere.

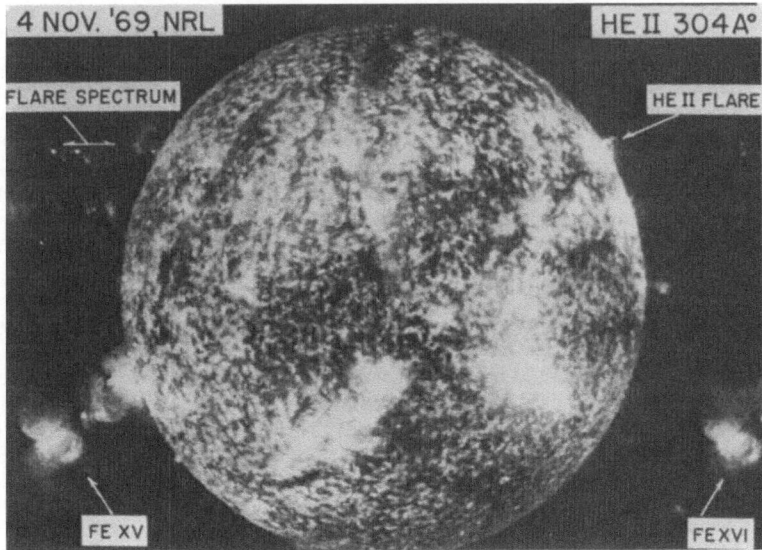

Fig. 21—A portion of a spectroheliogram made by NRL on November 4, 1969 during
a flare of importance 2N. The spatial resolution in these images is about 5 arc sec.

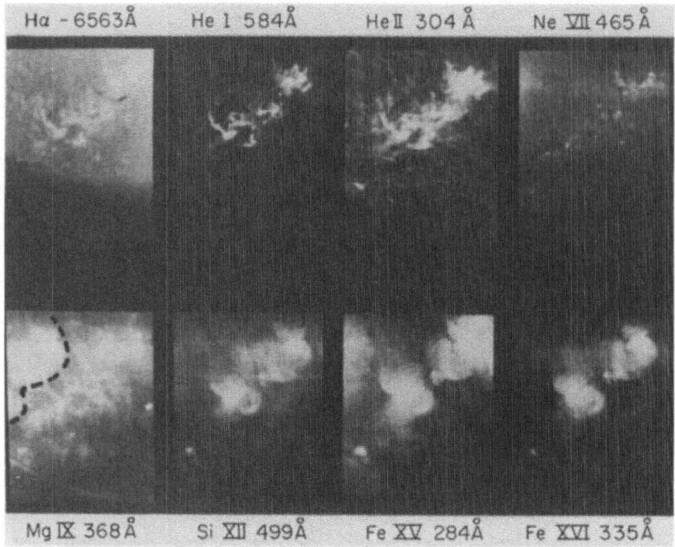

Fig. 22—The November 4, 1969 butterfly-like plage in 8 lines ascending
in altitude through the sun's chromosphere and corona.

butterfly-like active region is shown in Figure 22 in 8 spectral lines selected to show
how its appearance changes with temperature, and arranged in ascending order up
through the sun's atmosphere [26]. H-alpha of hydrogen is lowest in the chromosphere. Then follow the resonance lines of neutral helium and ionized helium, followed
by NeVII, which is in the transition region between the chromosphere and the corona.
The detail in the active region changes from line to line. The greatest change takes
place in NeVII; here the active region has almost disappeared and only a few bright
points remain. Ascending still higher we see MgIX from the very base of the corona,
SiXII as the corona is entered, and Fe XV and Fe XVI well up in the corona. In the
corona the active region changes in structure and is diffuse and cloudlike, completely
different from its character in the chromosphere. It is from photographs like these
that we hope, eventually, to understand what is going on in the sun's outer atmosphere.

But the most energetic and fascinating solar phenomenon of all is the solar flare.
In Figure 23 twelve images of the solar flare photographed from the 4 November 1969
rocket flight, are assembled [25]. These images show what the flare is like at different
levels in the sun's outer atmosphere. They are arranged, as before, in order of increasing altitude and temperature. The upper set shows the flare progressing up through the
chromosphere; here it shows as a bright ribbon-like object coming from a small somewhat brighter nucleus. By NeVII, the highest and hottest in the upper series, the
nucleus is much more intense and the ribbon weak, but great detail is still present.
The nucleus is only 6 arc seconds across; it covers something like the area of the North
American continent.

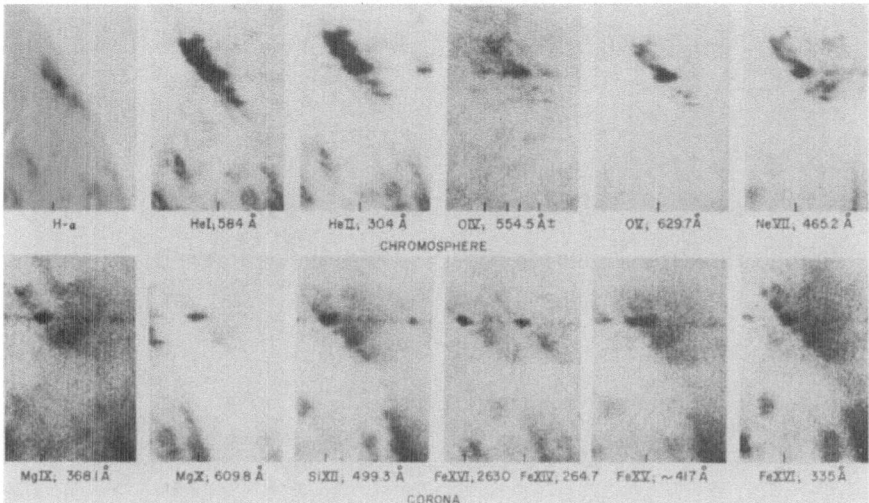

Fig. 23—The November 4, 1969 importance 2N flare shown in 12 emission lines, ascending through the chromosphere and corona. These images are taken from the complete set of spectroheliograms of which a section is shown in Figure 21.

The lower set shows the flare in the corona, again ascending to higher temperatures. Here the ribbon disappears completely, but the very hot nucleus is still present. In addition, the flare is surrounded by diffuse clouds of hot glowing gas.

The physics of the flare has not yet been explained. But through space research we are learning much more about it. The temperature of the plasma in the nucleus reaches many millions of degrees. It is a far hotter plasma than we can yet produce in the laboratory.

What is the very latest thing in solar space research? Perhaps I may be permitted to boast of a successful experiment in NASA's seventh orbiting solar observatory, OSO-7, an Anniversary present. Figure 24 shows two images of the sun's white light corona made one day apart [27]. In OSO-7 we have a white light coronagraph that records several times each day the sun's white light corona that from the ground can be seen only during the brief moments of a total solar eclipse. The images from natural solar eclipses never look alike. They show strange, weird formations, like helmets, arches and streamers, and they show us the solar wind of electrons accompanied by protons that streams forth from the sun and envelopes the earth. For the first time, from OSO-7, we are able to see how the corona changes with time and how rapid the changes may be. These images were produced with the aid of a Westinghouse SEC Vidicon [28] that records the image of the solar corona produced by an externally occulted coronagraph orbiting the earth at an altitude such that the sky is completely dark. The SEC vidicon is exposed for two seconds. Then the stored image is read out into a tape recorder, a process that requires 44 minutes to transfer the 65,000 picture elements.

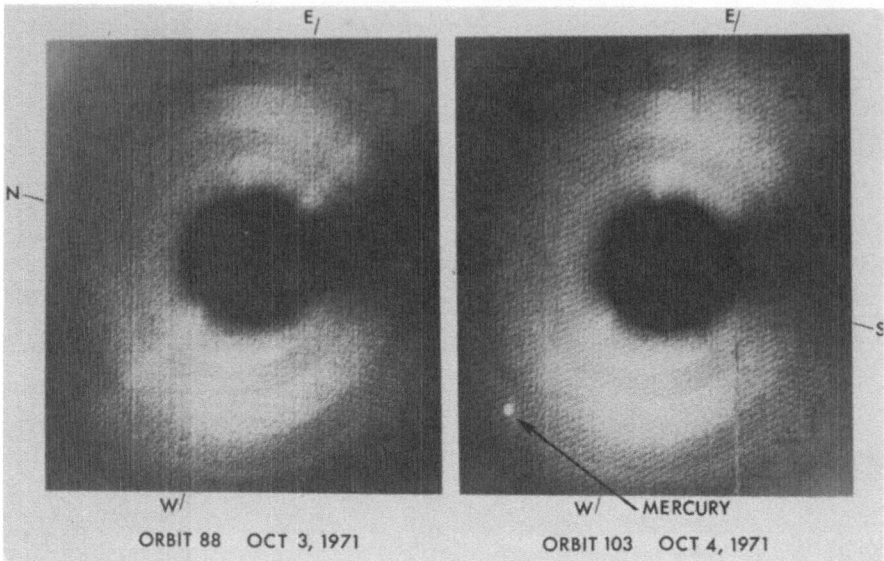

Fig. 24—The sun's white light corona in the region extending from 3 to 10 solar radii from the sun's center. These images were recorded on NASA's OSO-7 with a coronagraph designed by NRL, using an SEC vidicon and television techniques to record the image, which was sent to ground by telemetry.

The data stored in the tape recorder are dumped to ground by telemetry and then these images are reconstructed on a cathode ray tube and photographed.

This pair shown in Figure 24 is the first that we have reproduced. You can see that the corona has changed during the period of one day by more than can be attributed to the rotation of the sun. In the two weeks that we have been observing we have seen changes much greater than this, and if the sun were active, as it was a few years ago, I am sure that the changes would be far more spectacular. We were surprised to see the planet Mercury enter the picture on October 4. Powerful that the SEC vidicon is, however, it still cannot compete with direct photography.

For some years we have been building very large spectroscopic equipment for NASA's Skylab Mission [29] in the manned space program, currently scheduled for launch in 1973. A drawing of Skylab is shown in Figure 25. The cluster consists of several modules. The astronauts live in the orbital workshop, an empty fuel tank from a Saturn IVb booster fixed up for relatively comfortable living in zero g. Within this module various experiments will be performed. The central section is called the multiple docking adapter. Attached to one side of this is the Apollo Telescope Mount, or ATM, that contains the solar instrumentation. It is seven feet in diameter and eleven feet long. The scale can perhaps be realized by noting the size of an astronaut shown engaged in extra-vehicular activity, on his way out to the far end of the ATM in order to retrieve photographic film from the NRL spectrographs. The ATM is surrounded by

Fig. 25—Skylab, the USA's first manned space
station, to be placed in orbit on April 30, 1973.

its four solar panels that furnish it power. At the other end is the command module,
attached to the service module, and in which the astronauts are carried away from the
earth to dock with the cluster that had been placed in orbit previously. The entire
cluster is stabilized with control-moment gyros. In addition, the ATM is stabilized
with a servo system that keeps it pointed at the sun to ±2.5 arc sec in yaw and pitch,
and 10 arc minutes in roll. A photograph of the prototype ATM being assembled in
the large clean room at the Marshall Space Flight Center, Huntsville, Ala., is shown in
Figure 26. Two workmen at the right indicate the size.

Within the ATM are six major solar instruments [29]. There are two x-ray tele-
scopes, one constructed by the American Science and Engineering Corporation and
including a spectroscopic attachment, another designed and built by the Goddard and
Marshall Space Flight Centers. Another type of instrument is a white light coronagraph; this was designed by the High Altitude Observatory of Boulder, Colorado and
will record the white light corona by photography much as it is seen in Figure 24 from
television. There are three extreme ultraviolet instruments; a scanning XUV spectrometer produced by the Harvard College Observatory, and two photographic instruments
by NRL [30]. One NRL instrument is a large XUV spectroheliograph, intended to
produce hundreds of large solar images in various wavelengths in the XUV, of a character
similar to those shown in Figure 21 but twice the size. The other is a very high-resolution spectrograph that can be aimed at any point on the sun to obtain the spectrum of
any of the many special features present on the disk, for example, active regions, dark
filaments, flares, and the change in character of the spectrum from place to place as
the sun's limb is scanned. This spectrograph is expected to obtain over 6000 exposures

Fig. 26—The Apollo Telescope Mount (ATM) containing instrumentation for studying the sun from Skylab, shown during assembly in
the clean room at MSFC.

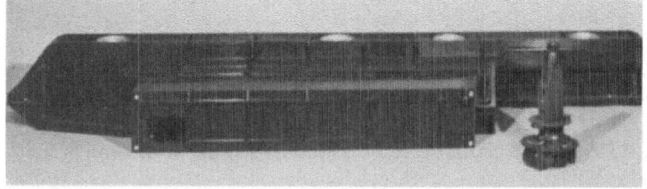

Fig. 27—The 9-foot-long NRL XUV spectrograph for Skylab's Apollo Telescope Mount, compared with the very first rocket spectrograph, flown by NRL in a V-2 rocket in 1946.

during the period when Skylab is in operation; 28 days for the first visitation, 56 days
for the second, and 56 for the third. These extreme ultraviolet instruments are being
manufactured by the Ball Brothers Research Corporation of Boulder, Colorado.

I should like to end this review with a photograph that dramatically illustrates the
progress of 25 years. Figure 27 compares the very first rocket solar spectrograph with
the largest that has ever been built; the little 1946 NRL V-2 rocket spectrograph of
1946 and the new high-dispersion spectrograph for Skylab's ATM.

REFERENCES

1. Baum, W. A., Johnson, F. S., Oberly, J. J., Rockwood, C. C., Strain, C. V., and Tousey, R., *Phys. Rev.* **70**, 781 (1946).

2. Johnson, F. S., Purcell, J. D., Tousey, R., and Watanabe, K., *J. Geophys. Res.* **57**, 157 (1952).

3. Johnston, H., *Science* **173**, 517 (1971).

4. Tousey, R., Purcell, J. D., and Garrett, D. L., *Applied Optics* **6**, 365 (1967).

5. Johnson, F. S., Malitson, H. H., Purcell, J. D., and Tousey, R., *Astrophys. J.* **127**, 80 (1958)

6. McAllister, H. C., Thesis, University of Colorado, (1959).

7. Tousey, R., *Space Science Reviews* **2**, 3 (1963).

8. Brueckner, G. E., *Bull. Am. Astr. Soc.* **3**, 259 (1971).

9. Milman, A. S., *Bull. Am. Astron. Soc.* **3**, 448 (1971).

10. Purcell, J. D. and Tousey, R., *Mém. Soc. Roy. Sci. Liège,* 5th Ser. IV, 274, 283 (1961).

11. Morton, D. C. and Widing, K. G., *Astrophys. J.* **133**, 596 (1961).

12. Jefferies, J. T. and Thomas, R. N., *Astrophys. J.* **129**, 401 (1959).
    Jefferies, J. T. and Thomas, R. N., *Astrophys. J.* **131**, 695 (1960).

13. Purcell, J. D. and Tousey, R., *J. Geophys. Res.* **65**, 370 (1960).

14. Lyot, M. B., *Monthly Notices Roy. Astron. Soc.* **99**, 580 (1939).

15. Edlèn, B., *Z. Astrophy.* **20**, 30 (1942).

16. Grotrian, W., *Naturwissenschaften* **27**, 214 (1939).

17. Austin, W. E., Purcell, J. D., Tousey, R., and Widing, K. G., *Astrophys. J.* **145**, 373 (1966).

18. Tousey, R., *Space Age Astronomy,* ed. by A. J. Deutsch and W. B. Klemperer, Academic Press, New York, (1962), p. 104.

19. Fawcett, B. C., Jones, B. B., and Wilson, R., *Proc. Phys. Soc. Lond.* **78**, 1223 (1961).

20. Purcell, J. D., Packer, D. M., and Tousey, R., *Nature* **184**, 8 (1959).

21. Purcell, J. D., Tousey, R., and Koomen, M. J., *Space Research VIII,* ed. A. P. Mitra, L. G. Jacchia and W. S. Newman, North-Holland Publ. Co., Amsterdam (1968), p. 450.

22. Hunter, W. R., and Tousey, R., *Journal de Physique* **25**, 148 (1964).

23. Tousey, R., *Space Research XI,* to be published.

24. Tousey, R., *Astrophys. J.* **149**, 239 (1967).
    Purcell, J. D., and Tousey, R., *Bull. Am. Astr. Soc.* **1**, 290 (1969).

25. Tousey, R., *Phil. Tran. Roy. Soc. Lond. A.* **270**, 59 (1971).

26. Tousey, R., *New Techniques in Space Astronomy,* ed. F. Labuhn and R. Lüst, D. Reidel Publ. Co. Dordrecht (1971), p. 233.

27. Koomen, M. J., Brueckner, G. E., and Tousey, R., *Bull. Am. Astron. Soc.* **3**, 440 (1971).

28. Brueckner, G. E., and Tucker, B. J., "Use of Television—Type Image Sensors" NASA SP-256 ed. V. R. Boscarino, pp. 55-75 (1971).

29. *Astronautics and Aeronautics,* June, 1971, pp. 20-77, various authors.

30. Winter, T. C., Jr., *Astronautics and Aeronautics,* March, 1969, pp. 64-71.

SPECTROSCOPIC TECHNIQUES IN X-RAY

ASTRONOMY

Herbert Gursky

Space Research Division
American Science & Engineering, Inc.
Cambridge, Massachusetts

INTRODUCTION

Discrete cosmic x-ray sources were first discovered in 1962
and observations made since that time indicate that x-rays
represent a non-negligible form of radiation in cosmos.   There
is a class of x-rays stars in which x-ray emission dominates
all other forms of radiation by at least a factor of 1000
and the x-ray emission from certain galaxies is comparable to
the sum of all the remaining radiated power. Overall, it appears
that about $1:10^4$ of all the radiation in the universe is in
the form of x-rays.

The first kind of observation an astronomer tries to make
is to take a picture and this is true in x-rays where much
of the effort has gone into localizing and determining
the structure of discrete sources. This has also been by far
the most significant kind of measurement made to date since
we have been able to determine that particular, sometimes
well-known, objects such as the Crab Nebula, certain peculiar
galaxies, etc., are emitting x-rays. The second kind of
measurement an astronomer will turn to is that of the spectral
content of the detected radiation. Historically, in radio
and optical astronomy, this has been an extremely fruitful
source of information. Certain fields of astronomy, such
as stellar evolution, are based almost entirely on spectral
measurements.

    Before actually discussing the spectroscopic techniques
that are being used or planned in x-ray astronomy, I will
describe some of the basic observations and ideas that have
evolved in this field. The subject has recently reviewed
by Oda and Matsuoka[1].

    There are two outstanding reatures of the x-ray sky. One is
the existence of a large number of discrete x-ray sources and
secondly, is the presence of a diffulse background of x-rays.
The discrete sources are found to be largely concentrated in
the Milky Way, thus indicating an origin in our own galaxy.
Furthermore, around the Milky Way the distribution is very
irregular. Figure I, which is a scan of the entire Milky
Way with a collimated proportional counter, shows the
distinctive clustering of sources within about 20° of $\ell = 0$
(the direction of the galactic center). The brightest x-ray
source, Sco X-I, which is about ten times stronger than any
of those shown lies 25° off the Milky Way and does not appear
in this scan.

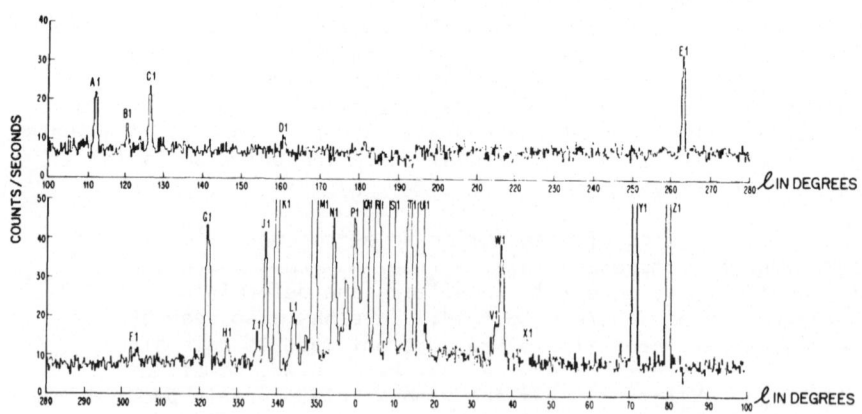

Figure I. Scan of the Milky Way in x-rays. Each peak is an x-ray
source. The zero of the coordinate system is the direction of
the galactic center.

Other concentrations of sources are found in the directions of known spiral arms. In addition a number of sources are located at high galactic latutude and several of these are found to coincide with known outstanding external galaxies.

The nature of the x-ray sources is still not clear. For only one class of objects, the supernova remnants, is x-ray emission believed to be a characteristic of that class. However, these represent only six of the approximately 50 or so galactic x-ray sources. The remainder of the sources apparently represent either an entirely new class of stellar objects such as black holes or an unexpected manifestation of some known object such as white dwarfs. Furthermore, it is quite possible that there are several different kinds of objects represented among these objects.

In the case of the extragalactic sources, we find a diverse set of objects among the presently observed sources. These include NGC 1275, a Seyfert Galaxy, the Coma Cluster, a distant cluster of about 10^3 galaxies, and Centaurus A, a giant radio galaxy.

Over and above the discrete sources, a diffuse background of x-rays is observed. Its intensity seems to be uniform around the sky and it is likely to originate well beyond the confines of our own galaxy, either from intergalactice space or as the effect of large numbers of unresolved sources.

The great bulk of the information described above has come from observations in the 1 - 10 keV energy range. Significant x-ray fluxes are found to about 0.2 keV but at these low energies, most of the sources are cut off due to absorption effects. At energies above about 10 - 20 keV, observations are limited by the fact that the spectra of the sources characteristically are found to fall off rapidly with increasing energy, and only a few objects are of sufficient strength to be seen with present levels of sensitivity. The spectrum of x-rays from Cyg X-1 is shown in Figure 2. These data reveal many typical features of the cosmic x-ray sources. The data points are actually the pulse height spectrum obtained from a proportional counter and the smooth curve is the best fit to those points of an assumed spectral shape. In spite of the fall-off with increasing energy, Cyg X-1 is one of the few objects observed at high energy - significant fluxes have been reported to about 200 keV. The two different spectra demonstrate another important characteristic of the x-ray sources, namely, variability of both intensity and spectral shape.

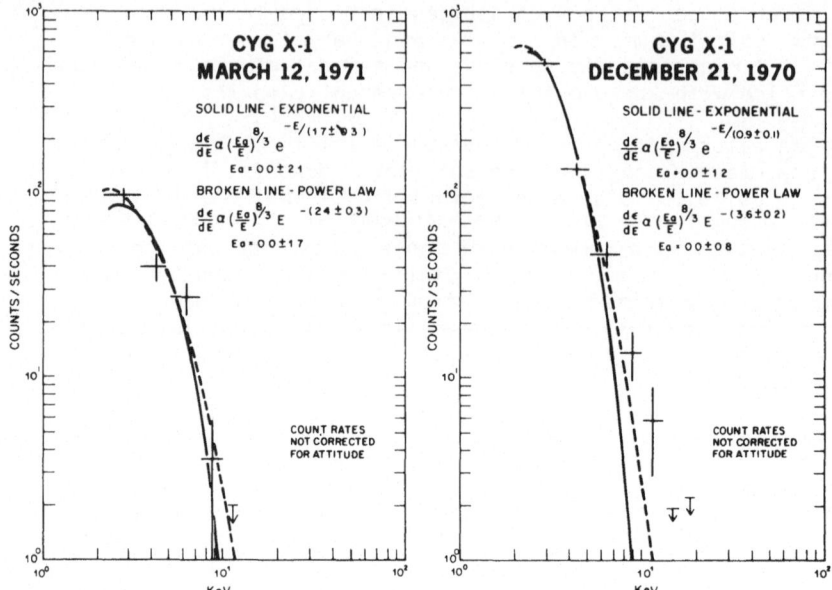

Figure 2.   X-ray spectrum observed by UHURU on two different
days.   These data show several important features of the x-ray
spectra; one is the rapid decrease with increasing energy, and
the other is the great variability from one day to the next.

    The integrated fluxes from the cosmic x-ray sources
is quite low.   The bright source, Sco X-I, has a flux of
about 100 photons/cm^2 -sec and the faintest source yet reported
are about 1/10,000 of this.

    There are three physical processes which are believed to
be the most likely to be operating in the cosmic x-ray
sources.   the first of these is thermal bremsstrahlung which
is just the radiation from a hot plasma.   If the plasma is
optically thin, the spectrum of the radiation contains a
strong continuum whose shape is

    $f\nu \sim exp(-h\nu/kT)$

Depending on the temperature (T) a significant fraction of the power can appear in spectral lines, particularly from O, Si, Ne and Fe which have fairly high cosmic abundances. To yield significant fluxes in the 1 - 10 keV energy range, T must be greater than 10^{7}K. At the other extreme from an optically thin plasma, the opacity can be such that the source appears as a black body, in which case, the continuum is given by

$$f\nu \sim \nu^{3} / (h\nu/kT) - 1)$$

and no spectral lines will be present. Of course, one can have any condition of opacity between the two extremes.

X-rays can also be generated efficiently by the synchrotron process in which radiation results from ultra-relativistic elec- trons gyrating in a magnetic field. This process occurs in a wide variety of objects and is known to be the source of much of the radio emission observed around the sky. Only a continuum is observed whose spectral shape is a function of the spectrum of the radiating electrons. If the electrons possess a power law, the x- ray flux also exhibits a power law spectrum.

The third process that is frequently discussed, especially in conjunction with extragalactic sources and the diffuse background is the inverse compton process which is the collision of low energy photons with high-energy electrons. Again only continuum radiation is observed and if the electrons possess a power law spectrum the x-rays will also exhibit a power law spectrum.

Finally, photoelectric absorption will always be present to modify any primary spectrum of radiation. Even if the source it- self is of low opacity, the interstellar medium will provide significant absorption of the x-rays up to some maximum energy. Two effects characterize photoelectric absorption. One is an abrupt cutoff of the x-ray spectrum at some low energy; the other is the introduction of sharp edges in the spectra which is the result of absorption edge discontinuities. Figure 3 shows the computed effect of interstellar absorption on a typical x-ray spectrum for several values of absorption along the path. The absorption is measured in units of atoms/cm^{2} of atomic hydrogen; in fact, the absorbing atoms are actually the small fraction of heavier elements present in the interstellar medium, particularly helium, oxygen and neon. The greatest absorption yet reported, in the source Cen X-3, extends to 4 keV and requires almost 10^{23} atoms/cm^{2} of hydrogen along the line of sight. This amount is greater than can be accounted for by the interstellar medium and must in part be intrinsic in the source.

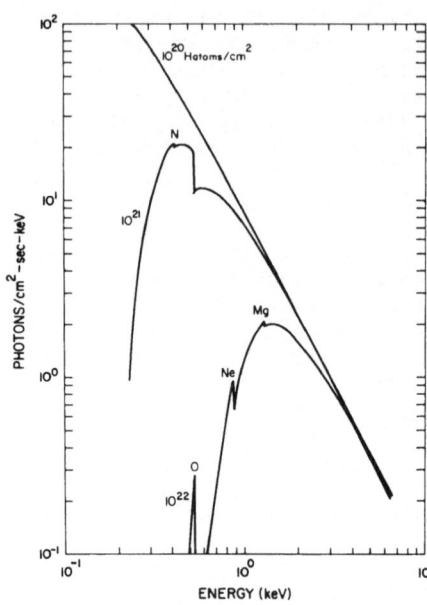

Figure 3. Calculated effect of interstellar medium on a typical x-ray spectrum (the Crab Nebula). The absorption unit (10^{20} - 10^{22} atoms/ cm^2) is the column density to the object along the line of sight.

Even based on such a limited presentation of the processes that give rise to x-rays, the significance of spectral data is clear. Typically, the shape of the continuum yeilds directly the following information:

- Temperature of the plasma
- Spectral index of electrons
- Column density of absorbing material

Unfortunately, cosmic objects are extremely complex; thus, continuum shape by itself cannot define a given physical process. For example, a plasma with a non-uniform temperature distribution can be made to emit a power law spectrum. Spectral lines on the other land are quite specific of thermal bremsstrahlung.

They also give data on electron temperature, velocity dispersion, red shifts, elemental abundances, etc. Absorption edges are also specific. If they can be examined with high spectral resolution, they can be used to determine the state of ionization of the absorbing material.

NON-DISPERSIVE SPECTROMETERS

Almost all spectral data obtained to date have been the re-sult of non-dispersive analysis of proportional counter signals. Although the design of counters have been dominated by consideration of maximizing area and minimizing the non-x-ray background, it has generally been possible to achieve close to the theoretical re-solving power of these devices. For example, the response to Fe[55] x-rays at best can be made to yeild a pulse height resolution of ~ 17%, the very large counters we have been flying yield at worst ~ 20% resolution. With these counters it is possible to disting-uish a power from an expotential continuum shape and to observe the effects of absorption ot low energy.

Figure 4. AS&E sounding rocket payload containing large area proportional counters and slit collimators used to study cosmic x-ray sources.

Figure 4 shows a sounding rocket payload developed and first flown in 1966 by AS&E that contains two banks of collimated proportional counters[2]. The net detector area exceeded 10^3 cm^2. Even large aperture instruments are routinely being flown, and for one future experiment the Naval Research Laboratory plans to fly an instrument with 40,000 cm^2 of detector area. The efficiency of proportional counter is determined by two factors; the entrance window which limits the low energy response by absorption of incident x-rays and the filling gas which becomes increasingly transparent at high energy. For observations in the 1-10 keV both sealed berylium window (down to 0.0001") counters and flow Mylar windows (down to 0.00015") have been employed. Thin organic films down to 1μ thickness have been utilized as windows for detectors which yield almost continuous sensitivity above 0.2 keV.

The analysis of spectral data from proportional counter is generally accomplished by a fitting technique rather than an inversion. The reason for this is that data is generally of poor statistical precision. By fitting technique I mean using a trial function, $f(E)$, to derive the pulse height spectrum, $N(h)$, via the relation,

$$N(h) = \int \epsilon(E)\, f(E)\, R(E, h)\, dE$$

where

$\epsilon(E)$ = counter efficiency

$R(E, h)$ = counter resolution function

then the chi square of the fit is used as a basis for distinguishing between various trial functions[3].

In principal it is possible to observe single, strong spectral lines with porportional counters, and there are several such reports particularly with the Fe lines from the strong source Sco X-1. If ScoX-1 were optically thin and of normal cosmic abundance, several experiments should have detected this line.

The present observations indicates that the line is of considerably smaller intensity than expected, probably as a result of
broadening by electron scattering.

Proportional counters are also used as broad-line filters by
using particular combinations of filters, windows and gas fillings.
For example, a thin organic film used as a counter provides a narrow
pass band, shortward of 0.2 keV, the carbon absorption edge.
Limilarly, windows of teflon are used to obtain a window at 0.7 keV
(Flourine edge) and aluminum window will yield an edge at 1.5 keV.
It is still necessary to use the pulse height relolution provided
by these counters in order to separate out the radiation that is
far outside the window.

DISPERSIVE SPECTROSCOPY

A. Bragg Spectrometry

It is natural to consider the application of Bragg Spectroscopy to the study of x-ray sources. In fact a number of instruments
have been constructed with this in mind, several have been successfully flown from sounding rockets and several others are being prepared as satellite experiments. In its simplest version a Bragg
spectrometer consists of a flat crystal with an appropriate pre-
collimator and detector. Since the x-ray sources are mostly of
small angular size, the incident radiation is parallel to a high
degree of precision. Furthermore, since only few sources exist
around the sky, radiation entering the spectrometers from odd
angles is negligible. An x-ray spectrum of the sun obtained with a
collimated Bragg crystal spectrometer is shown in Figure 5.
Because of the nature of the diffraction process in a crystal, these
spectrometers can only be used effectively to observe sharp spectral
features such as emission lines and edge discontinuities. Figure
5 which is data obtained from the OSO spacecraft[4], demonstrates
the richness of data that can be obtained using th Bragg technique
on an appropriate object and shows why observers have expended much
efforton developing Bragg spectrometers for x-ray astronomers
without any positive indication yet that spectral lines will be observable from the cosmic x-ray sources which are not at all like the
sun..

Typical crystals that are employed are LiF in the medium energy
(> 1.5 keV) and KAP (> 0.3 keV) in the low energy range. No really
satisfactory material exists for use at lower energies. The use
OHM has been discussed however it has yet to be used in a practical
spectrometer. The principle requirements for a crystal for use in
x-ray astronomy is that it by available in large pieces.(units
of hundreds of cm^2 are typical) and it be stable for long periods
of time ($\sim$ year) under laboratory and vacuum conditions.

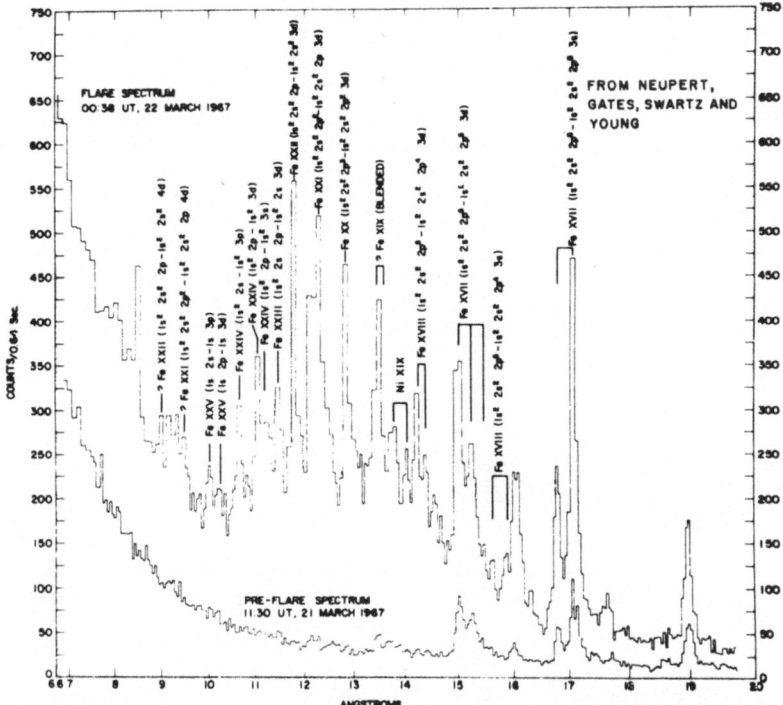

Figure 5. X-ray spectrum of a solar flare obtained using a flat
Bragg Crystal Spectometer.

The flat crystal spectrometer is used in a conventional manner
compared to the equivalent laboratory instrument. Typical of
instruments of this kind is a spectrometer flown by University
College, Leicester, to search for the iron line in Sco X-1.   The
Leicester group is preparing a similar instrument for UK-5 and
AS&E is preparing a spectrometer for measuring the silicon lines at
2 keV for the ANS satellite.

The flat crystal spectrometer has two serious drawbacks for use
in x-ray astronomy; firstly, the low fluxes require the use of large
crystals and secondly, the size of the detector must by made comparable to the size of the crystal. Because of the low incident
fluxes, the counting rate observed with these instruments is dominated
by the internal background of the detector, and the sensitivity of
such and instrument in terms of the faintest detectable line improves
only as the square root of the exposure (defined as the product of
the crystal area) and the observing time. Thus, much effort has gone
into devising instruments in which the crystal and detector area can
be reduced with little or no sacrifice in aperture.

If one simply curves the plane of the Bragg crystal it is possible to focus the reflected x-radiation to a point or a line. The detector then need only be large enough to compensate for the observations of the system and is substantially smaller than what is required for a flat crystal. For ordinary crystals and parallel radiation this doesn't work since each crystal segment then reflects a slightly different energy x-ray, and a line will reflect from only one portion of the crystal. However, if one is prepared to use focusing optics in conjunction with the crystal spectrometer then curved crystal instruments or their equivalent can be devised. It is not my intent here to discuss the status of x-ray focusing optics, howeverc as illustrated in Figure 6, which is an x-ray photograph of the sun taken from a sounding rocket, taken with a high resolution x-ray telescope[5], it is clear that such optics exist.

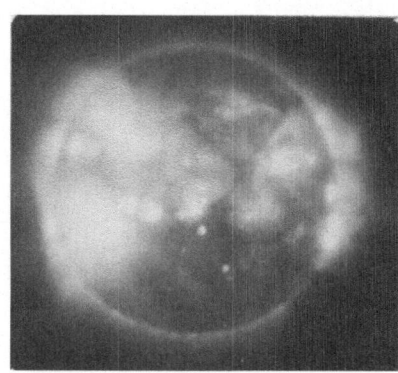

Figure 6. X-ray photograph of the sun obtained by the AS&E solar physics group from a sounding rocket.

One simple geometry which has been studied by Schnopper and Kalata[6] as shown in Figure 7 is to use focal spot of an x-ray telescope as the source of a spectrometer set up in a Johann Mount. This geometry is identical to that used in the laboratory with point sources of radiation. The focal spot and the crystal are are made to lie on a Rowland Curve. The curvature of the crystal compensates for the divergence of the x-rays beyond the focus and preserves the Bragg condition across the face of the crystal. In addition, the x-rays reflected off the crystal converge to a point on the Rowland Circle. If the crystal is curved in the form of a cylindrical section, a line focus results and if the crystal is made into a conical section, a point image results. One promising approach developed by the x-ray astronomy group at Columbia University[7] is the use of the newly developed mosaic crystals - especially graphite - as an objective crystal spectrometer. As shown in Figure 8 the crystal plane is flat and in front of the x-ray optics. The compacted crystals are composed of microcrystals, each defining a unique Bragg plane. However, the whole set of microcrystals comprising the surface constitute a spread in Bragg angles of order 15 - 30'. Thus, a single wavelength, λ, finds a set of crystals at the correctangle θ, to satisfy the Bragg condition, and a second wavelength λ_2 also finds a set of crystals at angle θ_2 that satisfy

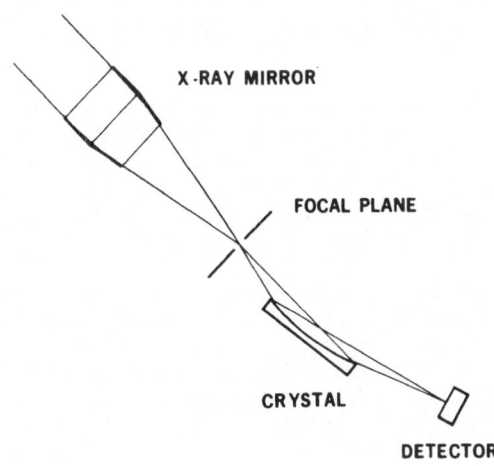

Figure 7. Schematic of curved Bragg Crystal Spectrometer used in the focal plane of an x-ray telescope.

the Bragg condition. The radiation comes off the crystal at angles that differ by $2(\theta_1-\theta_2)$ which then focusses to two different points in the focal plane of the optics. The power of this instrument is that a range of Bragg angles can be examined with high spectral resolution compared to a conventional instrument which must look independently at each increment of wavelength.

B. Diffraction Grating Spectrometry

It is possible to do x-ray spectroscopy with diffraction gratings as well as Bragg crystals. Reflection gratings are used as laboratory instruments in the very soft x-ray range and for application to x-ray astronomy, are competitive with the OHM

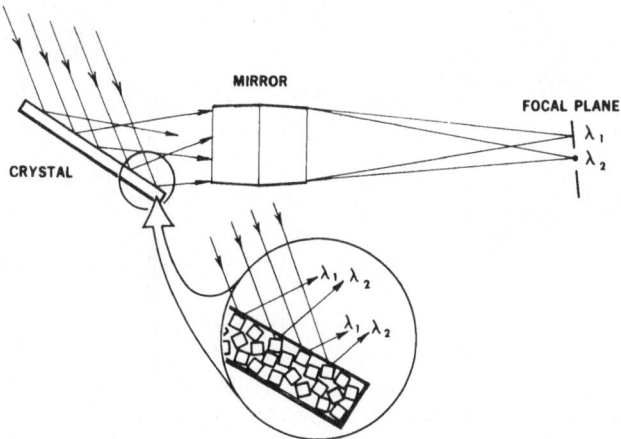

Figure 8. Schematic of an objective crystal spectrometer using a mosaic Bragg Crystal and an x-ray telescope.

crystals at energies below about 0.3 keV. The principal limitation
is that the instrument must be used at grazing incidence in order
to achieve high reflectivity.

Two possible geometries have been proposed. One is a curved
grating mounted on a Rowland Circle behind the focus of an x-ray
telescope. This geometry is identical with a Bragg Crystal mounted as shown in Figure 7. Alternatively, a curved grating can be
set to intercept a parallel beam of incident radiation. The grating is in the form of a section of a parabola.

There are two advantages of this type of spectrometer compared
to a Bragg crystal device when used at very low x-ray energies.
First it is possible to achieve higher spectral resolution with a
grating than with crystals such as OHM which is used at low energy.
Secondly, a range of energy can be observed simultaneously compared
to the point-by-point measurements possible with Bragg spectrometers.
However, one is faced with an awkward detector geometry with a grating spectrometer.

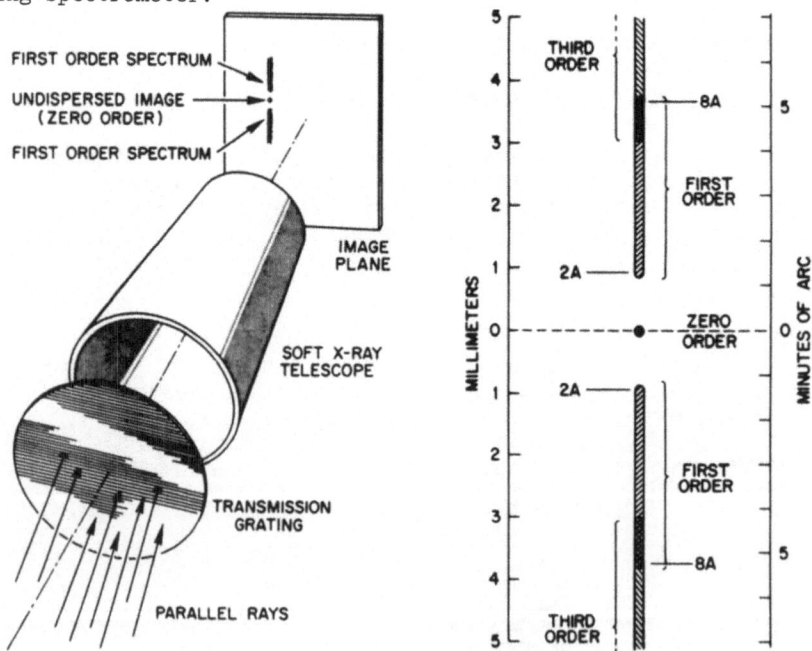

Figure 9. Schematic of an objective grating spectrometer which
utilizes a transmission diffraction grating at normal incidence
and an x-ray telescope.

A very different and novel grating spectrometer has been proposed by Gursky and Zehnpfennig[8] which is the use of a transmission grating as an objective grating spectrometer. As seen in Figure 9 the grating is placed normal to the beam of radiation and immediately in front or behind the x-ray lens. For a point source, the radiation is dispersed along a line normal to the rulings in the grating. Gratings have been constructed from replicas made on a thin organic film. Gold is then evaporated into the film at a shallow angle. The spectral resolution of this instrument is not great, of the order of 50 to 100, and is limited by the angular resolution of the mirror. However, the instrument has great speed. The entire dispersed spectrum can be examined simultaneously and in principle 40% of the beam diffracted into the first order.

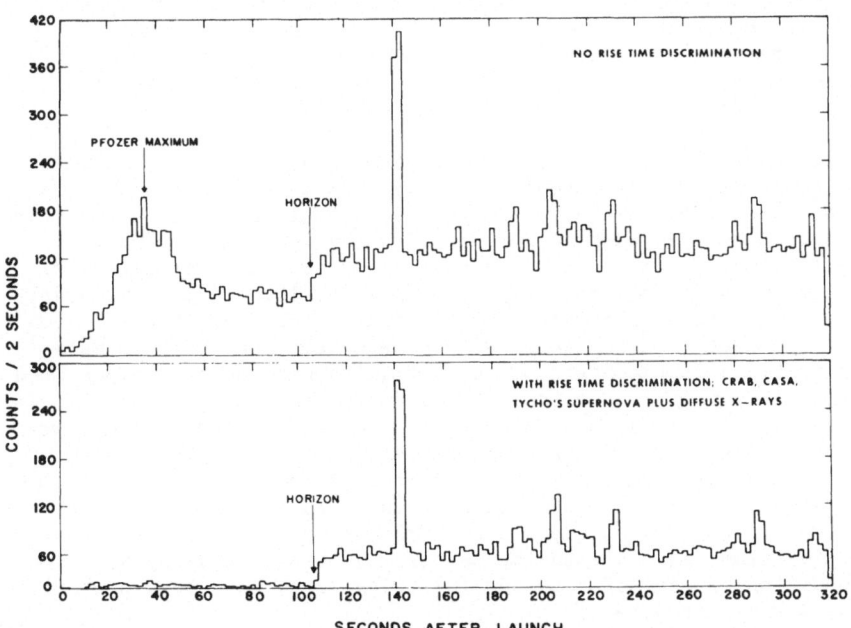

Figure 10. Discrimination against non-x-ray evernts in a proportional counter using rise time discrimination (PSD). Data is from a sounding rocket flight. Early data (< 100 sec) is while the vehicle is in the atmosphere and the observed counts are from cosmic ray secondaries. The Pfozer Maximum is a well known feature of the cosmic ray distribution.

IV. SUMMARY

     I can summarize as follows the spectroscopic techniques that
have been and will be applied to x-ray astronomy. First, the
principle source of spectral information will be the proportional
counter (or the scintillation counter at energies above about
20 keV) which can yeild resolving power of from 1 to 5 based on
pulse height analysis alone. This can be increased by about a
factor of 2 by using filters or particular combinations of counter
gases and windows. The principal application of proportional
counter spectoscopy is for the measurement of gross spectral features such as the continuum and the low-energy cutoff. Secondly
is the Bragg Crystal Spectrometer which can justifiably be used
for measuring spectral lines. Resolving power between several
hundred and several thousand can be achieved. In those cases where
spectral lines are in fact present, results of great consequence,
should be achievable. Between these is the objective grating
spectometer which has only modest performance but can be used to
provide a rapid survey of spectra for a variety of features including emission lines and absorption edges.

     I suspect that the x-ray spectroscopists will not learn much
new in the way of instrumentation from this talk. So far with one
or two exceptions the basic spectrometers that have been around for
50 years have been adopted for use in x-ray astronomy. I expect
it will continue to be the case that x-ray astronomers will rely
heavily on x-ray spectroscopists for new developments, particularly
in the same areas of crystals and gratings. However, there is at
least one area in which x-ray astronomers have been able to make
contributions to the state of the art which has not been emphasized
in the talk; that is, in the area of detectors. As an example,
Figure 10 illustrates how x-rays can be distinguished from x-rays
and charged particles in a proportional counter through the use of
rise time discrimination[9]. Exposed to a variety a radiation including x-rays, proprotional counters yield signals with a broad spread
in rise time; however, the signals from the x-rays yield the short-
est rise times which is the basis for rejecting non-x-ray events.
In addition, I can point to the following which have been accomplished by x-ray astronomers:

1. Development of large area sealed counters with windows down
    to 1 mil Be.
2. Development of large area flow counters with windows of the
    order of 1μ organic film.
3. Better understanding of count life limitations of proportional
    counters and photoelectric detectors.
4. Development of high spatial resolution detectors; namely,
    microchannel plates and multiwire proportional counters
    (adopted from nuclear physics).

3. Better understanding of count life limitations of proportional counters and photoelectric detectors.
4. Development of high spatial resolution detectors; namely, microchannel plates and multiwire proportional counters (adapted from nuclear physics).

In this area and also in the area of x-ray optics it is necessary to examine the literature of x-ray astronomy to establish the state of the art.

REFERENCES

1. Oda, M., and M. Matsuoka, Prog. in Elementary Particle and Cosmic Ray Physics X, Chapter IV (1971).
2. Gursky, H., P. Gorenstein, and R. Giacconi, Astrophys. J. 150 L75, (1967).
3. Gorenstein, P., H. Gursky, and G. Garmire, Astrophys. J. 153, 885,(1968).
4. Neupert, W. M., W. Gates, M. Swartz, and R. Young, Astrophys. J. 149, L79, (1967).
5. VanSpeybroeck, L. P., A. S. Krieger, and G.S. Vaiana, Nature 227, 818 (1970).
6. Schnopper, H., and K. Kalata, Appl. Phys. Lett. 15, 134, (1969).
7. Novick, R., and R. Angel, Columbia Astrophysical Laboratory, Columbia University, New York City.
8. Gursky, H., and T. F. Zehnpfennig, Applied Optics, Vol. 5, 875, (1966).
9. Gorenstein, P., and S. Mickiewicz, Rev. Sci. Inst. 39. 816, (1968).

SPECTROSCOPY IN BIOMEDICINE

SURFACE BIOANALYSIS OF HUMAN DENTAL ENAMEL USING INFRARED

INTERNAL REFLECTION SPECTROSCOPY

David J. Krutchkoff* and H.B. Mark, Jr.**

*Department of Oral Pathology, Health Sciences Center,
University of Louisville, Louisville, Kentucky   40201

**Department of Chemistry, University of Cincinnati,
Cincinnati, Ohio

INTRODUCTION

Dental enamel is the hard, protective tissue that normally surfaces the crowns of vertebrate teeth. Although enamel is the hardest of living tissues, (roughly 97 per cent inorganic mineral) it is relatively easily degraded (demineralized) in acid. Extensive degradation produces cavitation with subsequent rapid destruction of underlying more vulnerable dentin and eventual tooth loss. The disease, dental caries (in vivo, tooth destruction by acid demineralization) is in fact responsible for virtually all premature tooth loss prior to age thirty.

Recent dental research has in large part been concentrated on studies related to dental enamel in an effort to discover means by which teeth and, in particular, enamel surface could be made more resistant to the initial acid attack. However, most methods applied to enamel analysis fall short in that they have proven either insensitive to molecular characterization, destructive in nature or lack capabilities of surface analysis.

Infrared internal reflection spectroscopy (IRS) has been established as an effective method for non-destructive surface analysis of a variety of different samples.[1-4] Although the technique has found only limited application to biological research, it was thought that IRS was perhaps well suited to enamel analysis since it possessed critical analytical capabilities lacked by other methods. Accordingly, a long term investigation to determine the

utility of IRS as a method capable of dental enamel characterization
was begun.

Preliminary IRS studies were involved in determinations of
(1) Feasibility and sensitivity of IRS as a method for experimental
characterization of enamel surfaces; (2) Determination of IR spectral
features of normal dental tissues; (3) Surface transitions with
caries-like demineralization and therapeutic remineralization. All
studies were conceived in light of a long term objective; that being
the development of IRS to the stage of potential utility as a
clinical tool of diagnostic capability.

METHODS AND MATERIALS

Fresh extracted human teeth were selected for initial
studies. Attempts to analyze unaltered, natural enamel surfaces
using multiple reflection IRS techniques were unsuccessful due to
the limited prism-sample contact provided by the irregular morphology
of enamel surfaces. Thus samples were modified experimentally by
grinding flat surface facets on experimental slabs in vitro. This
technique (grinding procedure) admittedly destroys the natural
tooth surface. However, it provided a flat enamel surface with a
marked increase in prism-sample contact. Specimens so prepared
proved well suited to experimental IRS studies and yielded definitive IRS spectra with 2 mm. KRS-5 (n = 2.4) prisms in a Wilks
Micro ATR accessory.[5,6] Specimens were scanned from two to twenty
five microns.

INVESTIGATIVE PHASES

I.   Feasbility and Sensitivity Determinations:

Initial efforts devoted to IRS characterization involved a
number of reference minerals. In general, all apatitic phosphates
tested were transparent at higher energy levels (two to six microns)
while the lower energy region (six to twenty-five microns) exhibited
the characteristic IR "fingerprints" of minerals in question. Preliminary spectra results demonstrated the sensitivity of IRS for
enamel and other inorganic analyses.[6] Comparability to transmission
methods was also established. Spectra of a variety of compounds
(mostly inorganic phosphates and carbonates) were obtained and
specific band frequencies were recorded and compared to those subsequently recorded for dental enamel. It was found that each phosphate tested yielded similar spectral features in addition to unique
patterns which enabled each to be distinguished. For example,
synthetic hydroxyapatite (courtesy Monsanto Corporation, St. Louis,
Missouri) was characterized by a weak band at 630 cm^{-1} (the hydroxyl
librational mode), which was not observed with spectra of natural

fluorapatites. This feature was consistant with previously reported transmission data[7] and, in addition to demonstrating comparability of transmission and internal reflection methods, also allowed the two closely related apatites to be distinguished (Fig. 1).

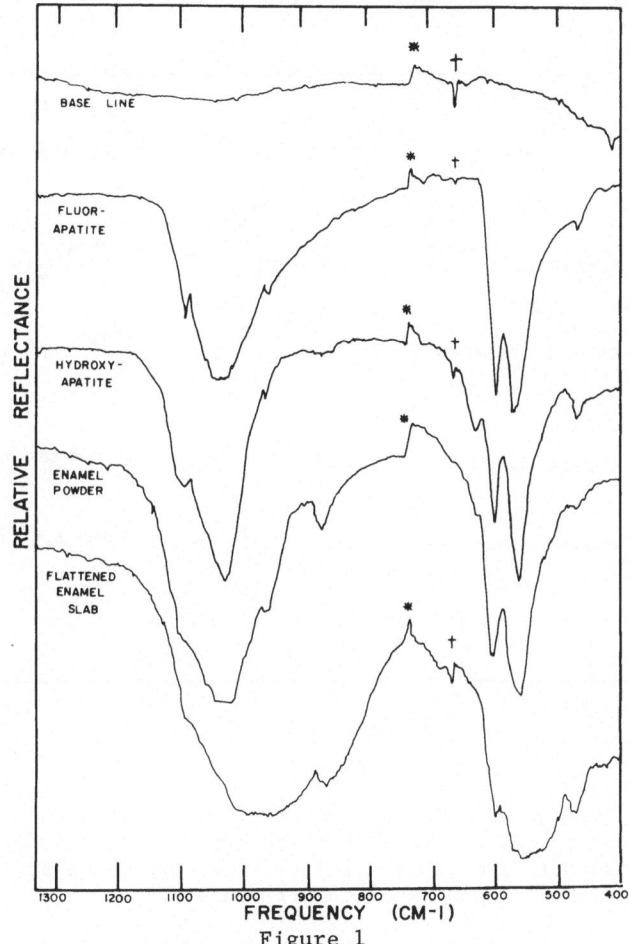

Figure 1

KRS-5 spectra of reference apatites and enamel (both powdered and flattened slab surface). Unique spectral features of each mineral provided specific characterization of all samples tested. (This figure was originally published in Reference #6).

II.   Surface Characterization of Human Enamel (Powder Plus Flattened Surface):

IRS spectra of powdered enamel specimens were obtained and found to exhibit significant differences from other phosphates tested.[6]   Although the γ_3 and γ_4 phosphate absorptions were similar, the carbonate bands (1405, 1450 and 870 cm^{-1}) plus sloping absorption from 700 to 600 cm^{-1} were characteristic and unique to enamel alone (Fig. 1, 2).

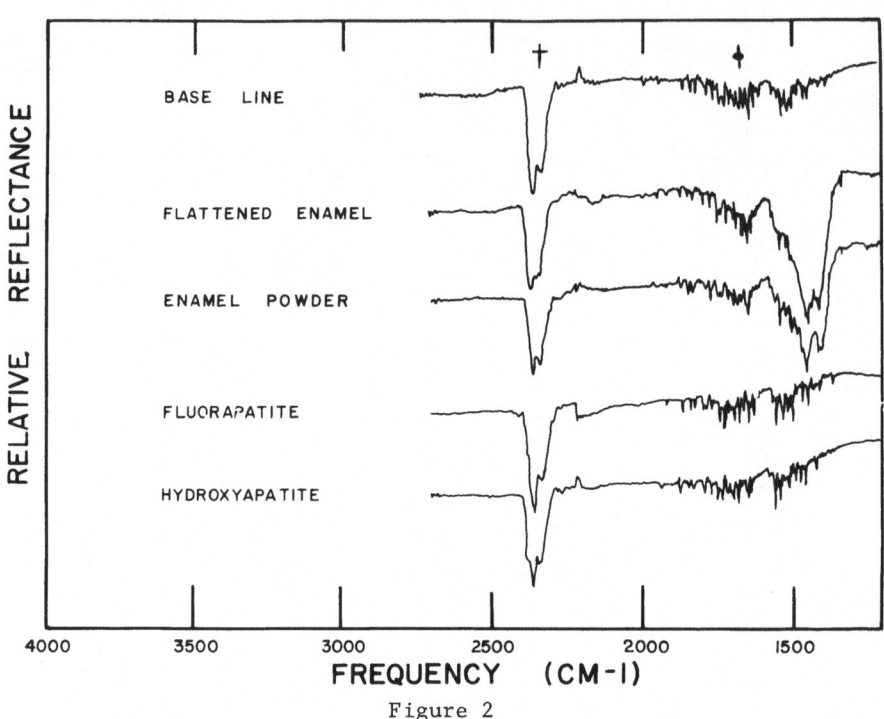

Figure 2

The distinct doublet at roughly 1405 and 1450 cm^{-1} is characteristic of carbonate.   This KRS-5 absorption was helpful in distinguishing enamel from closely related apatitic phosphates.   (This figure was originally published in Reference #6)

Spectra of flattened enamel surface were also run and compared to similarly obtained spectra of enamel powder[6]. Significant spectral differences were observed (peak broadening and shifts to lesser energy levels) and were originally attributed to chemical differences in the two phases (e.g., preferred distribution of an amorphous phosphate component limited to the surface of either flattened or unaltered enamel surface[8])(Fig. 1). Subsequent IRS analysis of identical samples using germanium (n = 4.0) demonstrated that our original interpretation was probably erroneous since

spectra of flattened enamel and enamel powder were virtually identical when germanium prisms (of higher refractive index) were used
(Fig. 3).

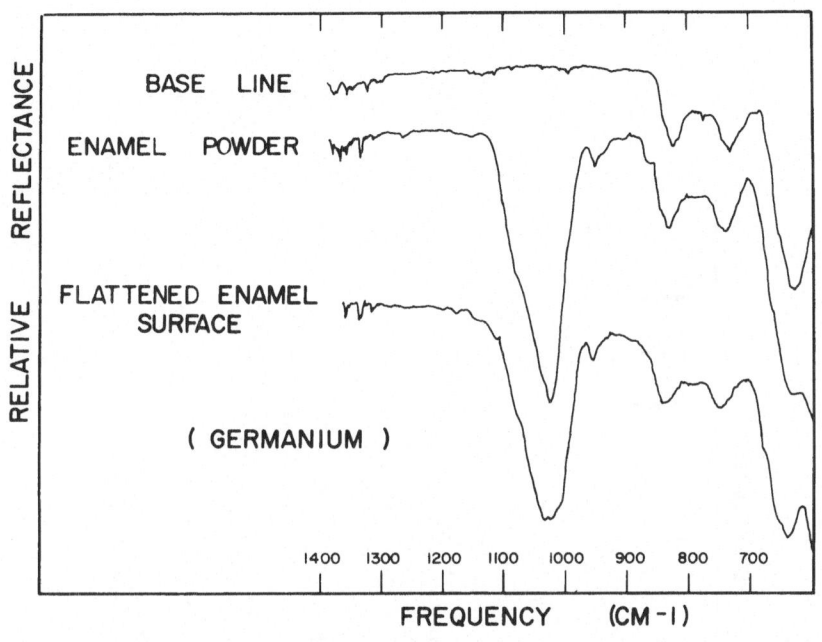

Figure 3

Germanium IRS analyses reveal essentially no difference
between flattened enamel surface and enamel powder. Thus, previously observed differences with KRS-5 (See Fig. 1) may now be attributed to refractive index dispersion.

Thus spectral differences in enamel powder and flattened
enamel surface observed with KRS-5 analyses are more properly
attributable to distortion produced by refractive index dispersion
during an absorption.[9]

III.   Transitions of Enamel Surface:

In Vitro Demineralization. Since in vivo dental caries is
essentially a process of weak acid demineralization, it was decided
to mimic the disease experimentally using flattened enamel slabs
and lactic acid-gelatin mixtures[10] to produce in vitro enamel
"lesions". Gradual etching of flattened enamel surface produced a
"honeycombed" pitting[11] similar to the pattern previously reported

for unaltered enamel surface.[12]   Spectral analysis before and after
such treatment revealed definitive changes in IRS spectra that
were characterized by improved peak resolution and shifts to higher
energy levels[8] (Fig. 4).

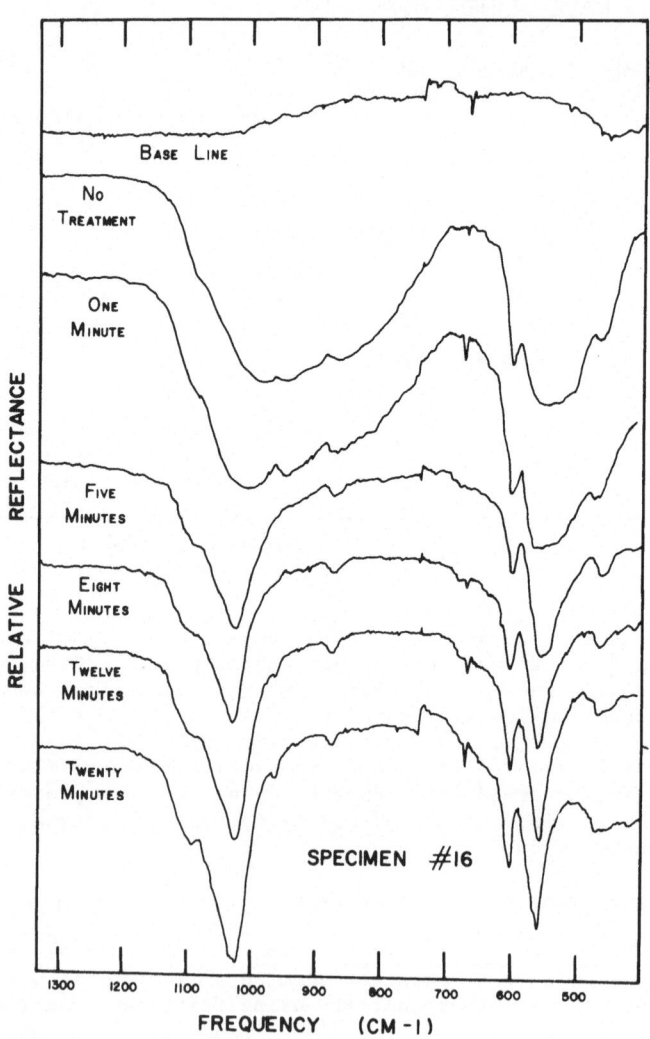

Figure 4

KRS-5 spectra reveal that the flattened enamel surface
appears to approach KRS-5 spectral identity with enamel powder
(see Fig. 1) with progressive acid demineralization. (This figure
was originally published in Reference #8)

It was proposed that the "honeycombed" pitting produced during
weak acid demineralization produced a change in refractive index
from that of solid enamel slab (1.62) to an intermediate level
between enamel and air (between 1.0 and 1.62). The diminished
critical angle resulting from this refractive index change would
provide sufficient margin between θ and $θ_c$ to avoid distortion that
would otherwise occur if the two were in close proximity.[9] With
the untreated enamel slab-KRS-5 system, θ exceeds $θ_c$ by small margin
thus providing a situation vulnerable to peak distortion.

Observed spectral shifts of flattened enamel slabs with
weak acid demineralization (Fig. 4) are in all probability indicative of surface change in refractive index rather than a strict
preferred dissolution of more soluble constituents as previously
reported.[8] Refractive index (n) of the slab surfaces probably
diminishes with progressive demineralization. A parallel decrease
in $θ_c$ accompanies a drop in sample refractive index. It is proposed
that a point is eventually reached when the change in (n) during
absorption would no longer be sufficient to cause artifactual
energy loss from failure to exceed the critical angle. At this
point, spectra of enamel powder and flattened slabs would become
indistinguishable.

Spectral shifts with demineralization are consistant and
experimentally reproducable even though they appear to be due to
refractive index phenomena rather than chemical differences. It
is conceivable that like changes of unaltered enamel surface could
be monitored by in vivo IRS methods thus providing a means for sub-
clinical detection of early "in situ" demineralization.

Surface Changes of Topical Fluoride Application. Various
types of "remineralizing" solutions have been advocated for treatment and/or prevention of dental caries. Substances are applied
topically to enamel surface in an effort to either make the surface
more resistant to carious breakdown or perhaps to partially replace
apatitic mineral lost in previous demineralization. Ideally,
mineral salts would precipitate from solution in areas of prior
demineralization. Such therapeutic precipitates would thus serve
as a form of mechanical "repair", and as such, pre-empt the necessity
of a mechanical restoration.

Many different solutions have been proposed by various
groups.[13-16] Variables of such include specific ionic constituents,
concentrations of such, and pH. In spite of the wide variety of
solutions, a nearly universal constituent of all remineralizing

solutions is the element, fluorine in any of several different
forms. In general, investigators have previously attempted to
achieve maximal incorporation of fluoride on or within treated tooth
structure reasoning that quantitative fluoride uptake could be
equated proportionally with caries preventive activity. Accordingly,
most early attempts to judge fluoride effectiveness have employed
quantitative fluoride analyses (either surface or bulk) with little
regard to the qualitative nature of the fluoride-enamel complex.
Recently, attention has been focused on the form of fluoride
incorporation rather than the earlier strict quantitative studies.
In an effort to refine and expand qualitative data already available, we employed infrared IRS in an attempt to characterize the
reaction products of several "remineralizing" solutions that are
currently in clinical use.[17]

Flattened enamel slabs were again used as the experimental
model. We found that reaction products differed considerably
depending upon the conditions of remineralization. Slabs previously
etched in lactic acid (pH 3.0) and subsequently remineralized in
acidulated fluorophosphate solution (AFP)[18] yielded IRS spectra
which indicated a surface precipitate of calcium fluoride rather
than fluorapatite which had previously been reported.[19] Spectra of
treated slabs revealed a diminished peak amplitude plus a "cut-off"
at 500 cm^{-1}. The apparent masking of surface phosphate absorption
after AFP treatment in addition to the proven transparency of CaF_2
to 500 cm^{-1} was taken as strong evidence for a surface precipitate
of CaF_2 following this form of treatment[17] (Fig. 5). The kinetics
of this reaction, specifically the fate of surface calcium fluoride
is under current investigation.

KRS-5 spectra of flattened enamel surface through etching
and subsequent AFP remineralization (Method II). Spectral data is
suggestive of surface acquisition of calcium fluoride following
AFP treatment. (This figure was originally published in Reference #17)

Previously demineralized slabs remineralized in neutral
(pH 7.3) fluoride containing saturated calcium phosphate "rehardening"
solutions[20] yielded entirely different reaction products as demonstrated by IRS spectra of slabs treated in this fashion.[17] The
shift in the γ_3 peak maximum plus the broadened irregular absorption
in the 1000-700 cm^{-1} range was interpreted as evidence of surface
acquisition of amorphous phosphate following treatment of this type
(Fig. 6).

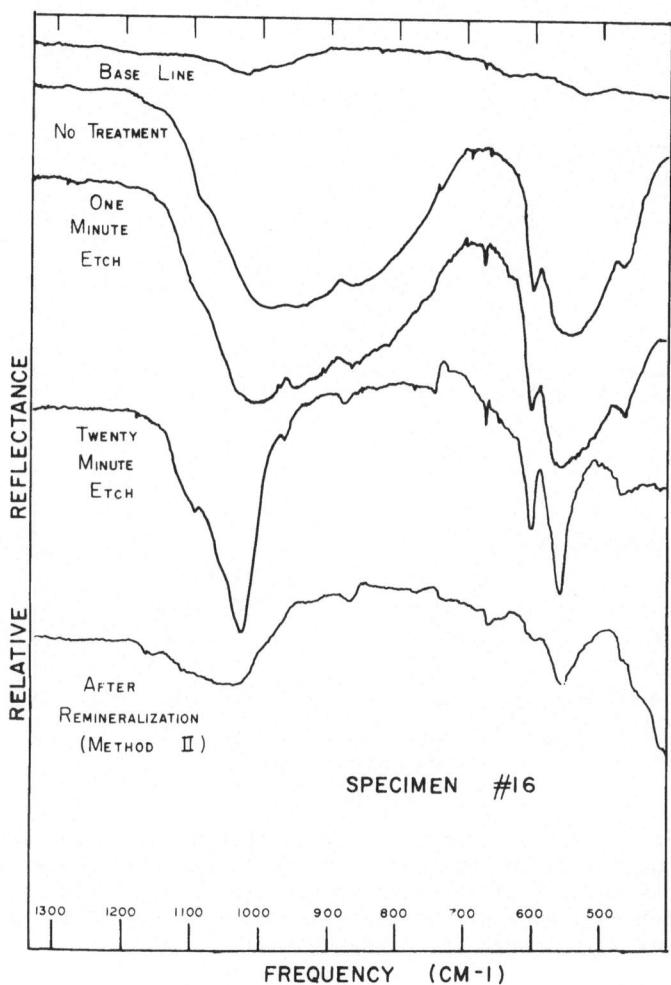

Figure 5

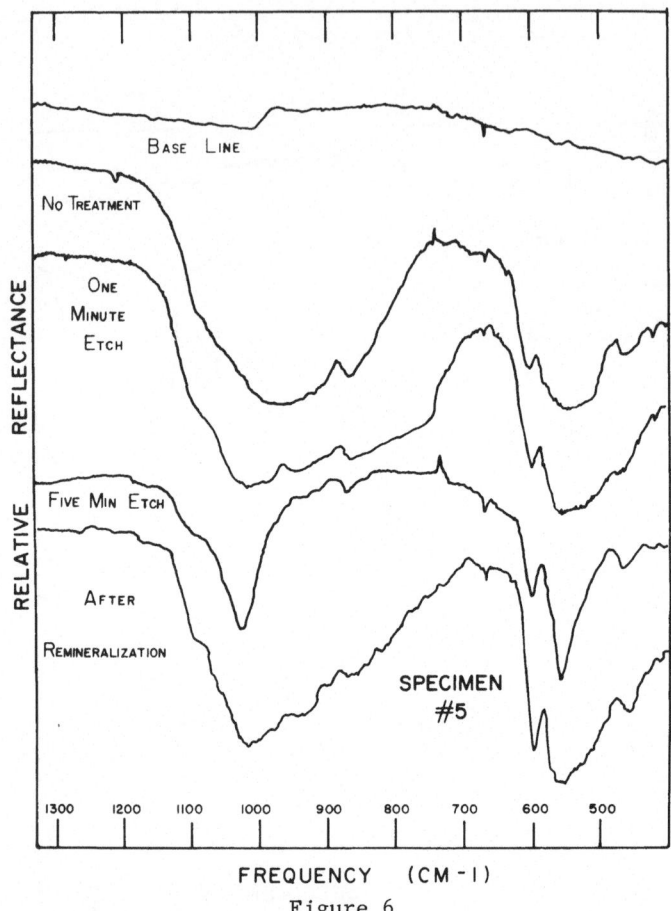

Figure 6

KRS-5 spectral monitoring of flattened enamel surface through experimental etching and subsequent neutral remineralization.[20] IRS spectra of slabs after treatment were interpreted as evidence for amorphous phosphate remineralization. (This figure was originally published in Reference #17)

The stannous fluoride-enamel interaction has also been investigated with IRS in an effort to characterize specific surface reaction products.[21] It was found that specimens of both enamel powder and flattened enamel slabs subjected to prolonged 10 per cent SnF_2 exposure exhibited definitive shifts in peak frequency such that post-treatment surface absorption maxima coincided precisely with those of pure $Sn_3F_3PO_4$ (Fig. 7). Thus IRS served to substantiate a previous hypothesis[22] that $Sn_3F_3PO_4$ was the principal reaction product of the SnF_2 — enamel interaction.

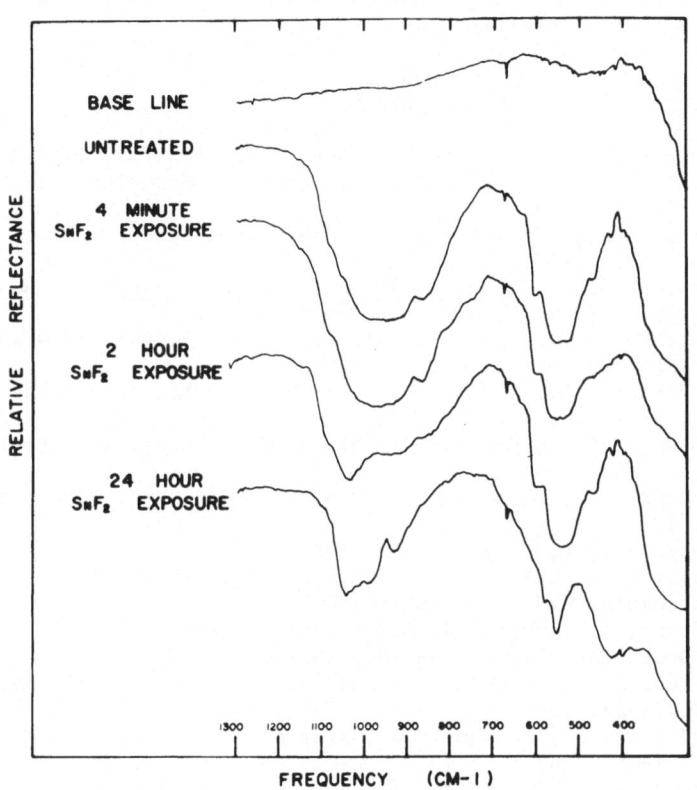

Figure 7

    KRS-5 spectra of flattened enamel surface following sequential exposure to ten per cent SnF_2. Gradual peak shifts with increasing exposure was indicative of surface precipitation of $Sn_3F_3PO_4$. (This figure was reproduced from an unpublished manuscript previously submitted for publication in Arch. Oral Biol.[21])

CONCLUSIONS

    It is clear that IRS has proven to be a useful device for experimental characterization of flattened enamel slabs in vitro.

Work is currently in progress to develop and modify the basic IRS apparatus to the point where enamel slabs may be analyzed without the creation of experimental surface facets. Hopefully, the apparatus could then be further modified for in vivo clinical analysis. For this to come to pass, other prism materials must be utilized due to the toxic properties of KRS-5. Efforts along these lines are presently under investigation.

REFERENCES

1.   Fahrenfort, J. Spectrochim. Acta. 17; 698-709, 1961.
2.   Mattson, J.S., Mark, H.B., Jr. Environ. Sci. Technol. 3, 161-64, 1969.
3.   Wilks, P.A., Hirschfield, T. Appl. Spectroscopy Rev. 1, 99-130, 1967.
4.   Harrick, N.J. Internal Reflection Spectroscopy. Interscience. (Wiley and Sons), New York, 1967.
5.   Krutchkoff, D.J. M.S. Thesis, Rackham Institute of Graduate Studies, University of Michigan, 1970.
6.   Krutchkoff, D.J., Rowe, N.H. and Mark, H.B., Jr. Arch. Oral Biol. 16, 161-75, 1971.
7.   Fowler, B.O., Moreno, E.C., Brown, W.E. Arch. Oral Biol. 11, 477-92, 1966.
8.   Krutchkoff, D.J., and Rowe, N.H. J. Dent. Res. 50, 1589-94, 1971.
9.   Krutchkoff, D.J. IN PREPARATION
10.   Silverstone, L.M. Brit. Dent. J. 125, 145-57, 1968.
11.   Krutchkoff, D.J. (Unpublished)
12.   Hoffman, S., Rovelstad, G., McEwan, S. and Drew, C.M. J. Dent. Res. 48, 1296-1307, 1969.
13.   Koulourides, T., Cueto, H. and Pigman, W. Nature (London) 189-226-27, 1961.
14.   Brudevold, F., McCann. H.G., Nilson, R. J. Dent. Res. 46, 37-45, 1967.
15.   Muhler, J.C. J. Dent. Res. 27, 51-54, 1960.
16.   Francis, M.D., Meckel, A.H. Arch. Oral Biol. 8, 1-7, 1963.
17.   Krutchkoff, D.J., Rowe, N.H. J. Dent. Res. 50, 1621-26, 1971.
18.   Pigman, W., Cueto, H. and Baugh, D. J. Dent. Res. 43, 1187-95, 1964.
19.   Brudevold, F., Savory, A., Gardner, D.E., Spinelli, M. and Spiers, R. Arch. Oral Biol. 8, 167-77, 1963.
20.   Aasenden, R., Brudevold, F. and McCann, H.G. Arch. Oral Biol. 13, 543-52, 1968.
21.   Krutchkoff, D.J., Jordan. T.H., Wei, S.H.Y., Nordquist, W.D. Submitted to Arch. Oral Biol., October 1971.
22.   Jordan, T.H., Wei, S.H.Y., Bromberger, S.H. and King, J.C. Arch. Oral Biol. 16-241-46, 1971.

NUCLEAR PARTICLE SPECTROSCOPY: APPLICATIONS OF NEUTRON ACTIVATION

ANALYSIS AND GAMMA-RAY SPECTROSCOPY IN BIOMEDICINE

David M. McKown and James R. Vogt

Research Reactor, University of Missouri
Columbia, Missouri 65201

INTRODUCTION

Nuclear Particle Spectroscopy embodies a large variety of physical methods for measuring the energy or time distribution of sub-atomic particles or gamma-ray photons associated with the decay of radionuclides. The principles of nuclear particle spectroscopy are materialized as selective radiation detection instruments which may be employed as refinements to radioisotope and radioactivation methods. This paper briefly reviews some biomedical applications of analytical techniques utilizing gamma-ray spectroscopy; however, rather than attempt an overview of their extensive applications, emphasis is given mainly to some general aspects of the impact of these techniques upon biomedical research that have been heightened by developments in gamma-ray spectroscopy.

The primary applications of gamma-ray spectroscopy in biomedical research have taken two main forms. First, neutron activation analysis techniques are used to determine the trace element composition of organisms and their environments. Secondly, radioactive tracers and stable isotope preparations in conjunction with activation analysis are used to study the dynamics of specific components in biomedical systems. The fact that activation analysis techniques are primarily applicable only to trace element studies does not imply that these techniques are of minor importance to biomedical research. Meticulous considerations of life processes have shown that a large number of trace elements are essential for vital functions; which is to say, the organism can neither grow nor complete its life cycle without trace amounts of certain elements.

TRACE ELEMENTS IN BIOMEDICINE

A comprehensive summary of the role of trace elements in human and animal nutrition has been given by Underwood (1), and the functions of trace elements in biochemistry have been discussed by Bowen (2).

The trace elements can be divided into four groups: dietary essential, possibly essential, non-essential, and toxic elements. At the present time 10 trace elements are known to be essential. These are iron, iodine, copper, zinc, manganese, cobalt, molybdenum, selenium, chromium, and tin. Several other elements, notably nickel, fluorine, bromine, arsenic, vanadium, cadmium, barium and strontium may be considered as possibly essential for certain vital functions on the basis of suggestive but incomplete evidence. Some 15 to 20 other elements, including aluminum, antimony, mercury, germanium, silicon, rubidium, silver, gold, lead, bismuth, and titanium are consistently present in biological tissue with no presently known biological role. The toxic classification is justified, perhaps, for a few elements, such as arsenic, lead, cadmium, mercury, plutonium, and tellurium which are highly toxic at extremely low levels. However, almost all the trace elements are toxic if ingested at sufficiently high levels.

Thus the trace element distributions, their dynamics, and essential functions are of vital importance to biomedicine; yet, the fundamental relationships between living organisms and most trace elements are not well known. This lack of understanding is largely due to the fact that until recently there were no analytical techniques available which could accurately measure such low elemental concentrations. If the presence of these micro-nutrients were detected, they were referred to only as "traces" -- a term which has remained in popular usage despite the fact that most of the elements can now be determined with a high degree of accuracy. Neutron activation analysis using gamma-ray spectroscopy is one of the methods presently available which is capable of measuring trace elements in biological systems.

NEUTRON ACTIVATION ANALYSIS

General Method

The theory of activation analysis as well as fundamental application considerations can be found in a number of books (3-5). Recent trends in the development and application of neutron activation analysis techniques were reviewed at the Third International Conference on Activation Analysis (6).

The two basic methods of neutron activation analysis are referred to as non-destructive or instrumental and destructive or
radiochemical activation analysis. Both methods involve irradiation of a sample with neutrons, usually in a nuclear reactor, to
produce radioactive nuclides from the stable elements in the
sample followed by quantitative assay of the radioactivities produced. In the radiochemical methods, the radionuclide or group
of radionuclides of interest are chemically separated from the
mixture of activities before counting. A weighable amount of a
carrier element isotopic with the radionuclide of interest may be
added to the sample after irradiation. Thus all of the standard
separation methods of analytical chemistry may be employed for
the radiochemical activation analysis of trace quantities. This
method provides maximum selectivity and sensitivity.

The instrumental activation analysis approach employs direct
assay of the irradiated sample without any post-irradiation processing. Thus the selectivity is determined primarily by the
choice of irradiation and decay conditions and the gamma-ray

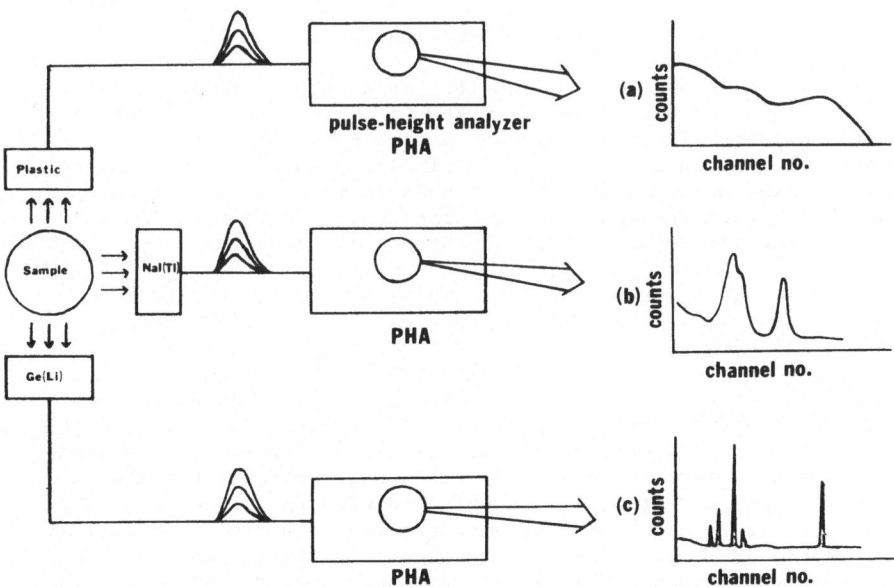

Figure 1.   Diagramatical Representation of Beta - and Gamma-Ray
            Spectroscopy. (a) Complex Beta-Ray Spectrum, (b)
            NaI(Tl) Gamma-Ray Spectrum, and (c) Ge(Li)
            Gamma-Ray Spectrum.

energy resolution of the gamma-ray spectrometer used to assay the irradiated sample. A diagramatical representation of gamma-ray spectroscopy is given in Figure 1. Gamma-rays interact with a detector which produces output signal pulses that are proportional to the energy deposited in the detector by each gamma-ray interaction. A pulse-height analyzer sorts the detector pulses according to pulse height and accumulates, in a channel storage memory bank, a gamma-ray spectrum. A gamma-ray spectrometer resolves the individual gamma-rays according to their energies and, thus, facilitates the selective detection and measurement of individual components of a mixture of radionuclides. As illustrated in Figure 1, a Ge(Li) gamma-ray spectrometer provides a much higher degree of energy resolution than a spectrometer employing a NaI(Tl) detector; however, the gamma-ray detection efficiency is much better using a NaI(Tl) detector. Also shown in Figure 1(a) is a beta-particle spectrum of a mixed radionuclide source to illustrate the poor energy resolution compared to gamma-ray spectroscopy. An excellent discussion of the principles and applications of gamma-ray spectrometry has recently been published by Crouthamel, Adams, and Dams (7).

Biomedical Applications

The full range of applications of neutron activation analysis is much too extensive to review in this paper. However, a very comprehensive bibliography indexed by author, technique used, element determined, and matrix analyzed has been prepared by the U.S. National Bureau of Standards (8). This bibliography, which includes the analysis of biomedical related materials, covers the literature available through January, 1971.

Estimated neutron activation analysis sensitivities are compared in Table 1 with the approximate levels of several essential elements in animal tissue and blood. These approximate elemental concentrations are taken from Bowen (2) and from data obtained in our own laboratory. The radiochemical neutron activation analysis sensitivities, which were estimated for assay of a radiochemically pure separated fraction using a NaI(Tl) gamma-ray spectrometer, represent typical sensitivities rather than the ultimate detection limits obtained with special experimental design. The instrumental neutron activation analysis sensitivities given in Table 1 assume a reasonable multi-element compton background using a Ge(Li) gamma-ray spectrometer. Note that the instrumental sensitivities are generally 10-50 times larger than the corresponding radiochemical sensitivities due mainly to the much lower detection efficiency of the Ge(Li) detector. Nevertheless, even the instrumental sensitivities are low enough to provide useful data for many types of biological samples.

Table 1.   Comparison of Estimated Neutron Activation Analysis
           Sensitivities With the Approximate Abundances of Some
           Essential Elements in Biological Materials.

ELEMENT	APPROX. ABUNDANCE (PPM)				APPROX. SENSITIVITY (μg)	
	BLOOD	LUNG	MUSCLE	HAIR	RNAA	INAA
B	0.1	0.2	0.3	--	NO	NO
Ca	60	480	105	200	0.2	10
Cl	2900	1200	2800	2000	0.001	0.1
Co	0.003	0.2	0.2	15	0.0002	0.05
Cu	1	6	3	80	0.0003	0.01
Fe	470	1300	140	130	0.1	10
I	0.06	0.001	0.1	2	0.0001	0.001
K	1690	8600	10500	1900	0.004	10
Mg	40	400	680	300	0.05	5
Mn	0.03	0.8	0.2	1	0.00005	0.001
Mo	0.004	<.2	<.2	--	0.005	0.1
Na	1900	1200	4000	200	0.0001	0.01
Se	0.3	--	2.5	--	0.002	0.1
Zn	6.5	62	180	170	0.0005	0.5

RNAA -- Radiochemical Activation Analysis Using a NaI(Tl) Detector
INAA -- Instrumental Activation Analysis Using a Ge(Li) Detector

     For the purpose of neutron activation analysis considerations,
biomedical related materials may be classified as three types:
(a) internal body pools and fluids, such as tissue, organs, blood,
and urine, (b) external body pools, such as hair, finger nails, and
bone, and (c) environmental reservoirs such as air, soil, water,
and feed. As illustrated in Figure 2, gamma-ray spectra obtained
for the external body pools and environmental samples are generally characterized by a relatively low multi-element background
interference so that a large number of trace elements can be
determined simultaneously using Ge(Li) gamma-ray spectrometry with
no post-irradiation chemical processing. From the particular hair
sample spectrum given in Figure 2, eleven elements can be determined. Utilizing a second irradiation of the same sample and
multiple counts, 15-20 elements can be determined instrumentally
in human hair. A number of examples of instrumental neutron activation analyses of biological materials are given in Table 2.
This list is not intended to be encyclopedic, nor does each example necessarily represent all of the elements that can be determined instrumentally.

     Sample types previously classified as internal body pools and
fluids are generally characterized by relatively high concentrations of certain elements such as Na, Cl, P, K, and Br. The acti-

Table 2. Examples of Instrumental Neutron Activation Analysis of
          Biological Materials.

MATRIX	ELEMENTS DETERMINED	AUTHOR
Lung & Muscle	K,Cl,Na,Mg,Fe,Al,Zn,Br,Rb, Se,Cr,Mn,Hg,Co,Cs,Sb,Sc	Rancitelli (9)
Hair	Mg,I,Cu,Na,Cl,Al,Ca,S,Sr,Mn, K,Au,Zn,Br,Hg,Sb	RRF (10)
Bone	Ca,V,Al,Mn,Na,Cl,Sr,I,La, Br,Sc,Fe,Co,U	RRF (10)
Blood	Rb,Cs,Fe,Zn,Co,Sb,Se,Hg, Na,K,Br	Haller (11)
Tobacco	La,Br,As,Cr,Se,Ag,Sc,Sb, Zn,Co,Fe,Hg,Eu,Au	Nadkarni (12)

vation products of these elements severely limit the determination of other trace elements using purely instrumental activation
analysis techniques. As illustrated by the gamma-ray spectrum of
an irradiated heart tissue, Figure 3, the extremely large ^{24}Na
activity precludes the determination of trace element activities
except those with half-lives much longer than ^{24}Na. Those elements whose activation products are relatively short-lived must
be determined by radiochemical neutron activation analysis procedures. Radiochemical procedures have been developed for the
individual determination of almost all of the naturally occuring
stable elements. These methods are generally very sensitive,
selective, and accurate but, on the other hand, require laborous,
time consuming laboratory operations.

    An important stimulus to the increasing utilization of
radiochemical neutron activation analysis has been the development of rapid group separation schemes which allow several elements to be determined simultaneously. These procedures involve
either separation of a group of radionuclides of interest by
utilizing a common chemical property, or selective removal of
specific radionuclide interferences without loss of the radionuclides of interest.

    For example, a simple group separation scheme for the deter-
mination of Cu, Co, Zn, and Mn in animal tissue might consist of
the following: (a) dissolution of irradiated sample and carriers
in a mixture of acids, (b) extraction of their ammine complexes
in ammonium hydroxide, and (c) precipitation as a mixed metal

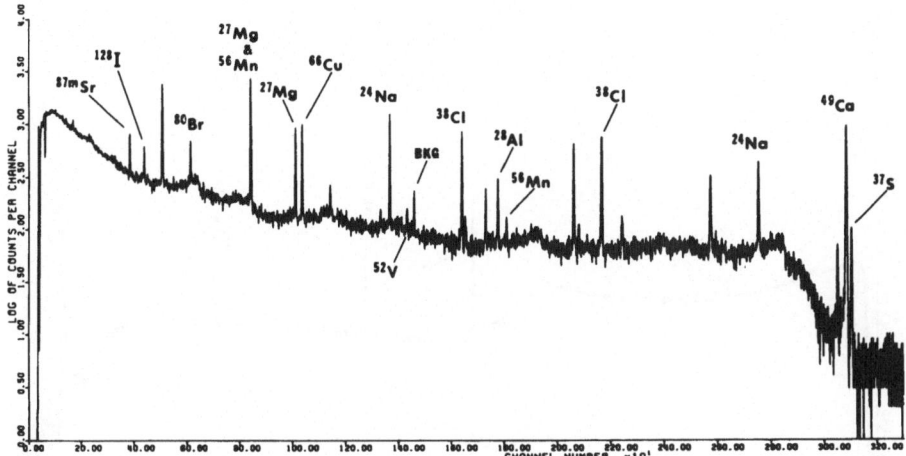

Figure 2.   Gamma-Ray Spectrum of a Human Hair Sample Following a
            1 Minute Irradiation and 5 Minute Decay Period.
            Spectrum Observed with a Ge(Li) Detector.

sulfide precipitate. Manganese will follow the procedure in the
absence of oxidizing materials, and cadmium will also separate.
The utility of this approach is illustrated in Figure 4, which
shows the gamma-ray spectrum obtained for the separated sulfide
fraction of the same irradiated heart tissue shown in Figure 3.
The major activities in the irradiated sample have been sufficiently reduced to permit the simultaneous determination of Cu,
Co, Zn, and Mn. This group separation procedure requires only
about 10 minutes per sample if 4 samples are processed simultaneously.

    More than 15 different separation schemes, excluding success-
ive modifications of a scheme, have been published which facilitate the separation of 10-50 elements into groups which can be
assayed by gamma-ray spectroscopy. The subject of radiochemical
separations has recently been reviewed by Girardi (13) with reference to these published procedures.

                        TRACER APPLICATIONS

                        Radioactive Tracers

    The radioactive tracer method has long been an important tool
in biomedical research. A wealth of basic knowledge of the re-

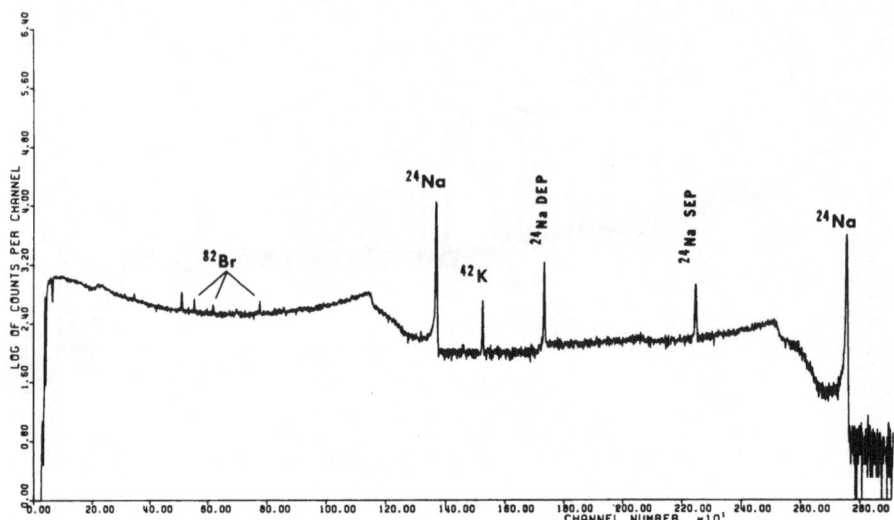

Figure 3. Gamma-Ray Spectrum of a Heart Tissue Following a 1 Hour
Irradiation and a 10 Hour Decay Period. Spectrum Observed
With a Ge(Li) Detector.

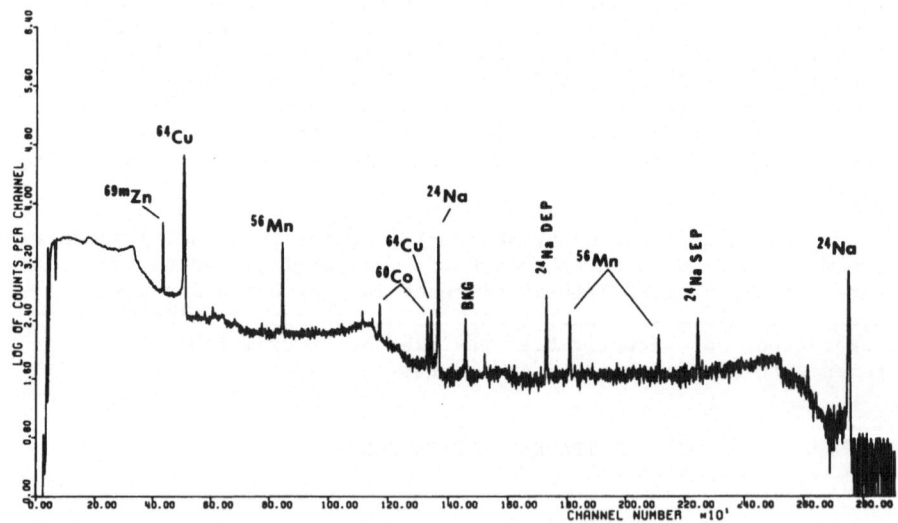

Figure 4. Gamma-Ray Spectrum of a Mixed Metal Sulfide Group
Separated From the Same Irradiated Heart Tissue Shown
in Figure 3 (See Text).

quirements, dynamics, and functions of specific components of bio-medical systems has been gained through radioactive tracer studies. From the information gained, a variety of more or less routine clinical diagnostic tests and therapeutic applications have evolved.

Many of these earlier radioactive tracer investigations employed a single radionuclide and a simple gross activity count-ing system. Such studies required considerable duplication of effort in order to study the relative behaviors of several differ-ent elements in the same system. An excellent example of the power and breadth of multiple isotope applications is the simul-taneous administration of radioactive sodium, potassium, chlorine, and tritium to determine in a single study (a) the exchangeable sodium, potassium, and chlorine in the body, (b) the plasma sodium, potassium, and chlorine, (c) the total body water, (d) the extracellular water, sodium, and potassium, and (e) the in-tracellular water and potassium (18).

With the present availability of gamma-ray spectroscopy equipment, a mixture of radioactive tracers may be assayed simultaneously with little or no additional effort. For example, a study was conducted in our laboratory to determine the uptake, accumulation, and excretion of radioactive fall-out by lactating animals. Feed samples were doped with a mixture of radionuclides. Collected urine, feces, blood, and milk samples were then assayed using high efficiency NaI(Tl) gamma-ray spectroscopy. Typical gamma-ray spectra are shown in Figure 5. The use of multiple tracers, gamma-ray spectroscopy, and computer assisted data re-duction made it practical to obtain 1300 individual analyses during the course of two feeding experiments involving 5 animals each. More important is the fact that the widely variable animal control parameters are not introduced into the relative behaviors of simultaneously injected multiple tracers.

Stable Isotope Tracers

There has been constant concern expressed by both professional and lay groups over the radiation exposure delivered by certain radioactive tracer methods. The question is especially germane when studies of healthy human subjects are contemplated. The use of stable isotope tracers in conjunction with activation analysis eliminates the radiation hazard to the system. Lowman and Krivet (14) have demonstrated the identical utilization of the stable and radioactive isotopes of iron for determining plasma clearance rates. Stable chromium has been successfully used as a tracer for determining the survival time of red cells in blood (15).

However, due to the previously discussed difficulties encount-ered in the activation analysis of certain biological materials,

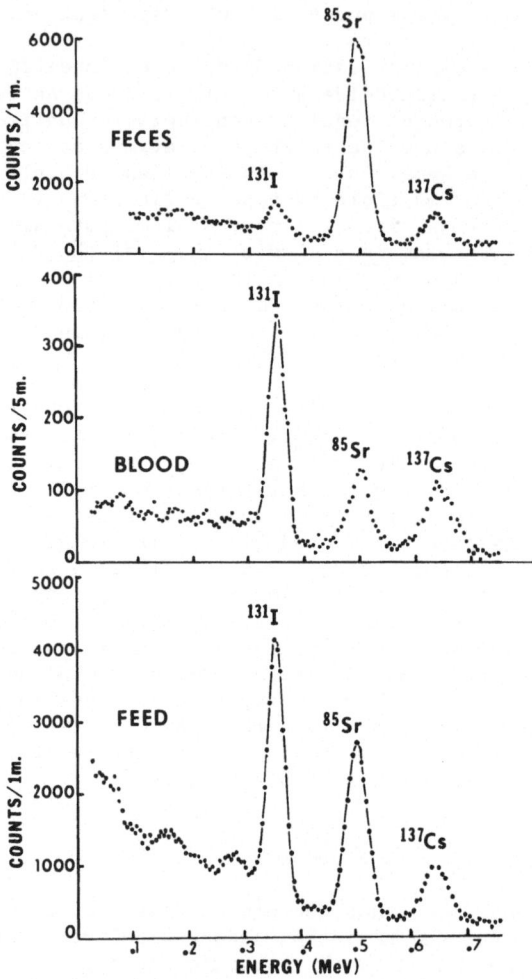

Figure 5. Gamma-Ray Spectrum of a Multiple Radionuclide Doped Feed
          Sample, and Gamma-Ray Spectra of Blood and Feces Samples
          Collected During a Radioisotope Feeding Experiment. All
          Spectra Observed With a Na(Tl) Well Detector. The Gamma-
          Ray Spectra Were Analyzed Using a Least Squares Computer
          Program.

the use of stable isotope tracers usually requires more time and effort than the corresponding use of radioactive tracers. The recent inovations in high resolution gamma-ray spectroscopy and radiochemical activation analysis promises to increase the utility of stable isotope tracers. For example, instrumental neutron activation analysis of a stable cerium inert marker has been used to determine ration digestibility (16).

A number of rare earth elements and other inert metals have been qualified as dietary markers. Radionuclides of these elements have very limited use in human nutritional studies, but their stable isotopes in conjunction with neutron activation analysis may be a feasible alternative. Preliminary studies in our laboratory (17) utilizing high resolution Ge(Li) gamma-ray spectroscopy indicate that at least five inert dietary markers can be determined instrumentally in a single irradiated fecal sample from an animal whose diet had been doped with the elements as markers for various nutrient qualities.

SUMMARY

The study of trace elements is an important aspect of current biomedical research. A wide variety of radiochemical and instrumental activation analysis techniques are available for these studies. In fact, with this array of related techniques, the general sensitivity and broad applicability of neutron activation analysis is probably unmatched by any other trace element analysis method. Utilization of sophisticated gamma-ray spectroscopy has also enhanced the capabilities of radioactive tracer methods and has opened new pathways for investigating the behavior of system components.

These radioisotope techniques have been developed and, for the most part, their biomedical applications have been demonstrated by the physical scientist. He has produced a dazzling array of analytical luxuries and will continue to make novel applications. But the physical scientist, helpful as he has been, cannot contribute more than superficially to biomedical problems without the guidance of biological and medical scientists. If there is to be a really significant contribution to medical technology, the physical scientist must be brought into the company of biomedical scientists and his techniques incorporated into meaningful large-scale biomedical research programs.

LITERATURE CITED

1. E.J. Underwood, "Trace Elements in Human and Animal Neutrition," 3rd ed., Academic Press, New York, 1971.

2. H.J.M. Bowen, "Trace Elements in Biochemistry," Academic Press, London, 1966.

3. J.M.A. Lenihan and S.J. Thompson, (eds), "Advances in Activation Analysis," Vol. 1, Academic Press, New York, 1969.

4. W.S. Lyon, Jr., "Guide to Activation Analysis," Van Nostrand, Princeton, N.J., 1964.

5. P. Kruger, "Principles of Activation Analysis," Wiley-Interscience, New York, 1971.

6. J.R. DeVoe, (ed.), "Modern Trends in Activation Analysis," Vols. I and II, U.S. National Bureau of Standards Special Publications No. 312, 1969.

7. C.E. Crouthamel, F. Adams, and R. Dams, "Applied Gamma-Ray Spectroscopy," 2nd ed., Pergamon Press, Oxford, 1970.

8. G.J. Lutz, R.J. Boreni, R.S. Maddock, and W.W. Meinke, (eds.), "Activation Analysis: A Bibliography," U.S. National Bureau of Standards Technical Note. No. 467, 1971.

9. L.A. Rancitelli, J.A. Cooper, and Richard W. Perkins, in: J.R. DeVoe, (ed.), "Modern Trends in Activation Analysis," U.S. National Bureau of Standards Special Publication No. 312, Vol. I, p. 101 (1969).

10. Unpublished work done at the Research Reactor Facility, University of Missouri, Columbia, Missouri.

11. W.A. Haller, R.H. Filby, and L.A. Rancitelli, Nucl. Appl., Vol. 6, pp. 365-370 (April 1969).

12. R.A. Nadkarni and W.D. Ehmann, Radiochem. Radioanal. Letters, Vol. 4, pp. 325-335 (1970).

13. F. Girardi, in: J.R. DeVoe, (ed), "Modern Trends in Activation Analysis," U.S. National Bureau of Standards Special Publication No. 312, Vol. I, p. 577 (1969).

14. J.T. Lowman and W. Krivet, J. Lab. Clin. Med., Vol. 61, pp. 1042-1048 (1963).

15. P.F. Johnson, P. Tothill, and G.W.K. Donaldson, Intern. J. Appl. Radiation Isotopes, Vol. 20, pp. 103-108 (1969).

16. S.E. Olbrich, F.A. Martz, J.R. Vogt, and E.S. Hilderbrand, J. Anim. Sci., Vol. 33, No. 4, pp. 899-902 (1971).

17. T.D. Luckey and J.R. Vogt, Unpublished work.

18. N. Veall and H. Vetter, "Radioisotope Techniques in Clinical Research and Diagnosis," Butterworth and Co., London (1956).

SPECTROSCOPIC WEAR METAL ANALYSIS IN LUBRICATING OILS

A CONTROLLED MAINTENANCE SYSTEM THROUGH USED OIL ANALYSIS

Edward J. Forgeron

Analysts, Inc.

Rolling Hills Estates, California

It is certainly a pleasure to speak to such a large group interested in oil analysis. It was not long ago that when two people gathered together to discuss this technology, it was considered a crowd. Today we have a room full of people interested in the application of oil analysis as a maintenance control tool.

I would like to speak primarily about our system of 'Controlled Maintenance' because I know it best. You notice that I use the word 'system.' By definition, a system is a regular, orderly way of doing something. A 'Controlled Maintenance' program links together all the people of the company concerned with maintenance in a coordinated effort. We are in the oil analysis business, but primarily we are in the management support business.

The basic management functions are:

1. Plan
2. Organize
3. Staff
4. Direct
5. Coordinate
6. Control

Without a dissertation on all these functions, a statement on the last two is in order. Management coordinates through good communication and controls through money, people, procedures, and techniques. 'Controlled Maintenance' through lube oil analysis is the method, the system, that helps accomplish this.

It is a tool by which management can monitor the exact condition of all the equipment, no matter where it is located. It is also a maintenance control tool which can pinpoint impending failure before breakdown or serious trouble occurs. Equally important, it can tell when the equipment is operating properly and requires no service. It is a quality control tool by which correct maintenance practices can be established. The lab report is a certificate as to the condition and quality of the equipment as delivered by the manufacturer or received from the overhaul facility. The report will also indicate whether the equipment has been serviced properly and completely in the field.

We call it a 'Controlled Maintenance' system because that is exactly what it does. The lab continuously provides accurate condition reports to management and to the maintenance department. Now everyone concerned with maintenance can continuously monitor the operation.

Oil analysis as a method of 'Controlled Maintenance' has come of age. Consider the diversity of the equipment controlled by a lube oil program at ANALYSTS, INC. They include stationary engines, off-highway and line-haul equipment, tractors, graders, shovels, oil rigs, airplanes, railroads, steamships, tugs, air conditioning units, compressors, and hydraulic systems.

The engine manufacturers recommend strongly that you use oil analysis to establish and monitor maintenance practices; the oil companies provide the service to many of their customers and advise the use of oil analysis as a norm for oil changes; the filter manufacturers suggest oil analysis as a means of demonstrating the efficiency of their filters. Everybody talks so well of oil analysis that - like sex -it's gotta be good!

The major problem facing anyone in maintenance is simple diagnosis. When something goes wrong with the equipment it can be repaired, but it would be infinitely less expensive if there were an indication of impending failure before catastrophic or serious trouble occurs. The mechanic usually must rely on concrete outward evidence of a mechanical malfunction because he cannot ask an engine what is wrong. However, with the use of spectrochemical oil analysis, the mechanic can look inside the engine to see what is happening without dismantling.

Trouble-shooting with oil analysis has the tremendous advantage that the equipment keeps working while the diagnosis is taking place. The used oil itself is the carrier and product that is required for the 'Controlled Maintenance' system.

The 'Controlled Maintenance' concept is based on the fact that oil which has been used to lubricate any moving mechanical apparatus for a period of time reflects the exact condition of

that assembly. Oil-wetted parts of an engine or mechanical component wear and deposit metallic-trace particles in the oil --
particles so small they remain in suspension and are not removed
by filtration. Many products of combustion and many contaminants
from the degradation process of the oil are present; thus the
lubricant becomes a working case history of the component.

    Our approach is considerably different from previous oil
analysis in that our tests are tailored to the component and include complete physical tests as well as spectrochemical tests.
We believe and know that all these tests are necessary for the
data analyst to give an accurate diagnosis of the condition of
the equipment. At least fifteen metallic elements such as iron,
copper, lead, etc., are identified and measured precisely in parts
per million for samples from every component. These elements reveal to specially trained analysts if any oil-wetted part is
wearing, and, equally important, if there is no wear. The physical
property tests reveal how the equipment is running. The choice
and selection of these tests is determined by the type and utilization of the equipment.

    Are all these tests necessary for a 'Controlled Maintenance'
program? Perhaps it might be good to use a medical examination as
an analogy. The physician checks the patient's temperature, blood
pressure, and pulse; then makes a diagnosis. This is at best a
cursory examination. The patient might have a serious ailment
that is undetected. The patient might have gall stones, kidney
stones, cancer, or other problems which would not be diagnosed.
However, if the doctor takes X-rays, blood and urine samples,
cardiograms, and makes other sophisticated tests, the chance of
missing a problem is lessened. The diagnosis is accurate. A
clean bill of health is meaningful. With incomplete data, the
surgeon might operate unnecessarily, treat the wrong ailment, or
overlook a serious problem.

    So it is with machinery. The more complete the data available
to the analyst, the more accurate is his diagnosis of the equipment
condition. He can attest that the equipment is in satisfactory
condition, and the oil is suitable for further use, that no service
is required. The machine can continue to operate until such a time
when there is indication that service is required or that it can be
programmed into the shop at the convenience of the operator. If an
abnormal or critical condition is noted, specific maintenance
recommendations are made and carried out by the service department
as soon as practical before serious damage or failure occurs.

    What makes a successful program is the completeness of the
histories and data bank stored in the laboratory files. It takes
many years to acquire the expertise to diagnose malfunctions and impending failures of equipment. The learning process is endless.

The oil is the product to be analyzed; therefore, it is extremely important to know as much about the oil as possible. There are over 3,000 petroleum lubricants analyzed, specified, and classified by brand and application in ANALYSTS' files. The metallurgy and specifications of the engines and components of over 500 manufacturers, gathered over twelve years, are stored in their files and data bank. The competent, analytical team at the laboratory assimilates this backup library with previous histories and compares the results of the sample analysis to provide specific and practical maintenance recommendations for 'Controlled Maintenance'.

One of the purposes of oil analysis is to provide a means of predicting incipient failure without dismantling equipment. This scientific approach to diagnosis is applicable to any diesel, gas, or jet engine, transmission, reduction gear, compressor, hydraulic, or closed lube system.

For each piece of equipment tested, the results are recorded on a permanent record card maintained in ANALYSTS' files or stored in a computer. Each subsequent test is recorded on this card and the composite results, along with specific maintenance recommendations, are airmailed the same day they are received.

Each piece of equipment has its own oil analysis record card held at the laboratory. The card contains all the data from these tests, the maintenance recommendations made by the laboratory, the feed-back information of maintenance findings, and action by the shop personnel.

How does it work? Samples are collected on a regularly scheduled basis or just prior to oil change and airmailed to the laboratory. The sample period is derived as the equipment's operating history is developed. The whole sampling procedure takes approximately one minute.

A truly representative sample is the key to a successful 'Controlled Maintenance' program. The best means of obtaining a representative sample is through a petcock or valve in the oil system located on a flowing line not at the cooler, nor in a gage line, nor downstream from the filter. A normal place to locate it would be somewhere between the pump and the inlet side of the filter or in the galley at a place which is readily accessible. The lube man will then take the sample easily, quickly, and in the same way each time. The oil sample is mailed to the laboratory as soon as possible because the efficiency and success of an oil analysis program is directly affected by the response time of the complete process from the time a sample is taken until it is reported back to the maintenance facility. Samples should be shipped as soon as possible and not gather dust on a shelf, bench, or desk.

Accompanying each sample is information pertinent to the equipment, including data listing time since last overhaul, actual miles or time on the oil, oil consumption, and any maintenance action taken since the last sampling. This information is important to assure a successful program. The lab is like a computer. The more information you provide and program, the more useful, specific, and practical are the maintenance recommendations.

At the laboratory, samples must be processed accurately and quickly. There is instrumentation presently available that provides repeatable analyses, whether it be for the wear elements or physical property tests.

We spoke earlier of determining at least fifteen wear elements and several physical property tests. How much research has been done by the instrument manufacturers on the analysis of the various makes and grades of new and used oils? How extensive have been these analyses on interelement interference? What effect does the presence of organic oil additives have on element determinations? Do the additives enhance and increase the concentration, or do they reduce this number?

These are not idle questions. New formulations and blending of oil packages are coming out monthly. An oil may have thousands of parts per million of barium, calcium, or other additive elements; another may have none. These oils are many times mixed in the same sump. There are also marked differences between synthetics and petroleum based oils.

The oil laboratory has need of more automated instruments to reduce the cost of analysis. New instrumentation should be developed to analyze for the different physical properties. Can we have a multiple fixed wave length infrared spectrophotometer? What applications and determinations are available in the u.v., X-ray, ultrasonic, spectrofluorimetry? How about a rapid, accurate method of measuring total acid number in used oil? These are some of the many challanges for the instrument manufacturers.

Enough about analyses and instruments. Does a laboratory need only good instrumentation and accurate data to offer a 'Controlled Maintenance' system?

Excellent salesmen do a marvelous job in selling their instruments. They convince the customer that all he has to do is to buy this terrific black box and he is in business. Gentlemen, it goes far beyond this. Many instruments may be required. More important, given all the necessary equipment to provide the accurate data, the easiest part of the program is accomplished. What do all these numbers mean? How do we translate these data into specific maintenance recommendations?

May I give you an example. Two used oil samples were taken
from a bulldozer and mailed to different laboratories. The analytical results for the wear metals were very close. High concentrations of silicon, iron, aluminum, and copper were present. The
physical property tests showed high solids.

One laboratory reported the high findings and blowby. They
recommended an immediate overhaul. Our laboratory, with similar
data, recommended a check of the entire air filter system, an oil
and filter change, and flushing the crankcase with new oil. Re-
sample in fifty hours. Our findings and recommendations were right.
The problem was corrected. The difference in cost to the customer
was fifty dollars for the service vs. four to five thousand dollars
for an overhaul.

There are many instances when a problem is undetected through
insufficient data or inexperience. The data analyst must know
machines and their various components. He must have a knowledge
of the different alloys that make up the components. Even with
the assistance of a computer, the diagnosis of unusual problems
still are made by the data analyst.

If you want to get a feeling of concern and humility, I suggest you visit a compressor plant or shipboard installation. Stand
next to an engine or reduction gear two or three stories high,
forty to fifty feet in length, valued at one million dollars or
more. Now look the engineer in the eye and say, "Let's take this
unit down." Or, perhaps, you might be in an airplane hangar when
they remove a JT-9 from a 747 at your recommendation.

Equally difficult is to tell a customer he must replace the
transmission on a used car he has just purchased. It may also be
your own automobile. There is a tremendous burden on the independent laboratory. You must be right if you are to stay in business,
or if your business is to grow.

I have made many points to put oil analysis and 'Controlled
Maintenance' in perspective. This is a technique that works and
can make money for the user. There is no short cut to good instrumentation, accurate, complete data, and, especially, experienced
laboratory personnel.

Lube oil analysis is not a panacea. It does not replace good
maintenance practice or management. If I may make a comment, I
recommend a complete independent laboratory specializing in oil
analysis. This means a whole team is working for the customer.
The laboratory is not selling equipment, parts, oil, or filters.
It is working for the man who pays the bill with no proprietary
interest other than his welfare. If the customer is at fault, he
will know it; if the vendor's product is other than specified or

the equipment defective within warranty, he will have the facts.

Am I sold on 'Controlled Maintenance'? You can bet your sweet duffy I am! I am excited and optimistic for the growth and acceptance of the technology. Your presence here attests to this. Thank you for your attention and interest.

A HIGH REPETITION RATE HIGH VOLTAGE SPARK SOURCE AND ITS

APPLICATION TO THE ANALYSIS OF WEAR METAL IN OILS

W.W. Schroeder, A. Strasheim and J.J. van Niekerk

National Physical Research Laboratory, CSIR, Pretoria

INTRODUCTION

Emission spectroscopy as well as atomic absorption techniques are generally used for the analysis of lubricants. Atomic absorption, being a sequentical method, is a relatively slow but accurate method whereas direct-reading emission techniques provide fast overall analysis of all elements required. This paper deals with a high speed spark source as an excitation medium and emission spectrometric techniques will be considered.

At present there is a demand for fast spark sources in the steel industry. The high repetition rate spark source used in the work discussed in this paper was primarily intended to serve as a source for analyses in the steel industry and it is therefore provided with an unipolar discharge.

Computerisation provides both high speed calculation and accuracy and is increasingly being employed in emission spectrometry. The reasons for speeding up oil analyses are varied. It is felt that the throughput of oil samples per instrument may increase to such proportions that a fast spark source in conjunction with a computer will be advantageous. There remains, of course, the problem of sample preparation which has to be solved if far-reaching automation of the process is to be achieved.

THE SOURCE

The reasons for deciding on a high voltage spark source of high repetition rate are as follows: The normal fast spark sources available at present work at a repetition rate of 400

cycles per second, allowing a vapour residence time of 2 milli-
seconds.   This is the domain of the medium or low voltage spark
sources.   On the other hand, a high voltage spark source with a
discharge capacitance of 5000 to 15 000 pF creates vapour clouds
with a life-time of the order of approximately 100 microseconds.
In this way a repetition rate of many thousands of sparks per
second can theoretically be obtained.

With respect to the analysis of lubricants, it seems that
both low voltage and high voltage sources produce similar detection limits, i.e. below one part per million for many elements.
The Enns-type spark source, in conjunction with the Michigan controlled spark, also provides a high repetition rate, but, as is
well-known, the number of sparks depends on the voltage across the
control gap and on the control gap-width, so this type of source
depends on two variables which can affect its reproducibility.

The pulse-type spark source described in this paper also
makes use of the Michigan control gap.   However, the number of
sparks per second is controlled by an oscillator.   The spark
source produces discharges of equal voltage irrespective of frequency, in sharp contrast to the Enns-type source.   Further, by
suitable design of the pulse transformer, the ignition circuit
and the discharge components, it is possible to produce discharges
ranging from the high voltage, low capacitance spark to the low
voltage high capacitance spark.   Figure 1 illustrates schematically the design of the source.

The design provides a variable frequency from 15 to 1000
sparks per second.   This allows comparisons to be made between
the various spark frequencies and, perhaps, at a later stage,
utilization of this "programmability" feature of the sparking
frequency.   A variable frequency necessitates an aperiodic
electronic design and designs employing swinging chokes etc., are
ruled out.   Larger losses are the penalty however.   In the present design, the three phase mains supply charges a capacitor of
900 μF (Cst) via rectification to a peak value.   This peak value
cannot be more than 450 volts because the capacitors are electro-
lytic.   Another capacitor (Cw) of 40 μF of the oil-filled type
is charged up from the power pool via a resistor of 6 ohms.   The
energy of this working capacitor (Cw) can be discharged with the
aid of a silicon controlled rectifier (Th) into the primary winding of a pulse power transformer (PT).   The combination of 40 μF
(Cw) and 6 ohms (Rw) has a time constant of 240 μsec, which is
short enough to allow a repetition rate of 1000 discharges per
second.   The discharge rate is determined by the uni-junction
oscillator (O) which drives the gate of the silicon controlled
rectifier.   The secondary winding will charge up the high voltage
capacitors (C_H).   The control gap (Cg) is set so that it will

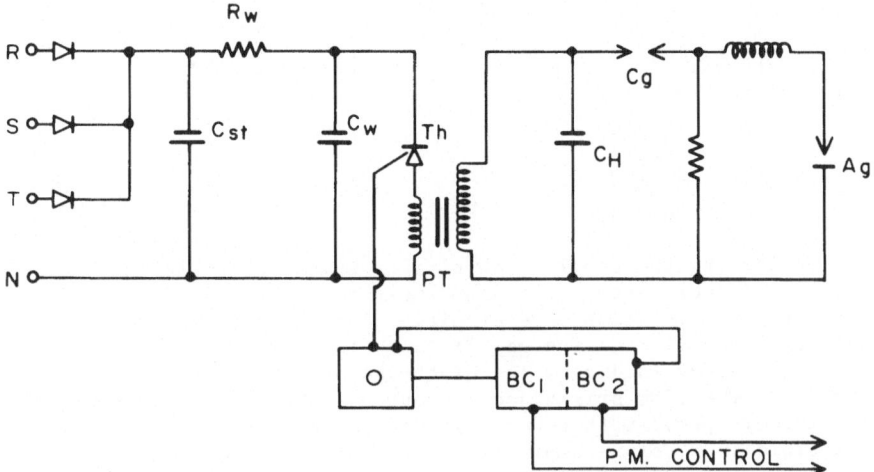

Fig. 1   Pulse-Type High Voltage Spark Source, Schematic.

break down before the crest value on the secondary capacitor is reached.

With the breakdown on the control gap, an ignition pulse is created on the analytical gap (Ag) and the energy will discharge through the inductance and both gaps.   A double counter (BC_1) and (BC_2) for spark counting is provided which is programmable in steps of 512 pulses with a total capacity of 7680 pulses on each counter.   Thus a given number of pre-sparks for sample pre-burn and a given number of sparks for annalytical integration purposes can be selected.

The output signals of the two counters control the photo-multipliers of a direct-reading spectrometer in such a way that during the pre-spark period the photocurrents are grounded and during the analytical period the currents are integrated.   In this way the high repetition spark source must be seen as part of a spectrometric system which is computer compatible and is also suitable for time-resolution if required.

1   W.W. Schroeder, J.J. van Niekerk, L. Dicks, A. Strasheim and H. v.d. Piepen. Spectrochimica Acta, Vol. 26B, pp. 331-340.

APPLICATION

In the experiments described, the following analytical lines were used throughout:

Cu 3247 Å
Al 3082 Å
Fe 2599 Å
C   2478 Å

Concerning the application of this spark source to the analysis of lubricants, it is customary to use a rotrode for oil transport from the sample reservoir. The counter electrode is made from graphite and the gap distance is 3 mm. This small gap is necessary to direct the discharge to the centre of the rotrode. The 1000 sparks/second source tends to produce a wider geometrical spread with repetitive discharges, which is especially noticeable in a point-to-plane configuration with a 6 mm gap. Figure 2 shows a discharge in a gap of 3 mm in oil at a repetition rate of 50 sparks/second.

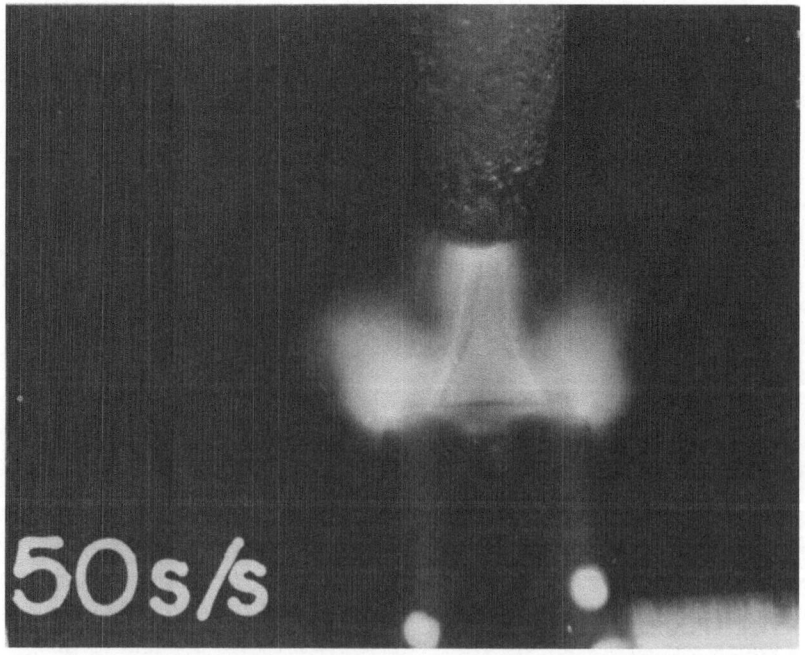

Fig. 2   High Voltage Discharge in Oil, 50 Sparks/Second.

The main channel is directed onto the centre of the wheel
and oil explosions take place which move upwards.   Figure 3
shows the same discharge at a repetition rate of 1000 sparks/
second.

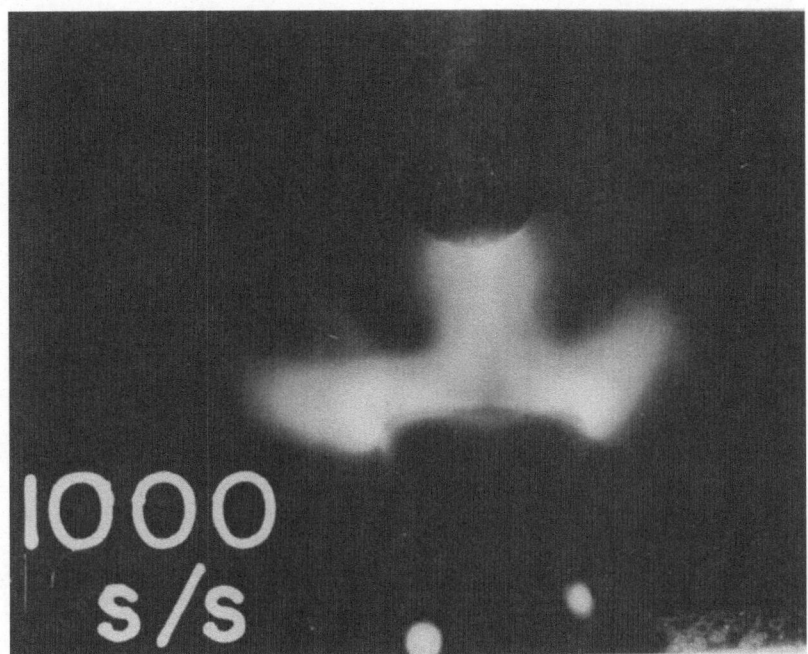

Fig. 3   High Voltage Discharge in Oil, 1000 Sparks/Second

The oil flares tend to move sideways but the centring of the
discharge is maintained.

Due to the greater power of a 1000 spark/second discharge,
the rotrode must turn faster.   With most samples the speed is
about 1 revolution/second.   Care must be exercised, however,
that no oil is thrown out by centrifugal forces.   As a sample
container, a relatively large aluminium block into which a trough
is cut is used.   This block provides sufficient cooling and pre-
vents the sample from igniting.

It must be mentioned that this kind of discharge is in
no way different to any other high voltage oscillating discharge.
Therefore, one cannot expect that there will be great deviations
from the usual parameters such as background reproducibility and
sensitivity for repetition rates from 50 to 1000 sparks/second.
The reason for this is that the single sparks are very short

compared to the repetition rate in the 1000 sparks/second case. The inductance is still the main parameter which determines the sensitivity and reproducibility of the spark source, as is shown in Table I.

TABLE I

Ratio Readings for Different Inductances
and Analytical Lines, Concentration 5 ppm

µH	$\dfrac{Fe}{C}$	$\dfrac{Al}{C}$	$\dfrac{Cu}{C}$
	019	127	180
55	202	111	172
	101	−30	080
	219	111	240
150	247	181	226
	349	246	225
	464	427	498
300	−165	−369	180
	218	−143	128

At a concentration of 5 ppm of iron, copper and aluminium, 6 exposures of 3000 sparks each have been made, using an inductance as indicated in Table I, at a repetition rate of 800 sparks/second.   Each of the figures listed in the table represent the average of two ratio readings and it is evident that at 55 µH aluminium failed once (the negative value), to show up above background.   At 300 µH both iron and aluminium failed (aluminium failing twice); only at 150 µH did no failure occur.   This indicates a property of the high voltage spark source which was previously reported for aluminium analyses.   The precision rapidly deteriorates with smaller or larger than optimum inductance relative to that of the discharge circuit. [2]

Having established the optimum discharge conditions in terms of inductance, the spark-off properties were investigated.   The three elements, iron, copper and aluminium, were recorded with respect to the carbon line as an internal standard.

Figure 4 shows spark-off curves as an average of three readings, each integrated over 1000 sparks for a repetition rate of 50 sparks/second.

---

2   A. Strasheim, F.R. Maritz and I. Kovacs.   XII Colloquium
     Spectroscopicum Internationale, Exeter 1965, p. 299-310.

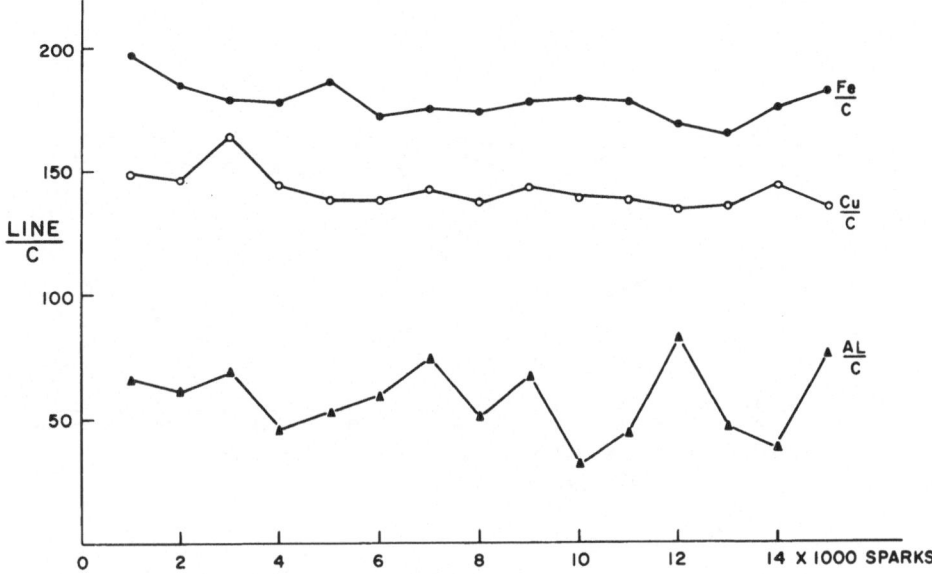

Fig. 4   Relative Spark-Off Curves, Oscillating Spark,
50 Sparks/second.

It is evident that there are large fluctuations, especially
in the beginning, which exceed 15% in the case of copper and iron.
For aluminium the variations are even worse and reach 100%.
Longer exposures - e.g. beyond 12 000 sparks - may seriously
affect the results.

Figure 5 shows spark-off curves for the same samples at a
repetition rate of 1000 sparks/second.

The curves are far smoother and the variations - especially
in the first few thousand sparks - are of the order of 5%.   Even
the aluminium shows far less fluctuation.   These results support
the choice of the region of 0-1500 sparks for spark-off purposes
and a region of 1500 to 3500 sparks for integration.

Table II illustrates the variance for 50 sparks/second and
for 1000 sparks/second at 10, 50 and 100 ppm concentrations for
18 repeat exposures with 500 pre-sparks and 3000 integrated
sparks.

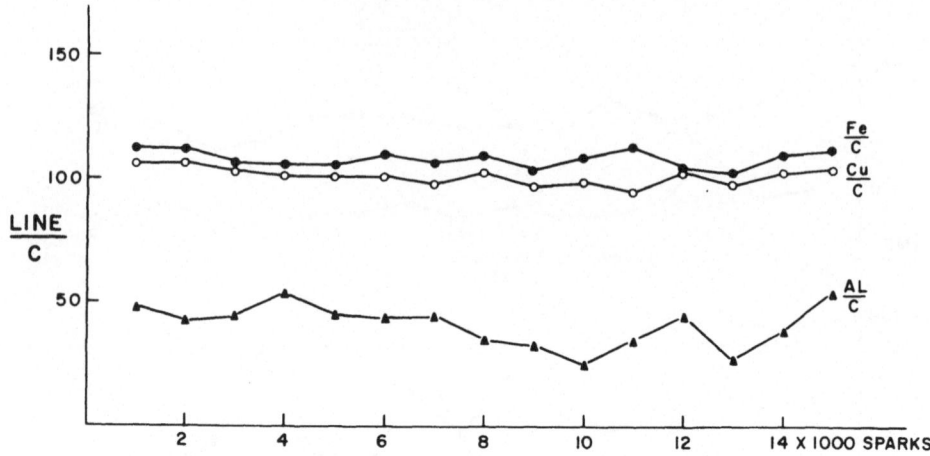

Fig. 5   Relative Spark-off Curves, Oscillating Spark,
         1000 Sparks/second.

TABLE   II

Variances of Different Line Ratios vs Sparking Frequencies
and Concentrations

	10 ppm		50 ppm		100 ppm	
	50 s/s	1000 s/s	50 s/s	1000 s/s	50 s/s	1000s/s
$\frac{Cu}{C}$	0,8	2,3	6,3	3,2	8,1	7,3
$\frac{Fe}{C}$	2.1	1.6	5,7	4,4	7,7	7,1
$\frac{Al}{C}$	5,2	6,9	3,1	6,1	4,4	2,6

There is no decisive difference between the two sets of
results. This confirms the prediction that no difference results
from using the two sparking frequencies.

Figure 6 shows a calibration curve at a sparking frequency
of 1000 sparks/second, a pre-spark period of 1500 sparks, and an
integration period of 2000 sparks.   The total exposure time is
3,5 seconds, being 20 times shorter than that normally used at

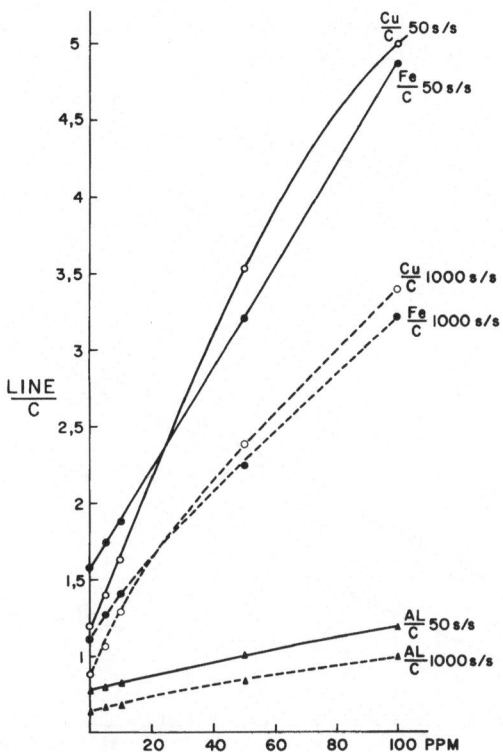

Fig. 6   Calibration Curves vs Sparking Frequency

50 sparks/second and 10 times shorter than required with an Enns
source.   There is, however, some difference in sensitivity for
all the elements concerned.   The sensitivity in the 50 sparks/
second case is higher due to less self-absorption, especially
for copper, although it will probably make very little practical
difference.   The real sensitivity is much more dependent on
other factors, e.g. the type of spectrometer, the size of exit
slits, the suitability of the line, and background reduction by
CO_2 injection.

Figure 7 shows the effect of blowing CO_2 into the gap.   A
significant enhancement of the sensitivity of the aluminium line
is achieved.

As mentioned earlier, this spark source is fitted with a
rectifier in order to produce unipolar discharges.   Figure 8
illustrates the method of introducing a high voltage stacked
rectifier into the source.

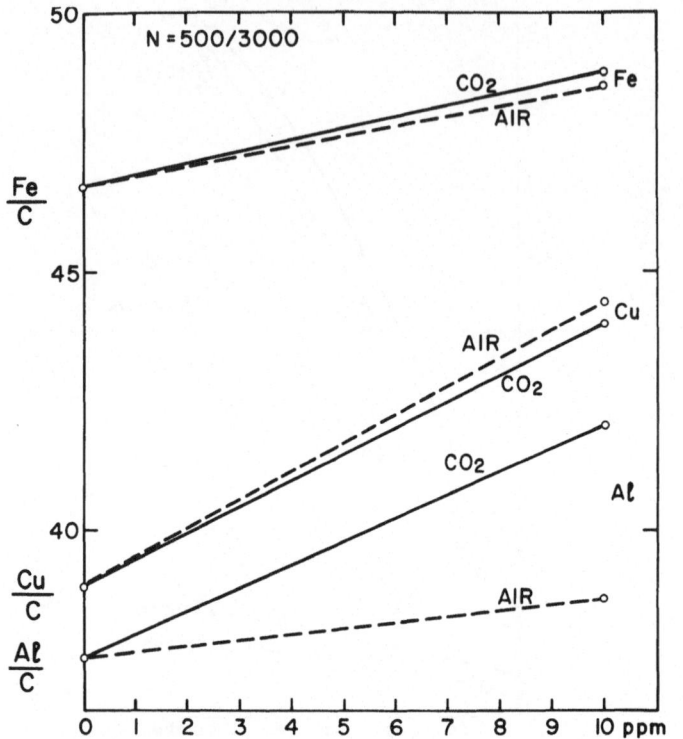

Fig. 7   Effect of CO_2 injection into the analytical gap.

For a series RLC circuit, the critically damped discharge requires –

$$\frac{R^2}{4L^2} = \frac{1}{LC}$$

or

$$R = 2\frac{L}{C}$$

and

$$L = \frac{R^2C}{4}$$

Two cases illustrate this –

(i)   C = 5 nF   (ii)   C = 5 nF

           R = 1Ω R = 200 Ω

           L = 1,25 10^{-3} μH   L = 50 μH

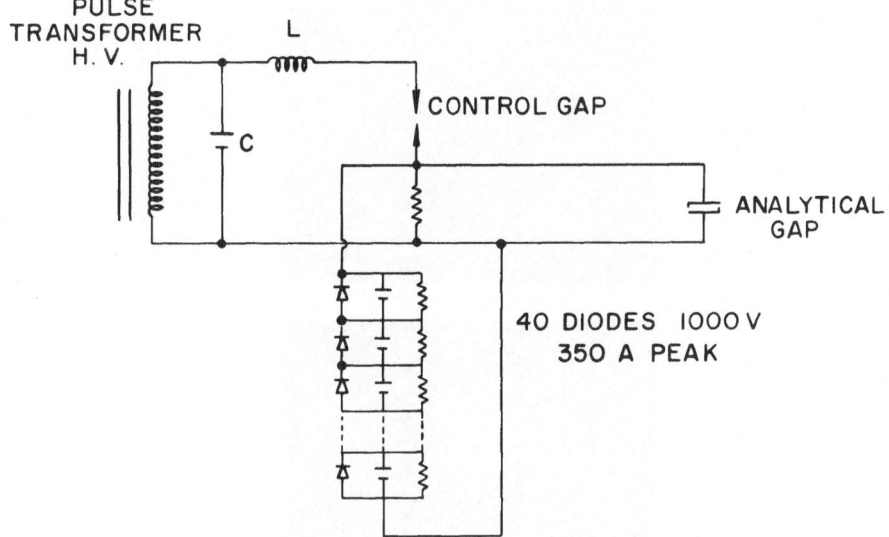

Fig. 8   Rectification of High Voltage Spark

In cases (i) and (ii) the durection of the discharges are only of the order of nanoseconds and 1,5 µsec,respectively. Therefore, we have to rectify the oscillating discharge in order to retain its time duration.

Walters[3] has introduced a rectifier stack in various configurations to the RLC circuit and obtained control over the waveform of the discharge. Some of these discharge forms are unipolar. In contrast, if the rectifier is placed across the analytical gap of this spark source, the current-time behaviour shown in Figure 9 is produced.

Two problems may be noted: The dynamic impedance of the spark gap competes with that of the rectifier stack, so that large currents in the gap bypass it and appear unrectified. Secondly, the gap's total voltage drop, which is again a function of the instantaneous current, competes with the 0,6 volt rectifier drop per rectifier unit, which is a property of a silicon rectifier.

3.   Walters, J.P.   Analytical Chemistry, 40, 1968, p. 1672-1682.

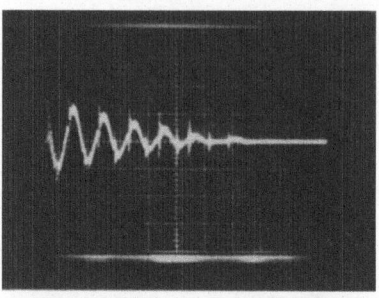

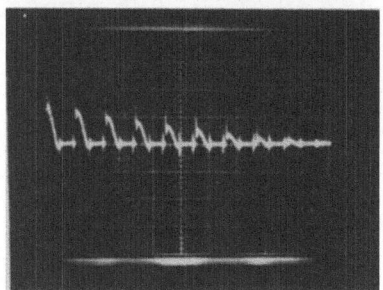

f = 255 Khz

Fig. 9   Rectified Spark-Current

The problem of safeguarding the rectifier against very high
voltage sparks necessitates some sort of bypass circuitry, as is
usual in mains rectification.   What has been done here in the
rectification of a high voltage spark can only be considered as
being a first experiment and further investigations are necessary.

It is interesting to note that some of the energy of the
suppressed half cycles is not lost.   In calculating the current
versus time integrals, one finds that, due to losses, the remaining area of the rectified spark is about 90% of the full oscillating spark area.   This effect is due to storage of the energy of
the short-circuited half-wave in the reactive elements.   This
system, therefore, works with less losses than the critically
damped spark.

It is well-known that the anodic discharge produces very
little vapour and this effect is used in the analysis of steel.
The counter electrode need not be replaced and lasts for many
hours.   This is a vital point when automation is considered.
Unfortunately, this is not the case in the analysis of oils.
Figure 10 shows a calibration curve for the rectified discharge,

again recorded for both the 50 sparks/second and the 1000 sparks/
second cases.

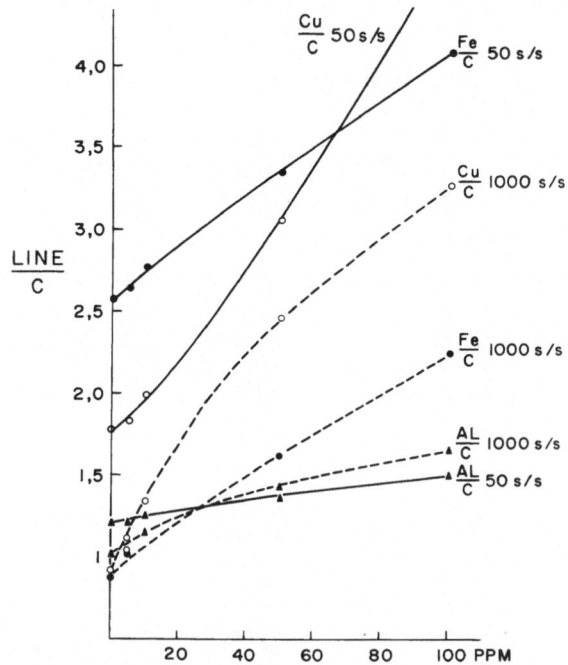

Fig. 10   Calibration curve of a rectified spark vs sparking
frequency.

It is quite evident that the rectified spark has a higher
background for copper and iron in the 50 sparks/second case.
Also, the sensitivity is less for copper due to enhanced self-
absorption.   On the other hand, there is some enhancement for
aluminium in the 1000 sparks/second mode.   Table III shows that
the rectified spark does not do away with contamination of the
counter electrode.

TABLE   III

Contamination of Counter Electrode Cu/c Ratio

	Oscillating Spark	Rectified Spark
After contamination with 10 ppm Cu 24 000 sparks	2,09	1,88
Blank sample, new rotrode	1,45	1,58
Blank sample, new rotrode and new counter electrode	1,35	1,43

The rectified spark shows en even higher reading of contamination than the oscillating spark. The reason for this is most probably the flooding of the counter electrode with oil and consequently with masking of the anodic effect.

A comparison of results obtained with the various sources available is shown in Table IV.

TABLE   IV

Comparison of Sources

High Repetition

	1000 s/s HV	Enns 700 s/s HV	Unipolar 25 s/s LV
Total time	6,5 sec	30 sec	60 sec
Spark-off	500 sparks	7000 sparks	250 sparks
Integration	6000 sparks	14000 sparks	1250 sparks
Intensity ratio variance			
Fe$_*^*$ 10 ppm	3,2%	Unknown	2,6%
Al$_*$ 10 ppm	8,4%	Unknown	6,2%
Cu 10 ppm	6,5%	Unknown	7%
Concentration variance			
Fe 10 ppm	10%	9%[**]	7.2%
Al 10 ppm	32%	3%[**]	11,4%
Cu 10 ppm	6%	6%[**]	10,6%

[*]   Analytical lines: Fe 2599 and Cu 3247 for all three sets of calculations; Al 3082 for the high repetition rate source, and Al 3961 for the remaining two.

[**]   Jarrell-Ash Information.

Regarding the results on aluminium, the comparison is not quite fair as different lines had to be used owing to the low dispersion of the spectrometer. If a better aluminium line were used, an overall exposure of a smaller number of sparks could be achieved, perhaps resulting in sufficient precision from from a total exposure time of 3 seconds.

CONCLUSION

These measurements were performed with a rather low energy source of experimental design and higher energies are certainly required for many practical applications. It has not been the aim of this paper to produce optimised analytical results, but rather to show that the pulse-type spark source can be used to advantage.   It is probably advantageous to extend the principle of the pulse-type source into the medium voltage and higher capacitance range.   This type of source can thus be made very flexible so that the available vapour residence time can be matched with the spark frequency for a desired analytical procedure.

Considering that a computer and a fast spark source can now be incorporated into a system which provides analysis times of only a few seconds, there is now a need to find methods of automating the process of sample introduction.   This would greatly increase the volume of work which could be handled on a single spectrometer.

WEAR METAL ANALYSIS

OF LUBRICATING OILS

Bernard B. Bond

Technical Support Center
Department of Defense Equipment Oil Analysis Program

"Wear Metal Analysis" is the process of monitoring items of operating equipment for mechanical condition by analyzing their fluid lubricants for trace metals content. In the narrow sense, this definition applies only to the analyses and judgments of the monitoring laboratory; in the broad sense, it applies to the complete cycle of operations involved in the monitoring process. Since it would not be possible to treat this subject completely, and in detail, in one short paper, this discussion will necessarily be limited to a brief summary of the essential requirements, plus a progress report on the Department of Defense Equipment Oil Analysis Program.

PRACTICAL APPLICATIONS OF WEAR METAL ANALYSIS

Wear metal analysis can be successfully applied to almost any kind of mechanical system having a recirculating, oily or oil-like lubricant. The original application was made to diesel-powered locomotives; the second to aircraft equipment, including reciprocating engines, jet engines and separately-lubricated gear boxes and transmissions. It is also being successfully applied to ships equipment, various kinds of ground equipment and radar installations.

DESCRIPTION OF METHOD

In theory, wear metal analysis is simple. The components of a mechanical system move against each other. The frictional

contact causes wear, regardless of lubrication, so that tiny metal
particles separate from the contacting surfaces and remain suspended in the surrounding lubricant. A spectrometric laboratory ana-
lyzes successive samples of the lubricant, identifies the metal
particles (appropriately called wear metals) and monitors their rate
of increase. When any abnormally high rate of increase is detected,
the laboratory relates it to the probable source within the equipment and warns the equipment operator. Having been warned, the
operator examines the equipment, finds the beginning failure and
takes appropriate maintenance action, thus avoiding the consequences
of ultimate equipment failure.

The most important wear metals are iron, aluminum, magnesium,
copper and silver. Of these, iron is by far the most important,
since it appears as an indicator in approximately 86 percent of all
failures detected by wear metal analysis. Iron is an indicator in
failures of anti-friction bearings, gears, shafts, piston rings and
cylinder walls. Aluminum usually relates to wear of oil pumps, cases,
housings, pistons and cylinder heads; magnesium to cases and housings;
and copper to bronze parts such as bushings and retainers. Silver and
copper, in combination, are frequently useful as preliminary indicators of anti-friction bearing failures. Silicon is useful only as an
indicator of lubricant contamination by air-borne sand or by too much
silicone additive.

APPLICATION OF METHOD

In practice, wear metal analysis is both workable and effective, but it is not simple. It consists of three sequences of
operations, each of which must be performed meticulously, and all
of which must be completed within a limited time frame if the
total operation is to be effective.

The sampling sequence, performed by the equipment operator,
consists of taking samples, identifying and documenting the samples
and transporting them to the laboratory. Each sample must be taken
in accordance with a prearranged schedule. It must be so taken that
it will be both representative of all the oil in the system and free
of outside contamination. It must be positively identified with the
equipment from which it is taken, and it must be documented with the
related operational history of the equipment. And finally, it must
be expeditiously delivered to the monitoring laboratory.

The analytical sequence, performed by the laboratory, consists of analyzing the samples, relating the analyses to previous
analyses to establish wear metal concentration trends, identifying
abnormal rates of wear metal increase and relating them to probable
sources within the equipment, and finally, advising the equipment

operator or customer of appropriate maintenance actions. All of
the analyses must be accurate enough so that abnormal can be distinguished from normal wear. The laboratory must have adequate
technical guidelines both to detect and assess abnormal wear and
to translate the assessment to meaningful advice to the customer.
And finally, the laboratory must be able to communicate quickly
with the customer.

The final sequence, performed by the customer, consists of
applying the laboratory advice, and of providing the laboratory
with pertinent feedback information. The customer must religiously
follow the maintenance advice offered by the laboratory, and the word
"religiously" is used advisedly since experience has clearly shown
that believers benefit; non-believers do not. Continuous feedback
information is necessary, both from the customer and from the overhaul activity, so that the evaluation criteria used by the laboratory can be continuously updated.

LABORATORY REQUIREMENTS

Since the laboratory is the *sine qua non* of the wear metal
analysis operation, each of the critical factors affecting its
ability to perform its function should be given individual attention. These factors, which consist of physical location and communications facilities, analytical instrumentation, calibration
standards, technical data relating to the equipment being monitored, record-keeping capability, and personnel, will be discussed
separately.

Accessibility

The first critical factor is the location and communications
ability of the laboratory. The laboratory must be so located in
relation to the customer, and must have such communications facilities, that an adequate sample response time can be provided. Sample response time is the total elapsed time between sample taking
and the receipt by the customer of advice related to the sample,
and the requirements vary with the type of equipment being monitored. For military aircraft engines and transmissions, for example,
the maximum response times considered necessary are 24 hours for
routine samples and 2-3 hours for emergency samples. This means
that the laboratory should be located not further than about two
hundred miles from each customer, and that both telephone and teletype communication facilities must be immediately available at all
times.

The Analytical Instrument

The second critical factor is the analytical instrument. The laboratory must have an instrument capable of analyzing for the metallic elements that are important wear metal indicators. It must have good enough accuracy, repeatability and range so that analyses of consecutive samples from the same equipment will reliably indicate minor changes in the rate of wear metal generation. Also, for efficiency and economy of operations, it should provide rapid simultaneous analyses of all elements, with automatic printout of the results. In addition, to provide for statistical use in a large program, the results should be automatically recorded in punched cards or tape. So far, of the several possible instrumental approaches to wear metal analysis, only the emission spectrometer and the atomic absorption spectrophotometer have had large scale application.

The emission spectrometer provides adequate accuracy and repeatability. It has the advantages of relatively good stability, of providing rapid simultaneous analyses of all required elements and of allowing automatic transcription of the results to paper and to punch cards or tape. It has the disadvantages of high relative cost, very complicated electronics systems, and until very recently, a requirement for close environmental control.

The atomic absorption spectrophotometer has the advantages of relative simplicity, relatively low cost and superior accuracy and repeatability. At this point, it cannot provide analyses of all elements simultaneously, and consequently, requires manual assembly and recording of the results. It requires frequent standardization, and it has an explosion hazard that makes it unsuitable for use at some laboratory locations.

A newcomer to the field is the atomic fluorescence spectrometer. This instrument, based on research by Dr. J. D. Winefordner of the University of Florida, promises to combine the rapid simultaneous analyses and automatic data transcription of emission with the accuracy and precision of AA. Still to be evaluated in service, this may prove to be the instrument of choice in the future.

Calibration Standards

The third critical factor is calibration standards. Calibration standards are very important since they provide the only means of assuring the comparability of sample analyses, whether from the same instrument at different times or from different instruments. The characteristics of good calibration standards are as follows: (a) they must be accurate, i.e., they must contain all the elements

of interest within specified tolerances; (b) they must cover the
concentration ranges of interest; (c) they must be stable, so that
they should not contain volatile, unstable or mutually reactive
components; (d) they must be uniform as to viscosity and minimum
flash point; and (e) they must be continuingly available from a
single source of supply. At this time there do not appear to be
any calibration standards that completely meet all of these
qualifications.

Technical Data

The fourth critical factor is technical data relating each
model of equipment being monitored to each kind of wear metal.
These data should include normal concentration or threshold limits,
rate-of-increase limits per standard sampling period, the most
probable sources of wear metals within the equipment and information concerning failure trends of the equipment.

Record Keeping

The fifth critical factor is record-keeping capability. The
laboratory must maintain such files that up-to-date historical and
analytical data will be instantly available for each item of operating equipment being monitored.

Personnel

The sixth and last critical factor, but not the least important,
is personnel. The laboratory must be staffed with personnel having
several skills, although not all of the skills need be combined in
any one person. An essential skill, which should be possessed by
at least two persons in any laboratory, is the ability to evalu-
ate the analytical results and to translate them into meaningful
advice to the customer. Another essential skill is that of operating the spectrometer and analyzing the samples. Other skills,
which may or may not be necessary, are those of operating filing
equipment, typewriter, card punch and teletype. And finally, the
laboratory must have enough personnel to handle the assigned
workload.

THE DOD EQUIPMENT OIL ANALYSIS PROGRAM

The Department of Defense Equipment Oil Analysis Program is
composed of three programs operated by the Army, the Navy and the
Air Force, and known individually as "ASOAP" for Army, "NOAP" for

Navy and "SOAP" for Air Force. The three programs currently operate
114 oil analysis laboratories; all of which are shore-based, and
most of which serve customers of only one military service. Under
the present DoD plan, there will eventually be 109 laboratories, 81
shore-based, 25 ship-based and 3 installed in self-contained, de-
ployable trailer vans. The ship-based laboratories will necessarily
serve only ship-based equipment, and the vans are intended primarily
for Army field units, but the shore-based laboratories will serve
customers assigned on the basis of sample response time instead of
military service affiliation.

                            Program Progress

     The DoD program is a relatively recent development. It was
superimposed on the individual Service programs by the Department
of Defense for the purpose of increasing effectiveness and decreasing costs; and these results were to be attained by coordinating effort and by standardizing both equipment and operations to
the maximum extent possible. The following progress has been made
to date in implementing the DoD operation:

     (1) A Program Office has been established in the Naval Material
Command in Washington, with the function of coordinating program
policy and of administering the entire DoD program.

     (2) The nucleus of a Technical Support Center has been estab-
lished at the Naval Air Station, Pensacola, Florida, with the function of providing all technical information and services required
for the DoD program.

     (3) A Logistics Support Activity has been established in San
Antonio, Texas. Administered by the Air Force, this activity will
provide spare parts for the new spectrometers now under contract.

     (4) A Training Activity for spectrometer operator/evaluators
has been established, and is now operating at Fort Lee, Virginia.
Administered by the Army, this school currently operates on a two-
shift basis, and is turning out graduates at the rate of six every
two weeks.

     (5) Detailed procedures and instructions have been prepared by
the Technical Support Center to cover most aspects of the DoD
program.

     (6) Minimum requirements have been established for laboratory
certification. These requirements apply to instrument capability
in terms of analytical matrix, analytical range, accuracy and

repeatability, and to personnel capability in terms of evaluator training and experience.

(7) Sites have been selected for the 81 shore-based laboratories.

(8) Emission spectrometers conforming to specification MIL-S-83129 have been contracted for, and are now being delivered to laboratory sites at the rate of approximately six per month.

(9) Some progress is being made in readjusting laboratory work-loads on the basis of sample response time instead of military service affiliation. This transition is slow and will continue to be so, since continuous customer service must be maintained in spite of the many complicating factors involved.

OTHER PROGRAMS

Although the Department of Defense is the largest user of wear metal analysis, the method is being used by others on an increasing scale. In the United States, other government users are the Coast Guard and the National Aeronautics and Space Administration. In-dustrial users include most of the major railroads, most of the major airlines and some of the aircraft manufacturers. At least one heavy equipment manufacturer offers it as a continuing customer service, and the number of commercial laboratories offering the service is growing rapidly. Outside the United States, wear metal analysis is now being used in several countries, including Australia, Canada, Denmark, England, France, Italy, South Africa, South Vietnam and Turkey. Still experimental in some of these countries, it is far advanced in others.

CONCLUSION

In conclusion, it can be stated categorically that wear metal analysis is here to stay. Less than one generation from infancy, it is today a giant, and is still growing. Uniquely and universal-ly applicable to mechanical power equipment maintenance, it is ideally attuned to our mechanistic civilization. It has become one of our most powerful and profitable diagnostic tools, and, as such, can be counted as one of the major developments of applied spectroscopy.

AN IMPROVED COMPUTER-DIRECT READER SYSTEM FOR THE ANALYSIS OF WEAR METALS IN LUBRICATING OILS AND ITS APPLICATION TO OTHER TYPES OF SPECTROGRAPHIC ANALYSIS

P.D. Coulter, N.L. Bottone, and
H.W. Leggon
Union Carbide Corporation

Carbon Products Division, Parma Technical Center
Parma, Ohio 44130

In the early 1940's railroads started using wear metal analysis for maintenance control of diesel locomotives. By the mid-1950's the Navy was employing the technique for aircraft maintenance. During the 1960's equipment manufacturers started developing emission spectrographs specifically designed to analyze lubricating oils for wear metals. More recently a great deal of interest has been shown in developing emission spectrographs capable of performing very accurate and reproducible low level analysis of wear metals. Not surprising, the instrument manufacturers discovered as they sought to tighten the tolerances on the spectrographs that the quality and properties of the spectrographic electrodes employed affected the precision and accuracy of the equipment. During this time we became interested in evaluating spectrographic electrodes for their performance characteristics in wear metal analysis. We became further involved with this effort when the government required that each lot of electrodes sold be certified against a standard lot which they had selected as being particularly good.

The primary objective of the Analytical Group at the Parma Technical Center is to support a wide variety of research and development programs both for the Carbon Products Division and for the Consumer Products Division of Union Carbide. We perform very few repetitive or truly routine analyses. Our direct reading spectrograph is not reserved for the analysis of only one or two different matrices, but is employed to analyze a wide variety of materials. New materials are added at intervals to our repertoire. Therefore, we need a versatile direct reader, which can be switched from one program to another with ease. Some problems are encountered when switching the programs of the

direct reader because there is a tendency for the sensitivity of
photomultiplier tubes to drift slowly for several days after a
change of the dynode voltages, and likewise, because the source
and associated electronics show some instability for several days
after program changes. This results in either decreased flexibility while waiting for equilibrium to occur, or decreased accuracy and reproducibility if the instrument is used for analysis
before equilibrium has occurred. To circumvent these limitations and to facilitate the evaluation of electrodes, and also their
certification, a computer direct reader system was developed.
It was described previously in a preliminary form at the 1969
National SAS Meeting. Since that time we have refined, improved,
and added additional capabilities to the system.

    The spectrograph employed in these studies was a Jarrell-Ash
1.5 Meter Atom Counter shown in Figure 1. The scanner-coupler
readout system was modified to allow simultaneous punching of
paper tape and typewriter printing of the data. The spectrograph
is equipped with an automatic background corrector and a special
matrix box which allows the readout program to be changed
readily. The computer employed was an IBM 1800 equipped with
magnetic disks, magnetic tape, a paper tape reader/punch, a line
printer and a card reader/punch. The source conditions used for
oil analysis are shown in Figure 2. The analytical lines used for
the various elements are shown in Figure 3.

    The computer program is composed of a series of sub-pro-
grams which are called as indicated on a control card which pre-
cedes each set of data. Data can be read from paper tape in the
form they are punched by the spectrograph or they can be read
from cards. Paper tape is used whenever possible to avoid transcription errors; however, the ability to read cards allows the
computation of data obtained on other direct reading spectrographs
or densitometer readings from photographic plates made on our
Jarrell-Ash 3.4 Meter Ebert photographic instrument. Any data
sets read from paper tape which contained hardware errors are
automatically deleted and indicated on the printout. After data
input and any deletions have been accomplished, several sub-
programs may be called. If the data are calibration points, the
calibration program is called. A card containing the concentration
of the elements for each burn is read. A first, second, or third
order equation is fitted to the data. Each data set is systematically
removed and the change of the relative error is observed. Any
data set which reduces the relative error by a factor of three is
thrown out. This is illustrated in Figure 4. We thought that this
provision would be very useful but oil analysis data has been so
reproducible that no eliminations have occurred in several years.
A first order equation is preferred, but if going to a second or
higher order equation reduces the relative error by forty per cent,
the higher order equation is accepted. The sequence is performed

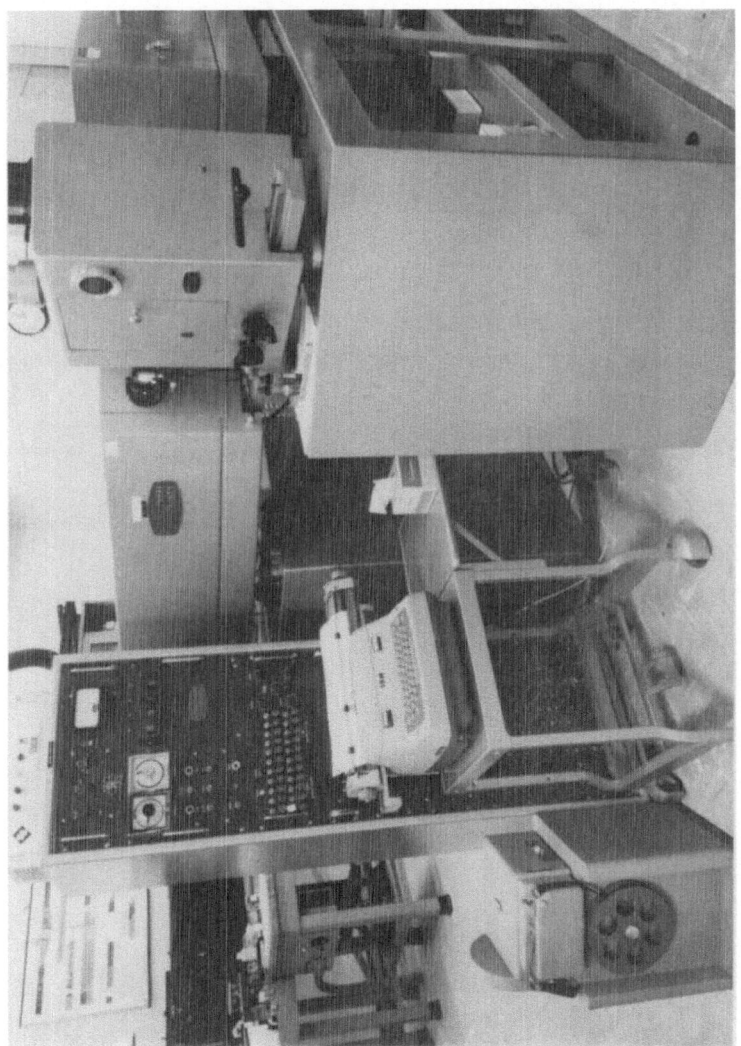

Figure 1. Spectrograph with Data Acquisition System.

Excitation Conditions	Exposure Conditions
Source: High Voltage Condensed AC Spark	12 second prespark
	Exposure cut off on
RF Current: 5 Amperes	4861.3 A H line (~40 seconds)
Capacitance: .00625 μf	
Inductance: 155 μh	25 μ entrance slit
Secondary resistance: Residual	
Breaks/half cycle: 5	Lines corrected for background
Analytical gap: 4 mm	
Auxiliary gap: ~3 mm	Draft - least amount possible
	Rotating disc attachment with tantalum holder for discs - 30 rpm

Figure 2. Excitation and Exposure Conditions for Oil Analysis.

Element	Wavelength (Å)
Fe	2599.39
Mg	2795.53
Al	3944.03
Cu	3273.96
Ni	3414.76
Sn	2839.99
Cr	4254.35
Pb	2833.07
Ag	3280.68
B	2496.78
Si	2881.58
Ti	3372.80
V	3102.30
Na	5895.92
H	4861.33

Figure 3. Wavelengths for the Various Elements in the Oil Analysis Program.

for each element. This statistical approach is patterned after
that developed by Margoshes while at the National Bureau of
Standards. In earlier variations of our program, we would try
fits up to the fifth order. However, we found that the time necessary to perform the curve fits, even on a high speed digital computer, was prohibitively long if there were very many calibration
points. Since we routinely make at least three burns for each concentration, and usually have five different concentrations resulting in a minimum of fifteen points for the calibration curve, we
found that we could not tolerate the luxury of checking fourth and
fifth order equations. As it is now, it takes the computer approximately a minute and a half per element to determine the best
calibration curve. These provisions for ease of calibration permit recalibration before running each batch of samples, insuring
greater accuracy while allowing a high degree of flexibility in our
programs.

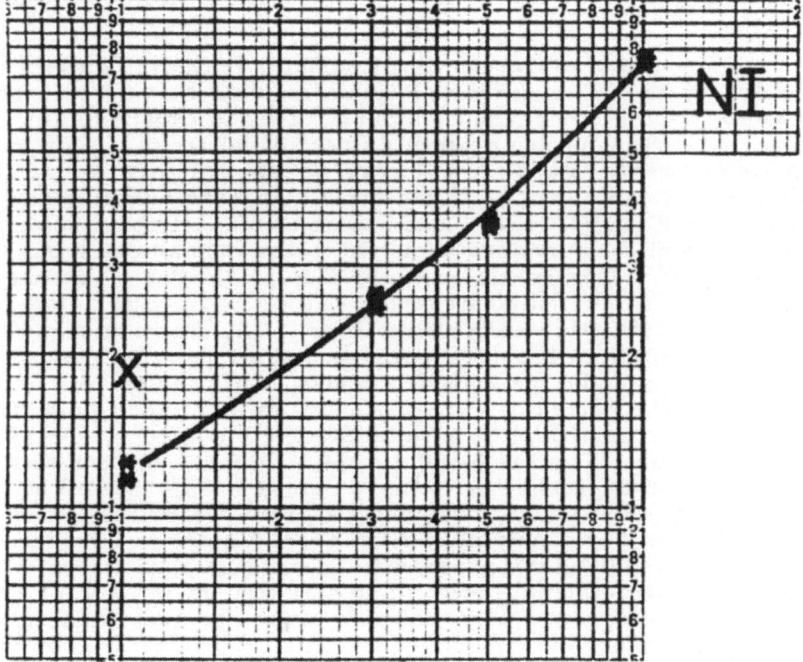

Figure 4. An Example of a Discarded Point in a Calibration Curve.

After the calibration equations for each element have been determined, the plotting of the data points and the least square fit can be made on a Calcomp plotter. An example is shown in Figure 5. We have found that this option can be an invaluable asset in establishing the proper calibration curves during method setup, and also for monitoring the performance of the spectrograph. If the shape of the curve starts to change, this indicates to us that our background correction could need readjusting. If the scatter of data points increases, this generally indicates that the auxiliary gap electrodes need to be replaced. We routinely use new auxiliary gap electrodes before each calibration run. Trends and small changes are more easily discerned from the plots than tabulated data. Consequently, we rarely calibrate the instrument without obtaining a plot of the calibration curves. Following the calibration, the calibration equations can be optionally stored in one of fifteen different disk files. One or more of these calibration curves may be used subsequently for calculating data. This feature has been very useful for studying variations arising from the preparation of calibration curves under different conditions.

If the input data are not calibration data, they can be calculated in several ways. The simplest is just a calculation of the concentration for each element for each burn. This is illustrated in Figure 6. If the data are grouped in replicate sets of two or more, they can be averaged and the standard deviation determined. The replicate sets can either be of the same size or varied in size as specified on a control card. It is possible to handle quintuplicates on some samples intermixed with triplicates or duplicates on other samples. This is illustrated in Figure 7. Paired or interleaved testing can be calculated when two or more materials are tested in a systematic and cyclic manner. The data are organized by the programs into groups according to their position in the cycle. The mean and the standard deviation of each group are calculated. The paired testing procedure is that specified by the government for certification of rotating disc electrodes. An example is shown in Figures 8 and 9. We have found the extension of this interleaving technique to more than two materials to be advantageous in the study of various electrode materials. It is possible to run, for example, five different materials in order many times, and have the programs separate the data into separate groups and calculate means and standard deviations for each element for each group. Slow changes in the spectrographic response do not affect the comparison. This option has allowed us to make comparisons between lots of electrode materials and likewise, comparisons within a single lot which has been separated according to some parameter of the electrode, thereby allowing us to investigate the influence of various electrode parameters on their behavior in wear metal analysis. These studies have not been completed but may be presented at a future meeting.

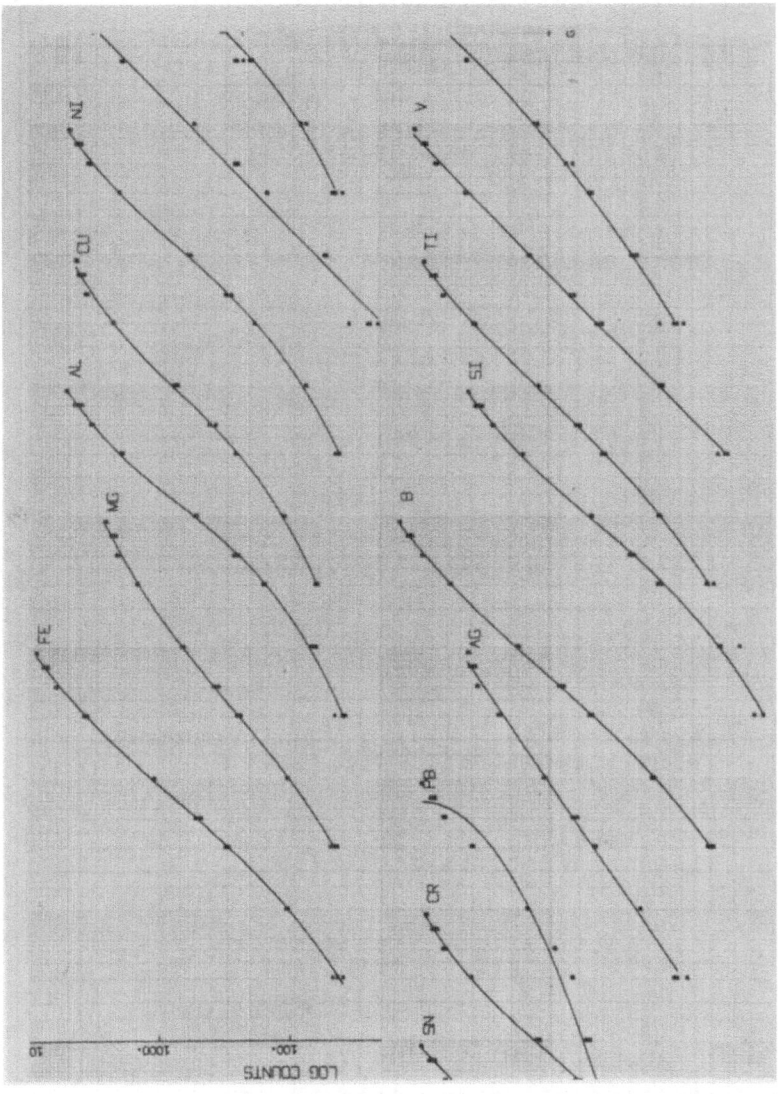

Figure 5. Fourteen-Element Oil Calibration Curve.

REFERENCE LOT

RUN NO.	FE	MG	AL	CU	NI	SN	CR	PB	AG	SI	N
2	30.5	34.3	32.2	31.4	32.1	30.6	32.1	30.7	31.5	31.2	30.0
4	30.5	34.7	32.6	31.8	32.3	30.2	31.0	30.3	31.6	31.6	31.4
6	29.1	33.2	28.4	29.4	29.5	30.4	28.6	32.5	29.1	30.0	24.4
8	28.0	31.3	28.6	30.6	30.3	30.4	30.1	31.6	30.5	30.5	
10	30.8	34.9	30.5	31.4	31.3	30.6	31.0	31.1	31.5	31.2	29.0
13	31.0	35.8	32.4	32.8	32.5	30.0	32.5	30.0	33.6	31.2	33.8
14	27.8	30.9	29.3	31.2	30.1	28.7	29.7	33.9	29.8	29.7	27.1
16	26.1	28.3	30.8	31.3	30.6	28.2	30.1	34.2		32.0	
18	26.9	28.5	31.2	31.8	32.2	29.6	30.1	35.6	29.1	28.3	27.8
20	28.4	27.8	31.3	30.1	31.5	29.5	30.0	36.0	28.0	32.7	
22	24.1	24.6	26.5	29.1	29.5	26.5	27.9	29.7	26.5	26.0	26.4
24	33.5	34.0	27.3	30.4	30.7	26.0	28.0	32.3	25.1		
26	26.5	27.2	29.6	32.0	31.9	28.7	30.7	30.7	29.1	27.7	29.8
28	26.2	27.7	27.7	30.1	30.6	30.1	28.8	33.1		31.3	
30	25.9	27.1	27.2	29.7	30.4	29.1	27.9	33.0	27.6	27.7	24.2
32	24.1		19.7	22.6	26.6	24.0	24.3		27.3		
34	26.1	27.0	24.1	31.3	30.3	27.1	28.0	27.9	30.2	27.7	35.3
36	25.8	26.0	23.3	30.3	31.2	25.8	28.0	26.3	29.0	28.0	
38	21.0	21.0	18.4	25.4	23.5	22.8	22.0	27.2	24.4	23.5	13.7
40	25.8	26.0	22.7	31.6	26.2	26.3	28.0	26.7	30.1	33.0	
42	25.6	25.3	29.1	31.5	32.2	27.8	31.5	27.4	28.8	27.4	36.1
44	26.3	27.4	25.6	30.9	33.2	27.0	28.1	34.1	26.3	27.2	
46	27.1	29.2	23.6	30.3	28.2	28.4	26.7	31.8	30.3	28.0	24.5
48	24.0	24.0	20.4	28.6	26.9	24.5	28.3	34.1	26.3	25.7	20.5
50	24.8	25.2	28.9	28.8	29.4	29.1	26.7	38.8	25.8	26.0	22.6
52	24.8	25.2	24.1	28.6	29.4	28.5	27.2	40.8	25.2	24.3	21.0
54	25.2	26.6	30.3	28.4	29.4	31.0	27.4	41.4	26.2	26.8	22.2
56	24.0	19.8	36.7	34.0	33.7	33.2	34.0	38.9	30.0	38.5	38.3
58	27.3	28.9	34.5	30.0	31.9	32.4	29.7	45.2	27.2	27.7	13.7
60	27.7	26.5	37.3	31.4	33.5	32.6	31.8	43.7	32.4	36.3	37.7

Figure 6. An Example of Calculated Concentration Data.

SPECTROGRAPHIC ANALYSIS REPORT

PAGE 0

9/29/71

CHARGE NUMBER 454-003
GROUPING OF DATA INTO REPLICATE SETS— TO CONTAIN AVE AND STD DEV   TAPE 71

RUN NO.	FE	MG	AL	CU	NI	SN	CR	PB	AG	B	SI
1	45.6	46.3	45.6	57.1	57.1	36.6	54.6	8.6	57.6	0.0	69.3
2	44.8	46.4	41.9	54.4	54.5	36.1	50.8	7.9	55.6	0.0	67.9
3	44.8	47.7	40.0	53.5	53.5	36.5	50.2	8.8	54.5	0.0	67.9
4	47.9	45.2	50.5	60.5	61.6	39.0	59.1	7.9	60.1	0.0	71.1
5	46.0	52.9	37.6	53.1	56.0	36.8	51.3	8.0	55.8	0.3	67.1
AVERAGE	45.84	49.33	43.16	55.75	56.57	37.06	53.25	8.30	56.78	0.10	68.71
STD DEV	1.2784	2.0825	5.0657	3.0991	3.1684	1.1327	3.7101	0.4489	2.2114	0.1204	1.5598
6	8.8	5.8	4.7	5.4	8.4	4.9	8.5	0.6	9.0	0.1	13.1
7	8.7	6.6	4.0	10.5	10.4	4.6	11.4	0.2	9.2	0.2	13.1
8	8.3	6.4	4.9	6.2	8.2	6.2	8.1	1.0	8.0	0.2	13.1
AVERAGE	8.60	5.72	5.52	9.41	9.07	5.25	9.41	0.66	8.80	0.22	13.15
STD DEV	0.3021	0.2262	1.8168	1.1570	1.1861	0.8571	1.7861	0.3977	0.6536	0.0388	0.0032
9	81.9	94.6	76.5	86.4	96.5	77.3	88.1	46.2	61.3	0.3	118.3
10	84.4	96.0	79.3	85.7	97.7	77.6	87.7	46.0	62.7	0.1	120.2
11	89.1	107.4	66.3	57.0	102.5	81.9	91.8	48.8	72.0	0.1	123.9
12	74.4	84.8	71.6	82.9	88.1	69.5	80.6	38.8	57.4	0.2	109.6
AVERAGE	82.51	96.20	77.73	89.80	96.26	76.63	87.10	44.99	63.38	0.22	118.03
STD DEV	6.1480	9.3215	4.0730	5.7535	5.9906	5.1591	4.6592	4.2823	6.1721	0.0680	6.0861
13	23.4	24.0	19.6	26.7	25.6	16.5	25.0	2.9	24.6	4.7	36.1
14	26.5	26.4	17.8	27.2	26.2	17.7	24.8	4.3	26.0	4.9	36.8
15	27.7	29.6	24.9	26.8	29.2	18.8	29.9	4.2	28.4	5.4	37.6
16	26.7	28.8	20.0	25.7	25.2	18.8	26.1	6.1	24.6	5.9	36.3
17	24.0	25.5	22.0	25.7	25.8	17.7	25.4	5.1	23.4	5.6	36.1
18	22.2	22.6	15.1	23.3	22.6	17.5	21.8	5.6	20.9	5.4	38.8
19	25.2	26.6	23.0	26.1	26.0	18.2	25.4	5.3	24.3	5.9	36.5
AVERAGE	25.16	26.74	20.55	26.31	25.83	17.92	25.39	4.82	24.67	5.43	35.08
STD DEV	1.9759	2.0548	2.4610	2.0051	1.0408	0.8142	2.4225	1.0939	2.2740	0.4489	1.6251

Figure 7. An Example of Interleaved Data.

	Fe	Mg	Al	Cu	Ni	Sn
Group I Avg. (ppm)	32.59	33.94	29.70	29.75	31.24	29.69
Group II Avg. (ppm)	32.55	33.81	29.95	29.91	31.37	29.43
Difference	0.04	0.13	0.25	0.16	0.13	0.26
%Difference	0.14	0.38	0.87	0.55	0.44	0.89

	Cr	Pb	Ag	Na	Si	Ti
Group I Avg. (ppm)	31.00	28.06	32.29	28.35	32.14	32.81
Group II Avg. (ppm)	31.38	26.68	32.47	28.88	32.36	32.97
Difference	0.38	1.38	0.18	0.53	0.22	0.16
%Difference	1.24	4.94	0.55	1.87	0.67	0.48

Figure 8. An Example of Paired Testing for the Average Concentration.

	Fe	Mg	Al	Cu	Ni	Sn
Group I σ (ppm)	2.73	3.05	2.55	1.88	1.89	2.15
Group II σ (ppm)	1.71	2.00	2.87	1.47	1.84	1.59
Difference	1.02	1.05	0.32	0.41	0.05	0.56
% Difference	3.12	3.10	1.07	1.38	0.18	1.89

	Cr	Pb	Ag	Na	Si	Ti
Group I σ (ppm)	2.13	3.43	2.06	3.10	1.45	2.26
Group II σ (ppm)	2.04	2.56	1.74	2.30	1.17	1.51
Difference	0.09	0.87	0.32	0.80	0.28	0.75
% Difference	0.29	3.10	0.99	2.83	0.87	2.29

Figure 9. An Example of Paired Testing for the Standard Deviation.

We have extended our programs to include analysis of solid materials by the DC arc technique. A calibration curve obtained employing a DC arc for the analysis of low level impurities in graphite materials is shown in Figure 10. The conditions used for this analysis are shown in Figure 11. We are also in the process of adding the "spark-in-spray" technique to our direct reader for the analysis of water samples. We feel that these computer programs which have been developed are useful for the evaluation of electrodes, of techniques, and for routine analyses by most emission spectrographic methods.

Our experience with electrode certification indicates there are several things that should be considered in determining future requirements for certification. It appears to us that it is not necessary to certify the electrodes against all the elements for which they will be used. Three to five elements with varying excitation potentials could be selected which would characterize the electrodes for certification purposes. This would facilitate the certification process, and in our opinion, no way reduce the stringentness of the test. We have observed that as the counter electrodes

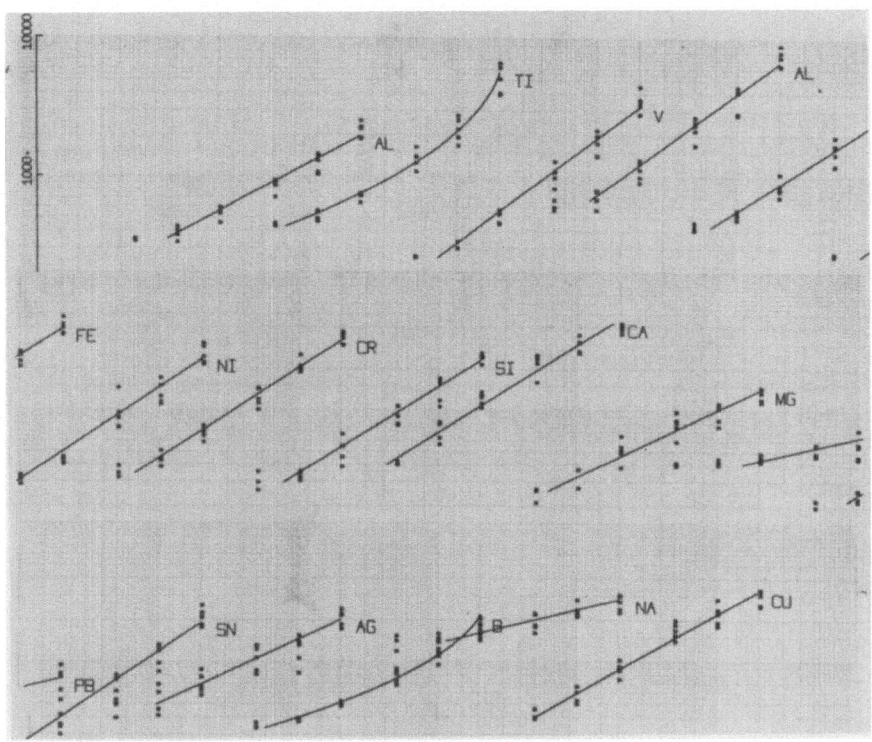

Figure 10. D.C. Arc Calibration Curve for Sixteen Elements.

Excitation Conditions	Exposure Conditions
Source: DC Arc	2 second preburn
DC Current: 10 Amperes	Exposure cut off on background at 3895 Å (~ 20 seconds)
	25 µ entrance slit Analytical gap: 4 mn
	Mask on grating
	Lines corrected for background
	Draft-least amount possible

Electrode System

Cupped sample electrode - Grade AGKSP - National L-4006

Hemispherical tip counter electrode - Grade AGKSP - National L-3954

Figure 11. D.C. Arc Excitation Conditions.

are used and resharpened, their behavior changes. After the
counter has been burned at each end there is some increase in the
standard deviation and variation in the mean which increases with
each burn. For certification analysis, we feel that a fresh counter electrode should be used for each burn. Current practice is
to use the counter repeatedly through certification. Some consideration should be given to using each end of the counter electrode
only once when certifying electrodes.

The stability of the oil standards has also been a problem.
Poor shelf life and variations in individual elemental concentrations within the same standard have been observed. Currently we
are evaluating CONOSTAN oil standards but have not yet determined their life expectancy under our laboratory conditions. More
work needs to be done in investigating the long term stability of
standards, and in assuring their accuracy. We believe that the
spectrographic method for the analysis of lubricating oil has a
promising future, and that the problems indicated can be overcome without too much difficulty.

X-RAY AND EMISSION SPECTROSCOPY

X-RAY DIFFRACTION STUDY OF NEW HAFNIUM COMPOUNDS

Frank L. Chan

Chemistry Research Laboratory
Aerospace Research Laboratories
Wright-Patterson Air Force Base
Dayton, Ohio 45433

and

G. G. Johnson, Jr.

Computer Science Department
The Pennsylvania State University
University Park, Pennsylvania 16802

The important use of x-ray powder diffraction as a
tool in the identification of unknown crystalline materials is dependent on the proper standards being in the
reference file. Such new standards are added yearly
to the Powder Diffraction File. The main purpose of
this paper is to present new data on several well characterized salts based primarily on hafnium both in the
hydrated and anhydrated form.

Specific compounds prepared in our laboratory and
studied in this paper include the sulfates, hydrated
sulfates, hydrated oxychlorides, as well as deuterium
oxide substituting for water in the hydrates. The
characterization of the phases has been made by means
of qualitative and quantitative chemical analysis.
Further verification of phase identification has been
made by comparison with isostructural compounds.

This work follows the recent work studying the change in water
of hydration and a similar work on deuterium oxide compounds
by the same authors.

INTRODUCTION

In the X-ray Sessions of the Tenth National Meeting of
Society of Applied Spectroscopy held at St. Louis, Missouri, on
18-22 October, 1971, the authors as well as other authors pointed
out that the element hafnium was discovered only several decades
ago. After some controversies Dirk Coster and George von
Hevesy were credited with the discovery of hafnium in honor of
Copenhagen, Denmark, in 1923. D. Coster and G. von Heversy
were the first to identify six prominent x-ray lines that con-
formed to the predictions made on the basis of electronic theory.
The fact that the chemical properties of hafnium are so much
similar to that of zirconium the date of discovery of the former
element could have been altered to a later date if the x-ray emission
method was not available at that time. In minerals, hafnium is
always associated with zirconium. In Brazil, it has been reported that baddelayite ores contain as high as 1.2 percent
hafnium oxide and 93 percent zirconium oxide. Large quantities
of nepheline-bearing rocks of Pocos de Caldas Plateau in Brazil
contain from one to three and one-half percent of hafnium. Other
zircons such as Alvite, Cyrtolite, Malacon, and Naegite contain
hafnium oxide ranging from 3.5 to 5.5 percent.

Hafnium has a good absorption cross-section for thermal
neutron, about six hundred times that of zirconium. Coupled
with its excellent mechanical properties and its corrosion
resistance, hafnium is used for reactor control rods. It is also
used in incandescent lamps. Hafnium carbide and hafnium
nitride are excellent refractory materials; the latter compound
has a melting point of $3310^\circ C$ and is the most refractory of all
known metal nitrides.

For the production of hafnium the Kroll process has been
adopted by the U.S. Bureau of Mines and by Wah Chang corporation, both located in Albany, Oregon. For the last three
decades, much research work has been carried out in the field
of analytical chemistry involving the precipitation methods by
Willard (1), Freund (1), and Hahn (2, 3). Determination of
hafnium in presence of zironium is a difficult task. For the wet

method, Hahn proposed the use of p-bromomandelic acid as a
weighing form in conjunction with the mixed oxide weight.
Because of the large gravimetric factor and greater selectivity,
the procedure using p-bromomandelic acid has been commented
on as the most attractive so far proposed.

The use of a radioisotopic source to excite the L_α and L_β
spectra of hafnium and the K_α and K_β spectra of zirconium in a
mixture containing these elements has been proposed by Chan
and Jones for the determination of either zirconium or hafnium
(4) The procedure for this method is simple and can be carried
out rapidly.

Closely connected with the qualitative and quantitative
determination of hafnium and zirconium is the x-ray diffraction
method for the identification of compounds of these elements.
In the present study a number of compounds have been prepared
and their diffraction powder patterns taken. These compounds
with their patterns described in this paper could be added to the
Powder Diffraction File compiled by the Joint Committee on
Powder Diffraction Standards.

EXPERIMENTAL

Preparation of Hafnium Compounds

For the present study, preparations of many compounds
have been attempted. A number of these compounds are not
pure enough to be included at this time. Others are amorphous
in nature and therefore no x-ray powder patterns can be obtained.
Preparations of these compounds, in general, fall into two
categories. The first category is by the fusion method and the
second category is by precipitation in aqueous solutions.

Six compounds have been successfully prepared and are
presented in this paper. These compounds are as follows:

(a) barium hafnate, $BaHfO_3$
(b) hafnium selenite, $Hf(SeO_3)_2$
(c) hafnium sulfate tetra deuterium oxide, $Hf(SO_4)_2 4D_2O$
(d) hafnium oxychloride hexa hydrate, $HfOCl_2 6H_2O$
(e) niobium hafnate, $Nb_2Hf_6O_{17}$
(f) tantalum hafnate, $Ta_2Hf_6O_{17}$

Materials used for the preparations of these compounds are the purest obtainable. A number of the compounds was used with other studies and was described in earlier papers. (5) Methods for the preparation of the six compounds are described here briefly. Barium hafnate was prepared by fusing equivalent quantities of barium carbonate and hafnium oxide at a temperature of 1600°C. for two hours. Hafnium selenite was prepared by reacting 0.0001 molar selenious acid with 0.001 molar hafnium sulfate in aqueous solution.

Hafnium sulfate tetra deuterium oxide was prepared by fuming pure metallic hafnium in concentrated sulfuric acid until all the gray metal disappeared. Traces of unreacted hafnium were removed by first dissolving the material in deuterium oxide and then filtering through an unglazed porcelain disc. Hafnium sulfate tetra deuterium oxide was obtained by re-crystallization in deuterium oxide. The procedure followed the general scheme described previously. (6) Hafnium oxychloride was prepared by grinding at 35°C. the octa hydrate and left in open air for several hours.

Niobium hafnate was prepared by fusing stoichemetric quantity of niobium pentoxide and hafnium oxide at 1600°C. for one and one-half hours. Tantalum hafnium was prepared by fusing stoichiometric quantity of tantalum pentoxide and hafnium oxide at 2800°C. for one-half hour.

Instrumentation and Procedures for Diffraction Patterns

The instruments used for this study have been reported in a number of papers by the authors. (6, 7) The arrangements with the necessary accessories have also been described and shown in an earlier paper. (8) These units with its heat exchanger (9), have been in working order for many years without breakdown. (6, 7, 8) The x-ray diffraction camera for taking powder patterns in the present study is the modified Philips camera having attachments designed by Chan. (10, 11) With these attachments it can take either powder patterns or single crystal rotation photographs of two different diameters using a special designed goniometer. Attachments for the G. E. powder camera have also been designed. (12) When the smaller diameter such as 57.35 mm. is being used, thirty to forty-five minutes are needed for taking a pattern. With an attachment 114.7 mm. in diameter, somewhat longer exposure is required. Provisions are being made to admit helium to the attachments to shorten the duration of x-ray exposure. Both the Straumanis and the Wilson techniques for loading the x-ray film can be used with modified cameras.

Figure 3, (A) to (F) are powder patterns of the six compounds prepared as described under preparation of samples. They were taken with the modified camera having an effective diameter of 114.7 mm. The exposure was two hours with 40 kV and 20 mA using copper radiation and nickel filter to remove the K_b radiation. Calibrations for the backing of the x-ray film were carried out similar to that described by Klug and Alexander. (13) It has been described in the previous paper. (7)

Computer Refinement

The computer techniques used in the refinements of the powder diffraction patterns of the new phases were three-fold -- first, a computerized identification system for the identification of multiphase unknown patterns. (Such a system can determine whether a single phase or a mixture of phases is present in a pattern). Second, an automatic indexing program for the determination of crystal system and trial cell parameters and third, a least squares refinement program for the determination of the best cell parameters for unit cell determination.

The diffraction patterns of the starting materials are in the Powder Diffraction File compiled by the Joint Committee on Powder Diffraction Standards. The x-ray powder diffraction patterns (obtained by Debye-Scherrer film techniques) were read and recorded (the series of $d.I_i$ representing the pattern). It is thus possible to see from the unknown chemical product if the reacting materials are still present and whether a new phase has been produced.

The recorded x-ray powder diffraction pattern of the product was placed in the Johnson-Vand (Search/Match Identification System ⌈Version 16⌉), (14, 15) and it was found that 3 of the lines of one of the product unknowns were from one of the starting materials. These 3 lines were very weak (thus showing a low concentration of the starter phase in the total amount of product material present). This program #1 (Search/Match) written completely in FORTRAN IV was used to identify which lines belong to each phase.

Since the lines of the unknown phase were now identified (determined) it should now be possible to find the crystallographic unit cell which will print all the observed interplanar spacings (d_i) which were observed in the unknown pattern. Such a procedure is called "indexing". We are fortunate indeed to have the PL/I version of the program, (16), running on our local IBM 360/67. The results of the program yields the crystal system, the cell parameters and the systematic extensions (i.e., the diffraction aspect).

The third program of the series was a FORTRAN IV least squares program to refine the cell parameters (a, b, c, α, β, γ) to minimize the $\Delta 2\theta$ residues of the calculated patterns (based on a, b, c, α, β, γ) and the spacing of the observed diffraction pattern. (17) Since 20 of the lines of the pattern were indexed in the indexing program, those 20 indexes were "fixed" in the refinement. With these 20 values of hkl to help determine the best cell parameters, the program automatically cycles 10 times, indexing more and more of the unknown pattern by modifying the cell parameters so that the $\Delta 2\theta$ is minimized. On each cycle of the least squares the tolerance on each of the $\Delta 2\theta$ residues is reduced to eliminate possible mis-matches with incorrect indexes. In the final refinement, all observed lines are approximately .02 2θ of the calculated positions.

The results of the cell refinement showed that all the materials were "indexed" and the cell parameters are given in this paper elsewhere.

RESULTS AND DISCUSSION

In this study attempts have been made to prepare a number of compounds by the fusion method as well as the aqueous precipitation method. Six powder patterns of the compounds prepared are shown in this study. They are not listed in the Joint Committee on Powder Diffraction Standards File.

Barium hafnate obtained by the fusion method appears cubic. TABLE I shows the x-ray diffraction data of this compound. The lattice constant has been found to be 4.16 Å, with one molecule per unit cell. The computer print-out for the 7th cycle and the 9th cycle with two-theta tolerances of 0.25882 and 0.10000 respectively are shown in Tables IX, and X.

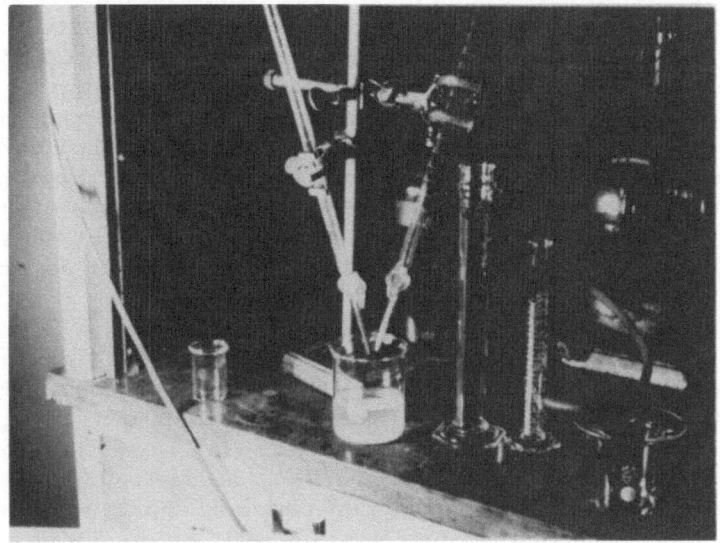

Figure 1.   Scheme for the precipitation of hafnium selenite
            from 0.001 molar solutions of the reactants.

Figure 2.   Inorganic compounds prepared for x-ray diffraction
            study.

Freshly prepared hafnium selenite from 0.001 molar selenious acid and 0.001 molar hafnium sulfate solutions shows the material to be amorphous in nature. The scheme of precipitation is shown in Figure 1. Precipitates prepared from dilute solutions of arsenic acid and hafnium sulfate; and precipitates prepared from dilute solutions of diammonium phosphate and hafnium sulfate likewise gave no well defined spectral lines. Upon repeated evaporation to dryness as many as four times of the freshly prepared hafnium selenite a number of spectra lines appeared as shown in Figure 3 (B). X-ray diffraction data are tabulated in Table II.

Hafnium sulfate tetra deuterium oxide prepared from the anhydrous sulfate and recrystallized from 99.5 percent purity deuterium oxide is shown in Figure 2. It follows the same general scheme described in an earlier paper. (6) It is orthorhombic and having the space group Fddd. After the least square refinement, 5th cycle, the lattice constants are shown in Table VIII. The observed interplanar spacings and relative intensities are shown in Table III. The indexes from the computer print-outs are likewise shown in Table III.

X-ray powder pattern for the hafnium oxychloride octa hydrate has been reported by the U.S. Bureau of Mines, Albany, Oregon. (18) It has been found to be tetragonal with lattice constants: a_o, 17.07 and c_o, 7.71. On exposure to the air at slightly elevated temperature the oct hydrate changes to the hexa hydrate variety. The x-ray powder pattern is shown in Figure 3 (D). Computer print-outs indicate that it is monoclinic with constants: a_o, 24.6; b_o, 11.7; c_o 23.3; and β, 96.0. X-ray diffraction data are shown in Table IV.

Both niobium and tantalum hafnates were prepared by the fusion method at different optimum temperatures as described. The least square refinement after the 7th cycle in the case of niobium hafnate gave lattice constants: a_o, 4.94; b_o, 5.10; and c_o, 5.26. It is orthorhombic and has a gram-molecular weight 1528.82. The least square refinement after the 6th cycle in the case of tantalum hafnate gave a_o, 4.91; b_o, 5.07; and c_o, 5.24. These constants were obtained by introducing to the computer the strongest lines, fourteen spectra lines in the case of niobium hafnate, and eleven in the case of tantalum hafnate, Table XI. The lattice constants so obtained were used to calculate the remainder of the indexes corresponding to the observed spectra

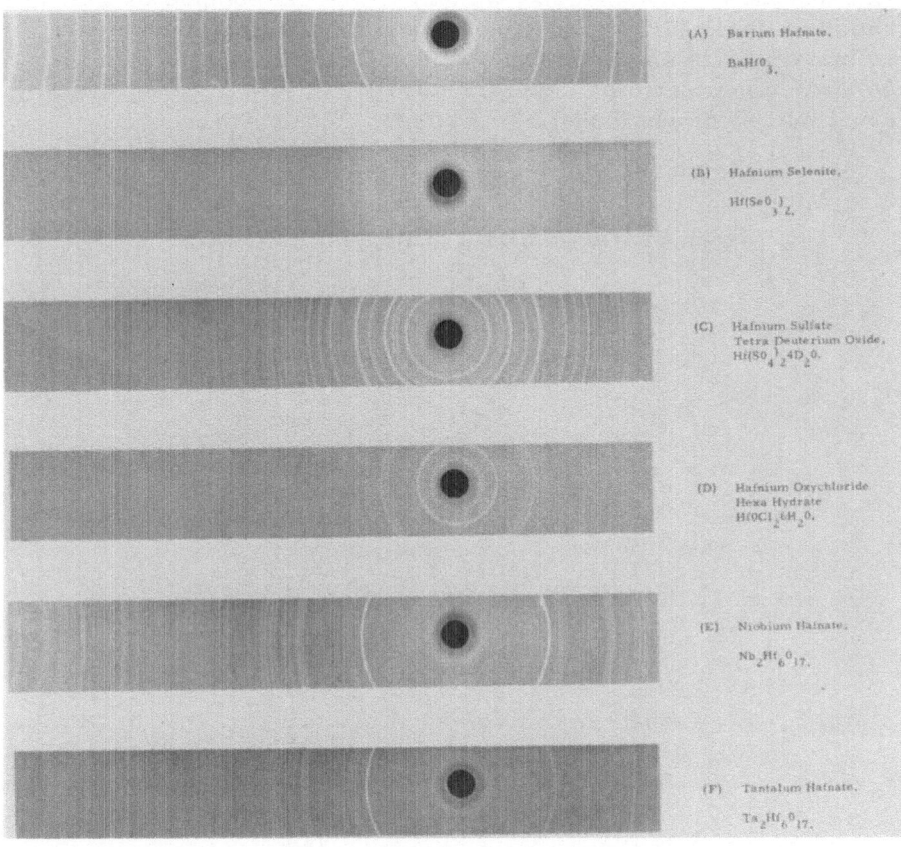

Figure 3
Some x-ray diffraction powder patterns of hafnium compounds.

lines, Table XII. In both the niobium hafnate and tantalum hafnate some two to three percent of the original substances were found to be unreacted. Faint spectra lines from the original substances were omitted from the final computation of the lattice constants. Some of the typical computor print-outs are shown in Tables IX to XII, inclusive. The crystallographic data and the gram-molecular weights of the six compounds are shown in Table VII. The lattice constants from the least square refinements are shown in Table VIII. X-ray diffraction data for niobium hafnate and for tantalum hafnate are shown in Table V and Table VI respectively.

TABLE I

X-ray Diffraction Data of Barium Hafnate

$BaHfO_3$

CuK_a, $\lambda = 1.5405$Å; temperature 20°C

d Å	I/I_o	hkl
2.925	95	110
2.075	35	200
1.861	5	210
1.701	100	211
1.474	40	220
1.389	5	221
1.318	45	310
1.204	20	222
1.1145	95	321
1.0420	10	400
1.0110	1	410
0.9883	25	411
0.9320	20	420
0.8900	10	332
0.8530	15	422
0.8175	80	431

TABLE II

X-ray Diffraction Data of Hafnium Selenite

$$Hf(SeO_3)_2$$

CuK_a, $= 1.5405$ Å; temperature $20^\circ C$

d Å	I/I$_o$	hkl
4.55	100B	110
3.75	50B	111
3.30	30B	200
2.65	90B	211

TABLE III

X-ray Diffraction Data of Hafnium Sulfate $4D_2O$

CuK_a, $= 1.5405$ Å; temperature $20^\circ C$

d Å	I/I$_o$	hkl
6.35	60	400
5.20	20	220
4.81	40	111, 11$\bar{1}$
4.24	90	311
3.55	10	$\bar{3}2\bar{1}$
3.42	65	620
3.11	10	131
2.93	100	331
2.86	25	040
2.62	10	1$\bar{1}$2
2.55	5	$\bar{2}\bar{1}$2
2.47	25	911
2.38	15	731
2.34	5	$\bar{3}$22
2.30	65	422, 602

continued

TABLE III cont'd

2.13	5	622, 1200
2.11	60	151, 931
2.06	15	351
1.961	65	551, 242
1.905	5	1040
1.872	25	1131, 1002
1.834	20	751
1.800	30	642, 113
1.770	35	$2\bar{1}3$
1.750	40	660
1.719	5	1240
1.702	5	951, 513
1.675	25	1331
1.643	10	$6\bar{1}3$
1.621	40	333, 713
1.605	5	1222
1.575	60	1600,
1.552	5	533
1.525	15	371

and 44 lines beyond 1.525

TABLE IV

X-ray Diffraction Data of Hafnium Oxychloride Hexa Hydrate

$HfOCl_2 6H_2O$

CuK_α = 1.5405 Å; temperature 20°C

d Å	I/I_o	hkl
12.4	100	200
10.2	50	102
8.00	20	202, $\bar{2}0\bar{2}$
6.85	90	203
6.19	5	113

TABLE IV cont'd

5.48	10	121
5.09	10	$\bar{2}0\bar{4}$
4.718	20	222
4.130	20	$\bar{2}0\bar{5}$
4.015	15	404
3.820	20	224
3.785	20	414, 131
3.637	1	$\bar{3}06$
3.568	50	116
3.405	5	701
3.287	5	207, 430
3.205	20	$\bar{5}0\bar{5}$
3.200	20	505
3.150	10	333, 622
3.100	5	515
2.990	1	$\bar{8}0\bar{1}$
2.950	1	334, $\bar{8}03$
2.710	10	$\bar{6}07$
2.625	5	507
2.530	5	109, 244, $\bar{7}07$
2.490	1	608
2.427	1	626
2.375	1	904
2.300	1	707
2.200	1	635
2.148	20	353

and 24 lines beyond 2.148

TABLE V

X-ray Diffraction Data of Niobium Hafnate

$$Nb_2Hf_6O_{17}$$

CuK_α, = 1.5405 $\overset{\circ}{A}$; temperature 20°C

d $\overset{\circ}{A}$	I/I_o	hkl
3.58	20	101
2.94	100	111
2.63	40	002
2.545	40	020
2.46	40	200
2.325	35	012
2.080	30	121
1.830	50	022
1.800	50	202
1.770	50	220
1.695	40	212
1.650	45	103
1.573	50	113
1.540	50	131
1.500	50	311
1.472	50	222
1.388	40	123
1.315	5	004
1.235	30	400
1.200	40	303, 410
1.185	40	133
1.170	40	024
1.155	40	331
1.134	50	240
1.115	25	402

continued

TABLE V cont'd

1.088	35	412
1.058	20	224
1.042	35	242
1.030	15	105
1.025	15	422
1.010	15	143
0.983	35	151
0.960	15	234
0.955	40	125

and 28 lines beyond 0.955

TABLE VI

X-ray Diffraction Data of Tantalum Hafnate

$$Ta_2Hf_6O_{17}$$

CuK_{α}, = 1.5405 Å; temperature 20°C

d Å	I/I_o	hkl	d Å	I/I_o	hkl
3.55	30	101	1.465	50	222
2.91	100	111	1.235	20	400
2.60	45	002	1.185	15	133
2.525	45	020	1.168	15	313
2.450	45	200	1.155	20	204
2.320	25	012	1.133	50	240
2.065	20	121	1.115	10	402
1.822	55	022	1.085	15	412
1.790	55	202	1.055	10	224
1.765	55	220	1.040	20	242
1.690	35	030	1.025	15	105
1.645	40	103	0.981	50	151
1.568	50	113	0.952	45	234
1.532	55	131	0.935	5	250
1.495	55	311	0.8875	45	252

and 7 lines beyond 0.8875

TABLE VII

X-ray Crystallographic Data

Formula	Crystal System	Space Group	Unit Cell Volume
$BaHfO_3$	Cubic	Pm3m	72.43
$Hf(SeO_3)_2$	Cubic		272.38
$Hf(SO_4)_2 4D_2O$	Orthorhombic	Fddd	1578.50
$HfOCl_2 6H_2O$	Monoclinic		6669.32
Nb_2HfO_{17}	Orthorhombic		132.47
$Ta_2Hf_6O_{17}$	Orthorhombic		130.72

TABLE VIII

Lattice Constants after Least Square Refinement

Inorganic Compounds	Formula	Gram Molecular Weight	Z	a_o Å	b_o Å	c_o Å	α_o	β_o	γ_o	Least Square Refinement
Barium Hafnate	$BaHfO_3$	263.86	1	4.16	--	--	--	--	--	9th cycle
Hafnium Selenite	$Hf(SeO_3)_2$	432.42	--	6.48	--	--	--	--	--	5th cycle
Hafnium Sulfate Tetra Deuterium Oxide	$Hf(SO_4)_2 4D_2O$	450.756	8	25.5	11.40	5.43	--	--	--	5th cycle
Hafnium Oxychloride Hexa Hydrate	$HfOCl_2 6H_2O$	373.510	--	24.6	11.7	23.3	--	$96.0°$	--	5th cycle
Niobium Hafnate	$Nb_2Hf_6O_{17}$	1528.82	--	4.94	5.10	5.26	--	--	--	7th cycle
Tantalum Hafnate	$Ta_2Hf_6O_{17}$	1704.90	--	4.91	5.07	5.24	--	--	--	6th cycle

Table IX. A Computer Print-outs on Least Square Refinement, Cycle 7, for Cell Constant and Unit Cell Volume, $BaHfO_3$

CYCLE 7   BAHFO3   2-THETA TOLERANCE = 0.25882   CUBIC

N	H	K	L	D CALC	D OBS	LAMBDA	2-THETA CALC	2-THETA OBS	2-THETA DIFF	WEIGHT
1	1	1	0	2.946838	2.924999	1.541780	30.32991	30.56186	-0.23195	1.00000
2	2	0	0	2.083729	2.075000	1.541780	43.42590	43.61792	-0.19202	1.00000
3	2	1	0	1.863745	1.860999	1.541780	48.86539	48.94220	-0.07681	1.00000
4	2	1	1	1.701358	1.700999	1.541780	53.88583	53.89813	-0.01230	1.00000
5	2	2	0	1.473419	1.474000	1.541780	63.09399	63.06625	0.02774	1.00000
6	2	2	1	1.389153	1.399000	1.541780	67.41248	67.42090	-0.00842	1.00000
7	3	1	0	1.317866	1.318000	1.541780	71.59930	71.59097	0.00833	1.00000
8	2	2	2	1.203042	1.204000	1.541780	79.70056	79.62451	0.07605	1.00000
9	3	2	1	1.113800	1.114499	1.541780	87.59723	87.52835	0.06888	1.00000
10	4	0	0	1.041864	1.042000	1.541780	95.44809	95.43173	0.01636	1.00000
11	4	1	0	1.010756	1.011000	1.541780	99.40308	99.37064	0.03244	1.00000
12	4	2	0	0.931872	0.932000	1.541780	111.63422	111.61102	0.02319	1.00000
13	4	3	1	0.817307	0.817500	1.541780	141.19464	141.11780	0.07684	1.00000

	A	B	C	ALPHA	BETA	GAMMA	VOLUME
RECIPROCAL CELL	0.23991942E 00	0.23991942E 00	0.23991942E 00	90 0.0	90 0.0	90 0.0	0.13810080E-01
R C CORRECTIONS	-0.34943543E-04	-0.34943543E-04	-0.34943543E-04	0.0	0.0	0.0	-0.60424209E-05
DIRECT CELL	0.41680660E 01	0.41680660E 01	0.41680660E 01	90 0.0	90 0.0	90 0.0	0.72410873E 02
D C CORRECTIONS	0.60749054E-03	0.60749054E-03	0.60749054E-03	0.0	0.0	0.0	0.31672466E-01

LARGEST RESIDUAL REDUCED TO UNIT WEIGHT 0.23195 OBS 1   STANDARD ERROR UNIT WT OBS 0.09632   DEGREES OF FREEDOM 12

Table X. A Computer Print-outs on Least Square Refinement, Cycle 9, for Cell Constant and Unit Cell Volume, BaHfO$_3$

CYCLE 9   BAHFO3   2-THETA TOLERANCE = 0.10000   CUBIC

N	H	K	L	D CALC	D OBS	LAMBDA	2-THETA CALC	2-THETA OBS	2-THETA DIFF	WEIGHT
1	2	1	0	1.864160	1.860399	1.541780	48.85382	48.94220	-0.08838	1.00000
2	2	1	1	1.701778	1.700399	1.541780	53.87280	53.89813	-0.02533	1.00000
3	2	2	0	1.473748	1.474000	1.541780	63.07828	63.06625	0.01202	1.00000
4	2	2	1	1.389463	1.389000	1.541780	67.39542	67.42090	-0.02548	1.00000
5	3	1	0	1.318160	1.318000	1.541780	71.58093	71.59097	-0.01004	1.00000
6	2	2	2	1.203310	1.204000	1.541780	79.67929	79.62451	0.05478	1.00000
7	3	2	1	1.114049	1.114000	1.541780	87.57266	87.52835	0.04431	1.00000
8	4	0	0	1.042096	1.042000	1.541780	95.42004	95.43173	-0.01169	1.00000
9	4	1	0	1.010983	1.011000	1.541780	99.37286	99.37064	-0.00223	1.00000
10	4	2	0	0.932080	0.932300	1.541780	111.59656	111.61102	-0.01447	1.00000
11	4	3	1	0.817435	0.817500	1.541780	141.12216	141.11780	0.00436	1.00000

	A	B	C	ALPHA	BETA	GAMMA	VOLUME
RECIPROCAL CELL	0.2399005E 00	0.2399005E 00	0.2399009E 00	90 0.0	90 0.0	90 0.0	0.13806887E-01
R C CORRECTIONS	0.10809924E-06	0.10809924E-06	0.10809924E-06	0.0	0.0	0.0	0.74505806E-08
DIRECT CELL	0.41683874E 01	0.41683874E 01	0.41683874E 01	90 0.0	90 0.0	90 0.0	0.72427612E 02
C C CORRECTIONS	0.0	0.0	0.0	0.0	0.0	0.0	-0.45776367E-04

LARGEST RESIDUAL REDUCED TO UNIT WEIGHT 0.08838   CBS   1   STANDARD ERROR UNIT WT OBS 0.03832

R C STNDRD ERRS   0.19930696E-04   0.19930696E-04   0.19930696E-04   DEGREES OF FREEDOM 10

DIRECT CELL VARIANCE-COVARIANCE MATRIX

						ROW
0.11992688E-06	0.11992638E-06	0.11992688E-06	0.0	0.0	0.0	1
0.11992688E-06	0.11992688E-06	0.11992688E-06	0.0	0.0	0.0	2
0.11992688E-06	0.11992688E-06	0.11992688E-06	0.0	0.0	0.0	3
0.0	0.0	0.0	0.0	0.0	0.0	4
0.0	0.0	0.0	0.0	0.0	0.0	5
0.0	0.0	0.0	0.0	0.0	0.0	6

C C STNDRD ERRS   0.34630462E-03   0.34630462E-03   0.34630462E-03   0.0   0.0   0.0   0.18051572E-01

Table XI. A Computer Print-outs on Least Square Refinement, Cycle 6, for Cell Constants and Unit Cell Volume, Ta_2Hf6O_{17}

CYCLE 6   TA2HF6017   STRONG LINES ONLY   2-THETA TOLERANCE = 0.25174,   ORTHO

N	H	K	L	D CALC	D OBS	LAMBDA	2-THETA CALC	2-THETA OBS	2-THETA DIFF	WEIGHT
1	1	1	0	3.579243	3.549999	1.541780	25.23355	25.08362	0.14994	1.00000
2	1	1	1	2.928055	2.910000	1.541780	30.52919	30.72327	-0.19407	1.00000
2	0	0	2 R	2.622403	2.599999	1.541780	34.19073	34.49454	-0.30380	1.00000
3	0	0		2.537049	2.525000	1.541780	35.37830	35.55276	-0.17447	1.00000
4	2	0		2.456024	2.450000	1.541780	36.58606	36.67923	-0.09317	1.00000
5	0	2		1.823394	1.822000	1.541780	50.02014	50.06104	-0.04089	1.00000
6	2	2		1.792606	1.790000	1.541780	50.94000	51.01947	-0.07947	1.00000
7	2	2		1.764662	1.764999	1.541780	51.80719	51.79529	0.01190	1.00000
8	1	1	3	1.566592	1.568000	1.541780	58.95486	58.89670	0.05817	1.00000
9	1	3	1	1.529689	1.532300	1.541780	60.52362	60.42282	0.10080	1.00000
10	3	1	1	1.493702	1.495000	1.541780	62.14111	62.08118	0.05994	1.00000
11	2	1	2	1.466027	1.464999	1.541780	63.54587	63.49878	0.04709	1.00000

	A	B	C	ALPHA	BETA	GAMMA	VOLUME
RECIPROCAL CELL	0.203581C4E 0C	C.197079366 00	C.19066477E 00	90	90	90	0.76497756E-02
R C CORRECTIONS	0.24883654E-07	C.87018520E-07	-0.63254447E-07	0.0	0.0	0.0	-0.37252903E-08
DIRECT CELL	0.49120512E 01	C.50740C995E 01	C.52448092E 01	90	90	90	0.13072278E 03
C C CORRECTIONS	0.2861C229E-C5	C.95367432E-06	0.38146973E-05	0.0	0.0	0.0	0.61035156E-04

LARGEST RESIDUAL REDUCED TO UNIT WEIGHT 0.19407   OBS   2   STANDARD ERRCR UNIT WT OBS 0.12587   DEGREES OF FREEDOM   8

R C STNDRD ERRS   0.32812241E-C3   0.33421908E-03   0.39051450E-03   0.0   0.0   0.0

DIRECT CELL VARIANCE-COVARIANCE MATRIX

						ROW
0.62679C86E-04	-0.16426115E-04	-0.22933542E-04	0.0	0.0	0.0	1
-0.16426115E-04	C.7404539RE-04	-0.24908106E-04	0.0	0.0	0.0	2
-0.22933542E-04	-C.24908120E-04	0.11539644E-03	0.0	0.0	0.0	3
0.0	C.0	0.0	0.0	0.0	0.0	4
0.0	C.0	0.0	0.0	0.0	0.0	5
0.0	0.0	0.0	0.0	0.0	0.0	6

C C STNDRD ERRS   0.7917C130E-02   C.86049624E-07   0.10742273E-01   0.28335041E 00

Table XII. A Computer Print-outs showing Interplanar Spacings Based on the Cell Constants, $Ta_2Hf_6O_{17}$, and the Various Indexes

TA2HF6O17   HKL LISTING - *** REFERS TO FIXED,R TO REJECTS ORTHO

N	H	K	L	D CALC	D OBS	LAMBDA	2-THETA CALC	2-THETA OBS	2-THETA DIFF	WEIGHT
1	0	0	1	5.224000		1.541780	16.97191			
2	0	1	0	5.073999		1.541780	17.47751			
3	1	0	0	4.912030		1.541780	18.05867			
4	0	1	1	3.639735		1.541780	24.45551			
5	1	0	1	3.578527		1.541780	24.88043			
6	1	1	0	3.529192		1.541780	25.23395			
7	1	1	1	2.924390		1.541780	30.56839			
8	0	0	2	2.612000		1.541780	34.33112			
9	2	0	0	2.537001		1.541780	35.37900			
10	0	2	0	2.455999		1.541780	36.58643			
11	1	0	2	2.322352		1.541780	38.77356			
12	0	2	1	2.306212		1.541780	39.05588			
13	2	0	1	2.282117		1.541780	39.48529			
14	1	2	0	2.254100		1.541780	39.99686			
15	2	1	0	2.222621		1.541780	40.58807			
16	2	2	1	7.710649		1.541780	40.81766			
195	5	0	4	0.785033		1.541780	158.17755			
196	4	5	0	0.782258		1.541780	160.44012			
197	6	0	2	0.781195		1.541780	161.36597			
198	3	3	5	0.781136		1.541780	161.37430			
199	2	2	6	0.780796		1.541780	161.72641			
200	6	2	0	0.779109		1.541780	163.34161			
201	5	4	0	0.776708		1.541780	165.96515			
202	5	1	4	0.775351		1.541780	167.03372			
203	0	3	6	0.774119		1.541780	169.53128			
204	4	5	1	0.773632		1.541780	170.34805			
205	3	5	3	0.772934		1.541780	171.66415			
206	6	1	2	0.772098		1.541780	173.58731			

TIME FOR THIS JOB WAS 8 SECONDS

SUMMARY AND CONCLUSION

Six inorganic compounds have been successfully prepared. Among these compounds, three of them were prepared by fusion at elevated temperatures, ranging from 1500° C. to 2800° C. The remainder were prepared in aqueous solutions. Six sets of x-ray diffraction spectra have been obtained for these compounds.

By the use of a computer the observed interplanar spacings were used by the least square refinement to determine the lattice constants, crystal systems, unit cell volumes and all possible indexes correspond to the observed and calculated interplanar spacings. The results presented could be used as one of the methods for the systematic identification of compounds. It is hoped that these results will be included in the X-ray Powder Diffraction File.

REFERENCES

1. H. H. Willard and H. Freund, "Fractional Separation of Hafnium and Zirconium by Means of Tiethylphosphate" Ind. Eng. Chem., Anal. Ed. 18, 195, 1946.

2. R. B. Hahn, "Determination of Zirconium - Hafnium Ratio with a p-Bromo Mandelic Acid", Anal. Chem. 23, 1259, 1951.

3. R. B. Hahn, "Determination of Zirconium - Hafnium Ratio and Its Applications in Geochemistry and Cosmochemistry", paper presented at the 18th Detroit Anachem Conference, held at Hilton Hotel, Detroit Michigan, 14 Oct., 1970.

4. Frank L. Chan and W. Barclay Jones, "Determination of Zirconium, Hafnium Niobium, Tantalum, Molybdenum and Tungsten in Aqueous Solutions by Radioisotopic - Excited X-ray Fluorescence", in K. F. J. Heinrich and C. O. Ruud Editiors, Advances in X-ray Analysis, Vol. 15, 1972, in press

5. Frank L. Chan, "Precipitation and Determination of Tantalum and Niobium from Homogeneous Solution with 3:3':4'5:7 -Pentahydroxy flavanone", Talanta, Vol. 7 pp. 253-263, 1961.

6. Frank L. Chan and G. G. Johnson, Jr., "Some X-ray Diffraction Spectra and Characteristic Properties of Deuterium Ocide Inorganic Compounds", in Developments in Applied Spectroscopy, E. L. Grove and A. J. Perkins, Editors.

7.   Frank L. Chan and G. G. Johnson, Jr., "A Study on the Change of Water of Crystallization by X-ray Diffraction Data Stored in ASTM Magnetic Tape", in Developments in Applied Spectroscopy", E. L. Grove and A. J. Perkins, Editors, Vol. 8, 53-75, 1970.

9.   Frank L. Chan, U. S. Patent No. 3, 384, 162

10. Frank L. Chan, U. S. Patent No. 3, 230, 367

11. Frank L. Chan, U. S. Patent No. 3, 079, 500

12. Frank L. Chan, U. S. Patent No. 3, 160, 748

13. Harold P. Klug and Leroy E. Alexander, "X-ray Diffraction Procedures", John Wiley and Sons, Inc. New York, 1954.

14. Johnson, G. G., Jr. Revised X-ray Powder Diffraction Technique. Ind. Engr. Chem. 61(5) p. 79 (1969).

15. Johnson, G. G. Jr. FORTRAN IV. Programs (Version 12) for the Identification of Multiphase Powder Diffraction Patterns. Joint Committee on Powder Diffraction Standards 154 pp (1970).

16. Visser, J. W. A Fully Automatic Program for Finding the Unit Cell from Powder Data. J. Appl. Cryst. 2, Part 3 pp. 89-95 (1969).

17. Evans, H. T., Jr., D. E. Appleman, D. S. Handwerker. The Least-Squares Refinement of Crystal Unit Cells with Powder Diffraction Data by an Automatic Computer Indexing Method. Abs. Amer. Cryst. Assoc., Cambridge, Mass., Program, p. 42-43.

18. Powder Diffraction File, Compiled by Joint Committee on Powder Diffraction Standards, Inorganic Index 1971.

THE ANNUAL CYCLE OF SOME MINOR ELEMENTS IN LINSLEY POND (NORTH

BRANFORD, CONNECTICUT) AS DETERMINED BY OPTICAL EMISSION

Ursula M. Cowgill

Department of Biology, University of Pittsburgh

Pittsburgh, Pennsylvania   15213

A procedure is described employing a Jarrell Ash direct reading optical emission unit. A special optical arrangement was
developed for this study. A 55 mm focal length, spherical quartz
lens was used in the arc stand to focus the electrodes' image on a
variable aperture located on a rider base on the optical bar. This
aperture is closed down to eliminate the incandescence of the electrodes entering the instrument. A 135 mm slit mounted lens was
employed to permit the light passed by the mask to completely fill
the grating.

Concentrated samples, collected weekly for a year, of the
column of water from Linsley Pond, were combined in a 1:1 ratio
with graphite and analysed in a pure argon atmosphere.

Beryllium, mercury, cobalt, boron, vanadium, silver, lithium,
bismuth, tin and molybdenum were present throughout the year of
study. Some aspects of the annual distribution of these elements
in Linsley Pond will be discussed.

INTRODUCTION

The purpose of this study was to develop a technique whereby
it would be possible to study the movement of all elements in a
natural thermally stratified lake. The elements which can easily
be detected by use of X-ray emission spectroscopy have already been
described (Cowgill, 1970). The present study is entirely confined
to those that can only be detected by optical emission spectroscopy
or those that can be more easily studied by this method either because of interference problems or because of low concentrations.

The elements considered here are beryllium, mercury, bismuth, tin,
cobalt, boron, sodium, molybdenum, vanadium, silver, and cadmium.
Cadmium, was previously determined by X-ray emission on the same
samples, but was again monitered on the optical emission spectrograph. The results obtained by examining cadmium served to provide some comparative measure for the other volatile elements in
the sample. As long as the concentrations of cadmium were essentially the same when obtained on both instruments, it was felt
that the problem of determining the other volatile elements was
surmounted. Because of the small quantity of water sample available, all detectable elements were studied on the same material.

                              PRIMARY OPTICS

     A special optical arrangement was made for this program. A
55 mm focal length spherical quartz lens was used in the arc stand
to focus the image of the electrodes on a variable aperture located
on a rider base on the optical bar. This aperture is closed down
to eliminate the incandescence of the electrodes entering the instrument. The slit mounted lens was changed to a 135 mm focal
length spherical quartz lens to allow the light passed by the mask
to completely fill the grating.

     The best reproducibility was obtained at the following exper-
imentally established settings: the width of the diaphragm of the
variable aperture was adjusted to 1.45 mm, the distance of this
aperture from the arc stand lens is 103 mm and from the exit slit
lens 148 mm, and the height of the aperture above the rider base
is 143.9 mm.

     The analytical gap is set at 4 mm or else incandescence gets
into the instrument causing higher background and lower signal.
A Stallwood jet is employed in a 100 percent argon atmosphere.
The spacing of the sample electrode above this jet is extremely
critical or else the argon blows the sample out. The sample must
be 6.5 mm above the jet to avoid this problem. The base of the
pointer electrode is placed exactly 10 mm into the upper jaw. As
long as this arrangement is adhered to exactly the reproducibility
between duplicate samples is better than would be expected.

     A problem arose with the mercury light. It interferred with
the mercury line at 2536 Å and with the molybdenum line at 3132 Å.
This situation made it impossible to have the mercury light on
during analyses and therefore line up could not be checked periodically, which is most desirable. The problem was solved by placing
a pyrex test tube over the light which eliminated the interference
with the mercury line and a piece of plastic of unknown composition

Figure 1. Assembled and unassembled electrode packer

over the test tube which eliminated the interference with the
molybdenum line.

OPERATING PROCEDURE

An electrode packer was designed to mechanize the filling of
electrodes. The empty electrode is placed in a slot that is
attached to an arm which when positioned puts the electrode directly under a funnel containing the sample. The sample is introduced
into the electrode by means of a plunger. The packing tension is
kept constant by use of a spring through which the plunger is used.
Figure I shows the electrode packer. The materials used in con-
structing the portions of the electrode packer that come in contact with the sample were selected and checked for elemental content to make sure that contamination problems were avoided. The
use of this packer has improved the reproducibility of duplicate
samples considerably. Sixty electrodes can be uniformly packed
with this small instrument in an hour.

An 1/8" Ultra Carbon No. U7-5440 crater type electrode is
packed, using the packer just described with the material to be
analyzed. The sample is then clamped in the lower jaw of the arc
stand and inserted through a Stallwood jet. The gas used is 100
percent Argon and the flow rate is set at 10 s.c.f.h. A counter
electrode No. U7-1964 is opposed to the sample electrode. The
sample is positive and the counter electrode is negative. The
sample is then arced in a controlled current d.c. arc of 6 amps
at an internal standard determined period of about 8 seconds. A
three second preburn period is employed to reduce the amount of
background getting into the instrument at ignition. The sample is
ignited by using a high voltage a.c. spark. Other instrument
specifications are given in Table I.

Optical bench specifications are given in Table II. All
photomultiplier tubes employed are extremely sensitive. The internal standard background set at 4016.1 is adjusted to terminate
the burning of the sample at 6000 counts. The second background
is not used as a background correction but only as a visual monitoring device the setting of which, at 3885.4 Å was empirically
determined to be the best position. Samples were considered
acceptable when they burned for 8 seconds + 0.5 of a second with a
background reading on the meter of 6 volts + 0.5 volts. Any
variation in the argon flow rate will cause drastic changes in the
burning time. Samples of high concentration tend to burn at less
than 8 seconds with a background reading of less than 6 volts but
nevertheless remain within the specifications set.

TABLE I

Optical Emission Apparatus and Operating Conditions

Item	Specifications
Spectograph	Jarrell Ash 1.5M direct reader, 22 channels
Grating	1180 grooves/mm
Angle of incidence	34°
Wavelength range	2000 - 8000 Å
Reciprocal linear dispersion	5.59 Å/mm
Resolution	0.2 Å
Entrance slit	25 μ
Capacitance	0.0025 μ F
Inductance	310 m H
dc arc	6 Amps
Electrodes	Graphite ultra carbon U7-1964
	Graphite ultra carbon U7-5440
Internal standard	Background 4016.1 Å
Analytical gap	4 mm
Atmosphere	100% Argon

TABLE II

Analytical Line Wavelengths and Exit Slits

Element	Wavelength Å	Order	Slit and Refractor Plate (μm)*
Be	2348.61	I	75 Q
Hg	2536.52	I	35 Q
Bi	3067.72	I	75 Q
Sn	3175.05	I	25 Q
Co	3453.50	I	25 Q
B	2497.73	II	50 C
Na	5895.92	I	75 G
Mo	3132.59	I	35 Q
V	4379.24	I	25 G
Ag	3280.68	II	50 C
Li	6707.84	I	75 G
Cd	2288.02	II	75 C
Hg Monitor	5460.74	I	25 G

continued

TABLE II - cont'd

| Internal Standard (background) | 4016.10 | 300 G |
| Background[+] | 3885.40 | 75 C |

* G = Glass, C = Corex, Q = Quartz. Mo has a prism behind slit
for beam deflection and Ag has a mirror in a similar position for
the same reason.
+ Background is set on CN band.

SAMPLE AND STANDARD PREPARATIONS

Samples of the water of Linsley Pond, North Branford,
Connecticut were taken weekly for a year. Twenty 1 each were
collected from the water column at the following depth ranges:
surface - 2.5 m, 2.5 - 5.0 m, 5.0 - 8.0 m, 8.0 - 11.0 m and
11.0 - 14.0 m. These samples were evaporated to dryness giving
in each case about 2 gr of solid material. The matrices of these
samples were determined initially by X-ray emission spectroscopy.
All detectable elements were analyzed by this method. Using the
concentrations of the major elements, a matrix was developed consisting primarily of $CaCO_3$, MgO, NaCl, SiO_2 and K_2SO_4. The
elements to be studied by optical emission were added to the basic
matrix in increments that would bracket the amounts of the elements
expected in the samples. The samples and the standards were treated identically at all stages of preparation. Samples were dried
at 80° C for 48 hr. They were mixed with high purity graphite in
a 1:1 ratio. The resulting mixture was mechanically ground in an
agate mortar and pestle for one hour. The size of the total sample
was one gram. Twenty duplicates of each sample were made. In most
cases 15 samples were acceptable, namely the time of burning was
about 8 seconds with a metered background of about 6 volts.

Samples of ice, plants growing in the lake, rock and soil
surrounding the lake were treated in the same fashion as the dried
water samples. In all cases the matrices of the samples were first
determined by X-ray emission.

The elements being studied by optical emission in this study
are not likely to suffer from external contamination problems with
the exception of sodium. The electrode packer is completely cleaned between samples with a high purity grade of acetone. This procedure appears to be quite acceptable. Each electrode is filled
by four successive additions which are added by depressing the
plunger four times. After each electrode is filled the base of

the funnel is cleared by using tefalon coated forcep tips. Other
forcep tips employed in this fashion have added cadmium to the
sample. No detectable contamination has resulted from the tefalon
coated forcep tips. The sodium contamination has proved proble-
matic. In order to avoid this situation, the best procedure proved
to be to wear disposable vinyl medical gloves. These were discarded
and replaced with each change of samples. This approach appeared
to be quite satisfactory. Sodium was determined in the plant
samples by optical emission. In all other samples it was analyzed
by X-ray emission since the concentration was too high to be able
to study it adequately by optical means.

Another difficult contamination problem which occurs early in
this procedure involves boron. The water samples were evaporated
to near dryness in pyrex beakers placed into heating mantles. The
same beakers were used each Sunday for the evaporation of the same
level from the lake. It was felt that this procedure would help
to avoid possible contamination. Initially new pyrex beakers and
watch glasses were used. These were aged in 6 N HCL for 48 hrs.
The beakers and watch glasses were then washed in a mixture of
2 1 HNO_3 and 2 1 H_2SO_4. They were then rinsed 18 times in tap
water, 18 times in single distilled water and 18 times in deionized
doubly distilled water. To check the purity of the deionized
doubly distilled water as well as the cleanliness of the dishes
used in the study, 120 1 were evaporated and no sediment on dryness
was obtained. The efficiency of the cleaning method was checked
once a month and always resulted in no detectable elements in the
evaporate. The beakers and watch glasses were washed after each
evaporation was completed. The same beakers were used throughout
the study. Apparently this procedure solved the boron contamination problem that would normally be expected to occur when pyrex
beakers were employed.

STATISTICAL CONSIDERATIONS

In the concentrated water samples, all elements determined
by optical emission were below 25 ppm in concentration with the
exception of sodium and boron. Table III shows the lowest and
the highest concentrations of the elements studied in the concentrated water samples and the relative standard deviation for
ten duplicates in each category. The limit of detection for all
these elements with the optical emission unit described is 0.06 ppm.
That is to say, with ten duplicates of a prepared standard it is
possible to distinguish between 0.06 ppm and 0.07 ppm. For the
variety of natural samples being studied neither sodium nor boron
would ever be encountered in such low amounts. For the low concentrations shown in Table III, bismuth, tin, molybdenum, vanadium
and silver were estimated by extending the lower portions of the
calibration curves. Since nothing is known about the behavior of

TABLE III

Highest and Lowest Concentration of Elements Found
in Concentrated Water Samples and the Relative
Standard Deviation for 10 Duplicates

	Concentration (PPM)		% Relative Standard Deviation	
Element	low	high	low	high
Be	.05	4.80	4.2	1.36
Hg	.65	10.61	3.1	0.62
Bi	.0011	1.30	2.4	1.33
Sn	.0000002	.60	2.2	3.50
Co	.11	1.40	2.9	2.46
B	760	9100	2.5	0.40
Mo	.044	7.50	0.4	7.10
V	.00008	5.40	2.1	4.41
Ag	.008	.80	1.9	3.12
Li	.142	1.90	4.5	1.24

most of these elements in natural waters, it was felt that even a
precarious estimate of amounts was better than no information at
all. For certain portions of the year studied the estimation of
the quantities of tin and vanadium was quite difficult. The rest
of the elements, with the exception of an occasional sample were
always present in sufficient amounts that determination of their
concentrations created no problem.

It is well to point out that the relative standard deviations for either range of concentration listed, tend to be smaller
than would be expected. The expected precision with a direct reader
would be of the order of 10 percent. It is felt that this precision
has greatly improved as a result of serious consideration given to
detail. First the mixing procedure which has been examined with
great scrutiny is probably the weakest link in any analytical chain.
If the mixing is not adequate, then all steps between this point and
that of burning the sample are in jeopardy. In order to study the
effectiveness of the mixing, a sample mixed with graphite was
ground for one hour. It was then pressed in a pellet suitable for
examination by X-ray emission. Calcium, which is present in high
concentration was examined in all positions of the sample. The
sample was then turned upside down and again the calcium concentration was examined in many positions of the sample. If the counts
per second obtained by this procedure were within one percent of
each other, it was assumed that the mixing was adequate for elements

of high concentration. An element that is present in low concentration must also be checked by this method. A sample was selected that contained 6 ppm cadmium. The same procedure as was used for calcium was repeated for cadmium. Mixing of a one gram sample in an automatic grinder for one hour proved to give adequate results by the criteria established. The next step is the filling of electrodes. Prior to this step all samples were dried at 80° C for 24 hours since the mixing procedure allowed for the readsorption of water. The filling of electrodes, using the electrode packer described, insures as uniform packing as possible. The extent of packing has an enormous affect on the characteristics and time of the burn. The position of the electrodes in the arc stand is extremely important as far as reproducibility is concerned. Any variation from the described procedure in this regard can destroy any chance of acceptable results between duplicates. Accepting burns that have lasted 8 seconds beyond the 3 second preburn time that give a metered background reading of 6 volts also improves the reproducibility. These criteria were established on the basis of carefully prepared standards. Any variations outside the stated range tended to produce unacceptable calibration curves.

COMPARATIVE DATA ON MERCURY

The following discussion is included merely to supply information on the optical emission results obtained and their validity in comparison with a far more sensitive method. At one point it was necessary to analyze some paint samples for their mercury content. The results appeared suspiciously high and so Dr. Frederick Breck of the Jarrell-Ash Division of Fisher Scientific Company kindly offered to examine aliquots of the same samples for their mercury content. A given amount of paint was digested and brought up to 50 ml. volume. A 5 ml aliquot was then diluted to 20 ml. Mercury was determined by a flameless method of atomic absorption which permits detection at the 0.2 ppb level. The method is based on the reduction of mercury salts in solution to elemental mercury. The mercury is aerated from solution as a vapor and is passed through an absorption cell. The absorption produced is measured by an atomic absorption spectrophotometer. The results of both methods are given below.

	Atomic Absorption	Optical Emission
Paint taken from kitchen crack	9.8 ppm	10.0
Paint taken from near sink	5.5	6.1
Paint taken from bedroom pipe	3.3	3.1
Fresh Paint	-	5.3

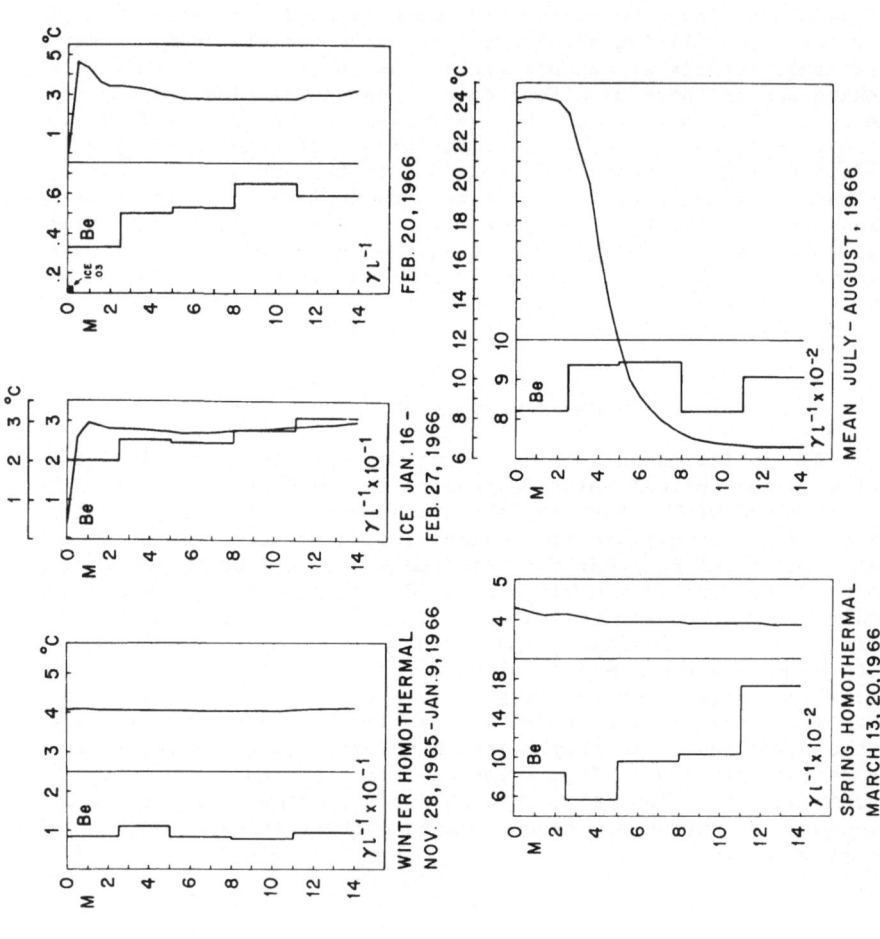

Figure 2.   The vertical distribution of beryllium at various times of the year

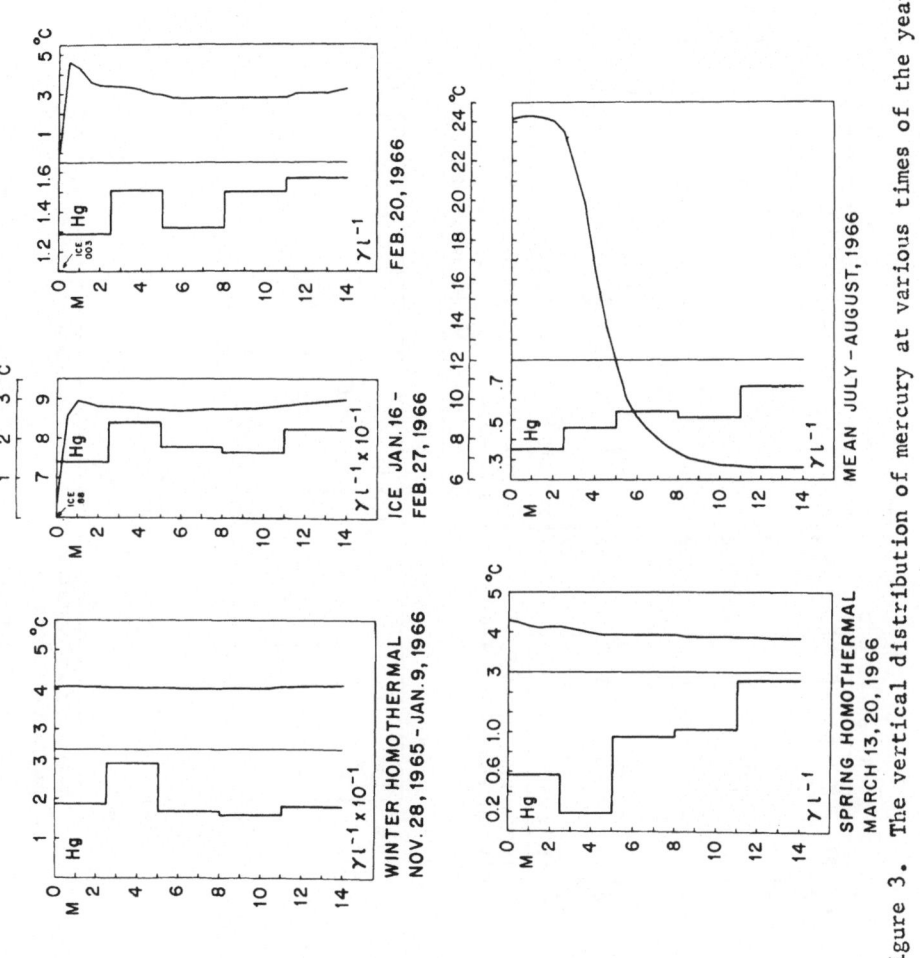

Figure 3. The vertical distribution of mercury at various times of the year

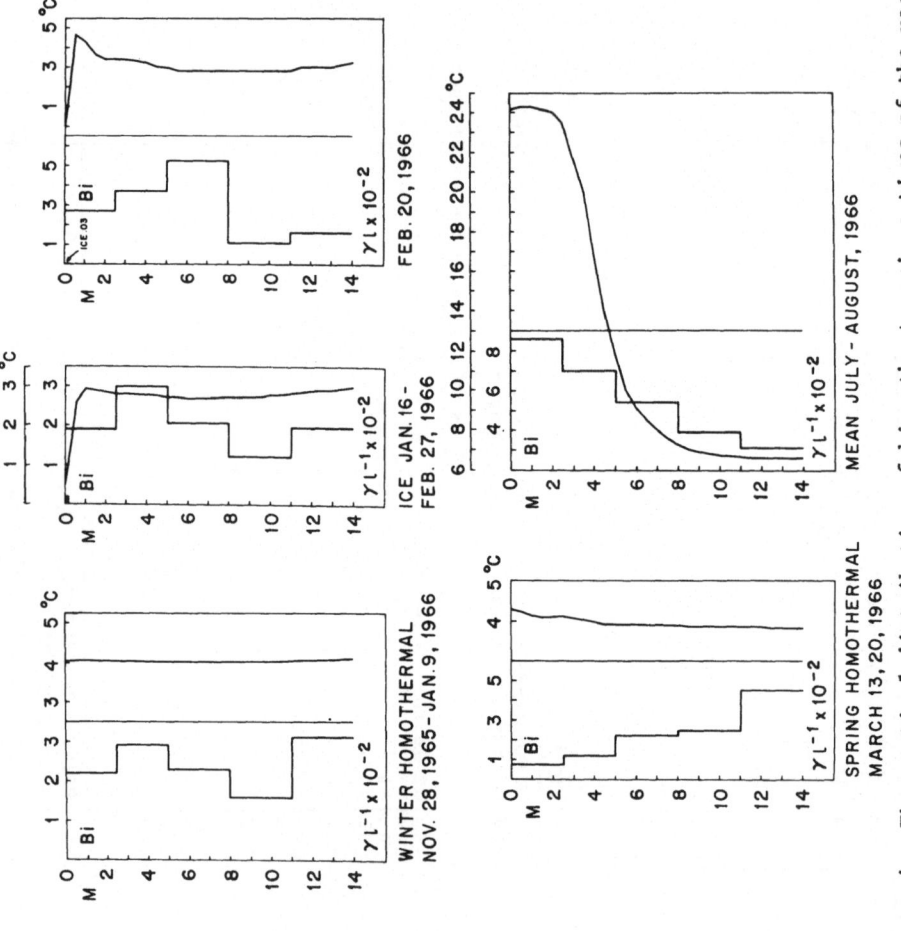

Figure 4.   The vertical distribution of bismuth at various times of the year

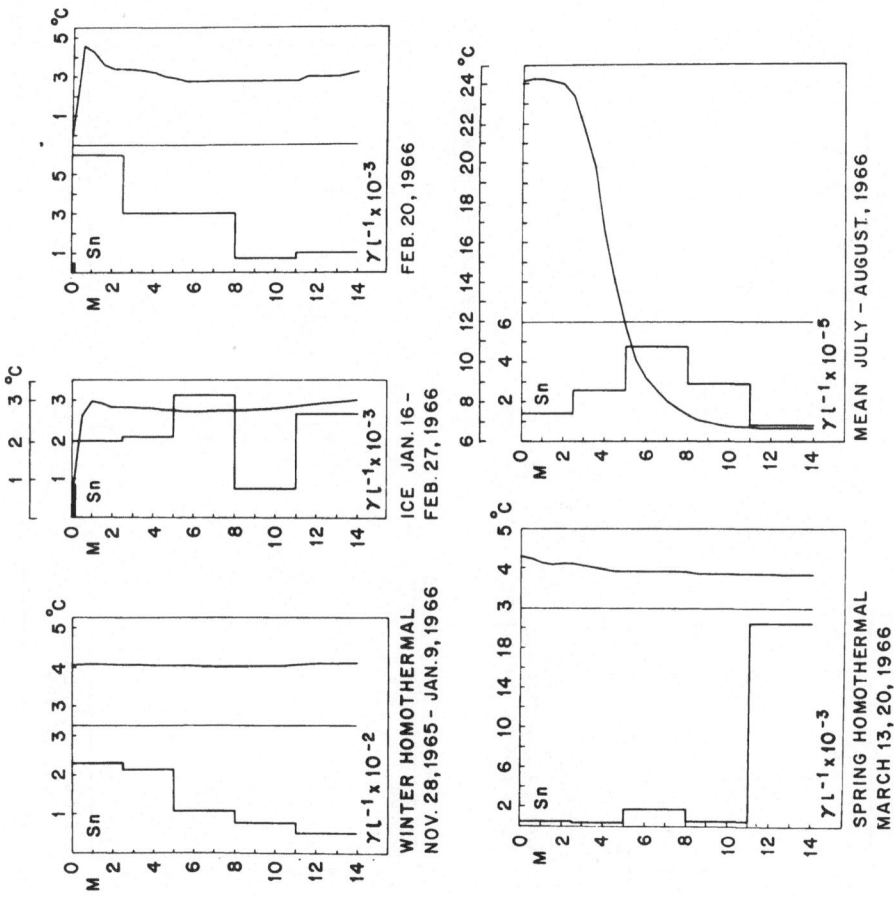

Figure 5. The vertical distribution of tin at various times of the year

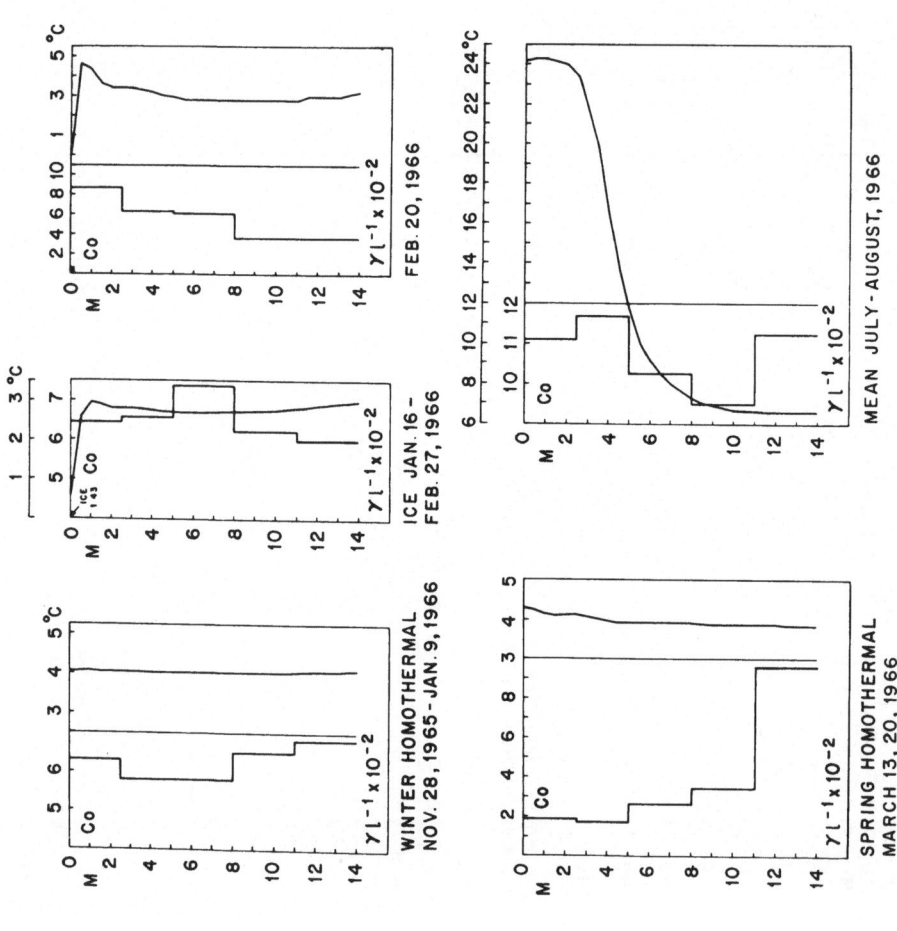

Figure 6. The vertical distribution of cobalt at various times of the year

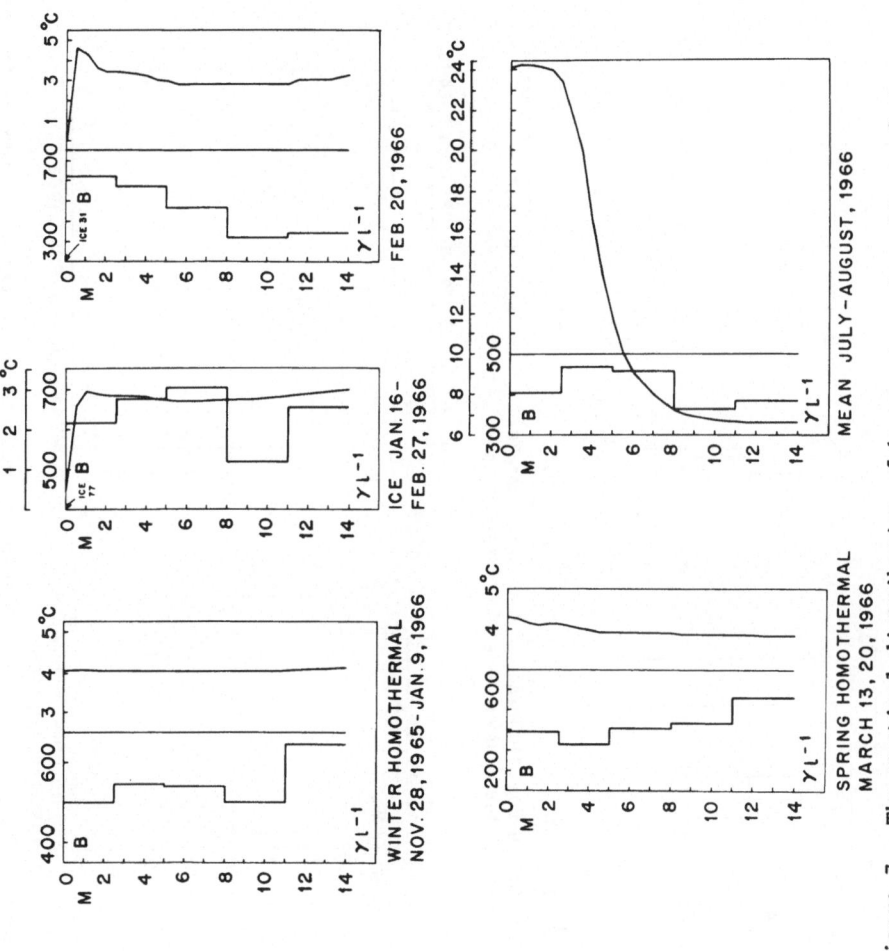

Figure 7. The vertical distribution of boron at various times of the year

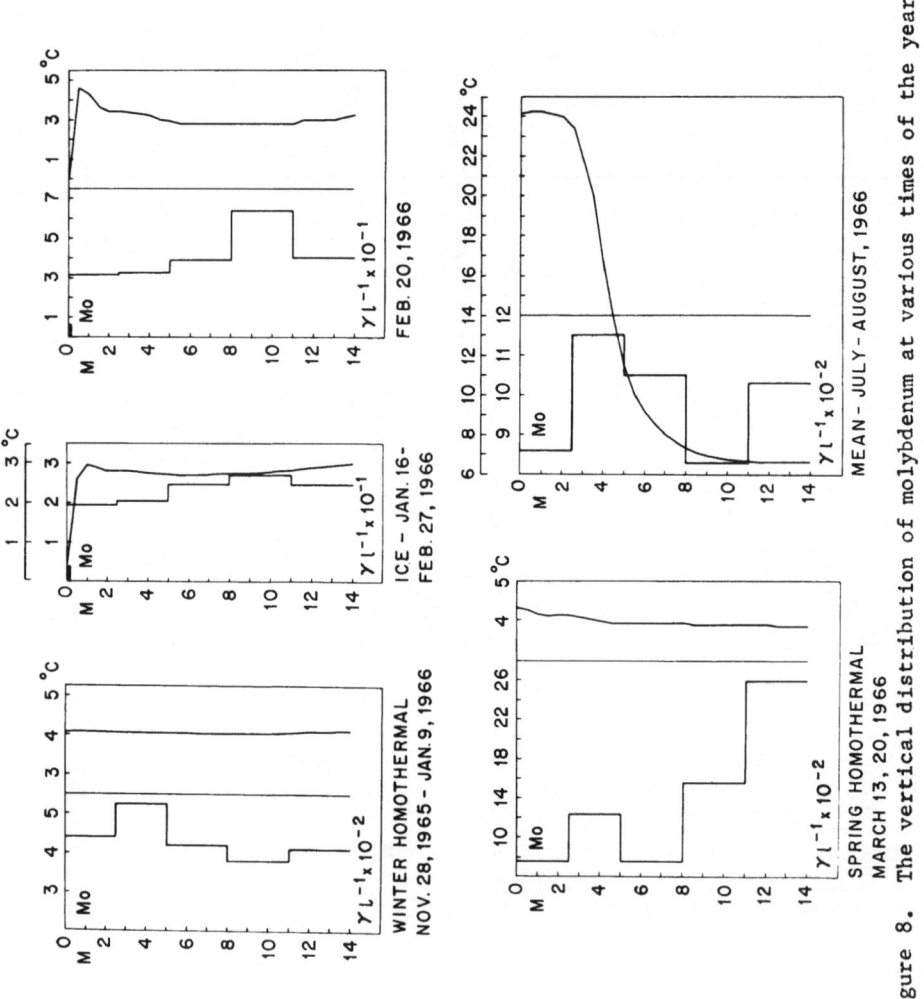

Figure 8. The vertical distribution of molybdenum at various times of the year

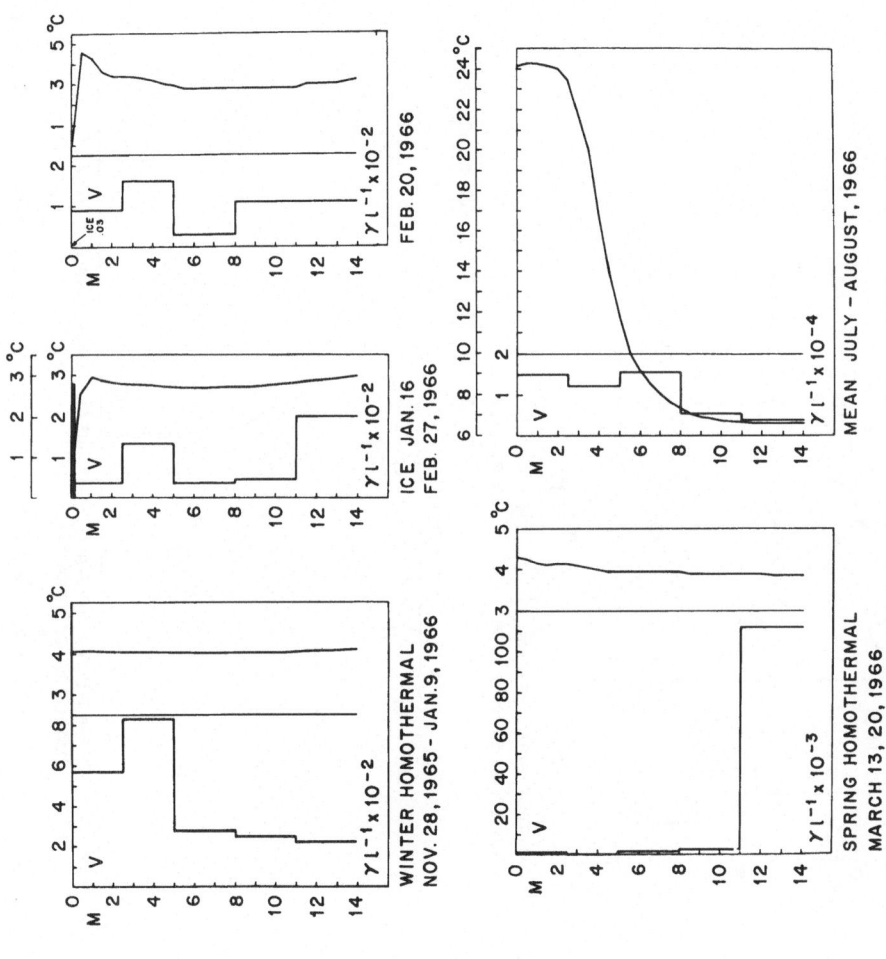

Figure 9. The vertical distribution of vanadium at various times of the year

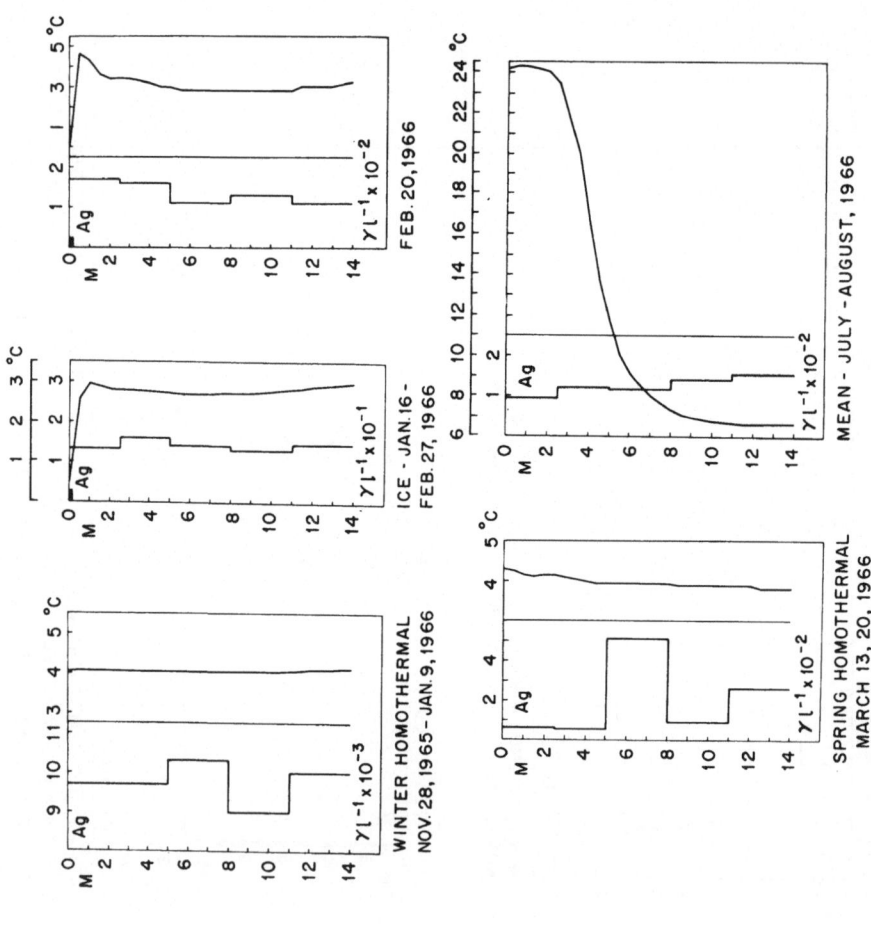

Figure 10. The vertical distribution of silver at various times of the year

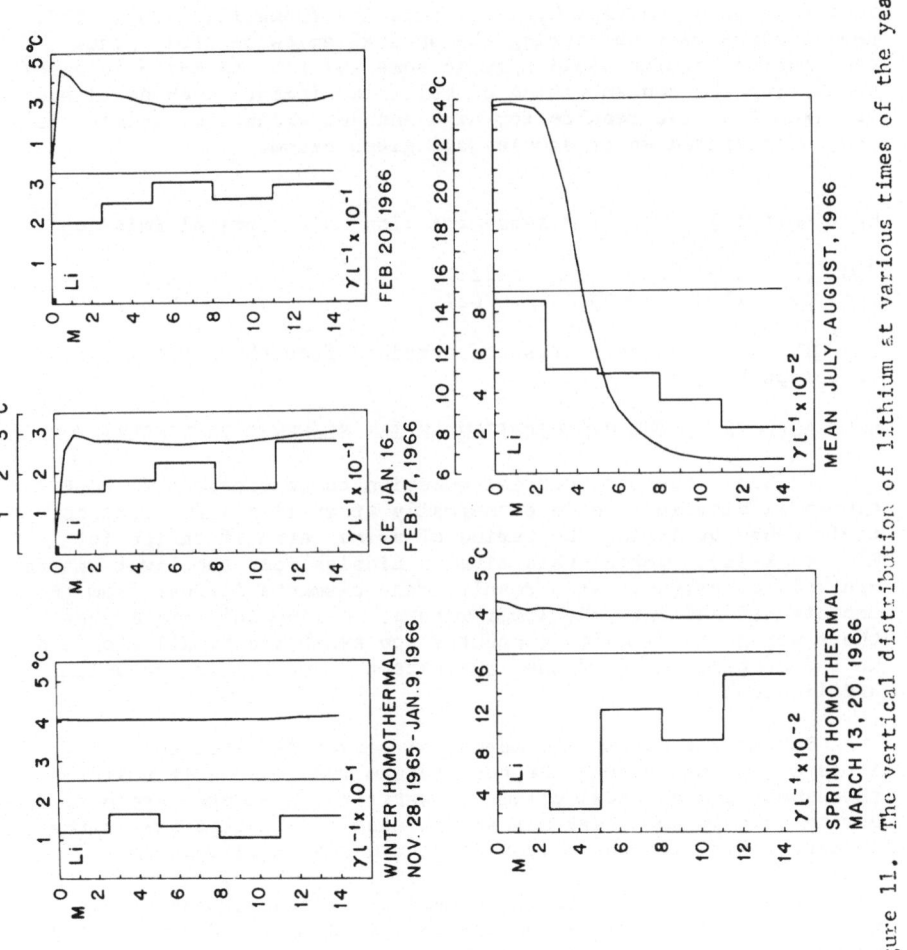

Figure 11. The vertical distribution of lithium at various times of the year

It can be seen from the results obtained that the two methods are amazingly concordant.

COMPARISON OF CADMIUM DATA OBTAINED BY TWO DIFFERENT METHODS

It was mentioned earlier in this paper that cadmium had initially been analyzed by X-ray emission (Cowgill, 1970). This analysis was repeated during the optical emission study since comparative results would provide some solution to the problem of estimating the concentration of volatile elements such as lithium for example. The results for high and low amounts of cadmium in the concentrated water samples are given below.

Cadmium (ppm)	X-ray emission	Optical Emission
Jan. 23 Ice	22.4	20.1
Oct. 17 8-11 M	6.1	6.3

The concordance of results obtained from these two methods is acceptable.

SOME ASPECTS OF THE DISTRIBUTION OF TEN ELEMENTS IN LINSLEY POND

Probably the most crucial question to be considered about elemental distribution in a thermally stratified lake concerns their behavior during the period of summer stratification (cf Figures 2-11). During this time in Linsley Pond the lower waters are void of oxygen. As a result, many elements diffuse from the unprotected mud into the lower waters. During July and August there was an increase in concentration exhibited by all elements in the deepest layer of the lake except bismuth, tin, vanadium and lithium.

During the autumnal homothermal period (cf Figures 2-11), it is clear that though the temperature was reasonably uniform throughout the vertical column of water the elements were not. During this period there was an increase of all elements in the deepest layer with the exception of tin and vanadium.

From January 16 through the end of February, Linsley Pond was ice covered. Of the ten elements studied (cf Figures 2-11), only vanadium concentrates in the ice, that is to say, the quantity in the ice is 7.4 times higher than it is in the surface to 2.5 M section from which the ice was made. Possibly the cause of this phenomenon may be that the vanadium in the atmosphere as a result of heating fuels (Israelyan, 1959) comes into the lake in the form of snow and rain. During the ice covered period, all the elements increased in concentration in the deepest layer except cobalt and molybdenum.

During the week prior to February 20 there was a thaw. Ice in
the littoral zone of the lake melted releasing less dense water
which slid under the ice throughout the lake carrying bulk chemical
material with it. In short, there was a density current. Beryllium,
mercury, boron, molybdenum, vanadium and lithium all increased in
their total quantity in the lake on this day. The rest of the
elements reached a peak in concentration the previous Sunday.

The next major incident that occurs in a lake in the temperate
zone is the spring homothermal period (cf Figures 2-11). At this
time the temperature is uniform throughout the vertical column of
the lake. However, this is not the case with the chemistry. During
this period all ten elements increased in the deepest layer of the
lake.

The coherence of various elements with each other and the
types of compounds which they make-up hopefully will be further
elucidated when all the data from this entire study of the distribution of all detectable elements in Linsley Pond will be subjected
to computer analyses.

SUMMARY

A method has been described employing an optical emission
direct reader for the analysis of ten elements. Comparative data
are offered for mercury and cadmium employing entirely different
analytical procedures. The results are quite concordant.

Some data are offered concerning the vertical movement of ten
elements in a thermally stratified lake throughout one year of study.

ACKNOWLEDGEMENT

The support of this work by the National Science Foundation is
gratefully acknowledged. Mr. Donald Crowley of the Instrument Shop
of Yale University made the electrode packer, without which this
study might still be in progress. I thank Mrs. V. Simon for her
care and speed in preparation of the illustrations.

REFERENCES

1. Cowgill, U. M. 1970 The hydrogeochemistry of Linsley Pond,
    North Branford, Connecticut. I. Introduction, field work
    and chemistry by X-ray emission spectroscopy. Arch. Hydrobiol.

68:1-95.
2. Israelyan, A. D. 1959 Trace elements in the ash of Maikop
    crude oil. Trudy Azerbaidzhan. Nauch. - Issledovatel. Inst.
    po Dobyche 8:274-281. (Chem. Abst. 55:18082L).
3. Jarrell-Ash Company 1970 Determination of mercury by flameless
    atomic absorption. Atomic absorption analytical method.
    No. Hg-1. pp. 1-6.

INDEX

INDEX